National Audubon Society
Field Guide to
North American Insects and
Spiders

A Chanticleer Press Edition

National Audubon Society Field Guide to North American Insects and Spiders

Lorus and Margery Milne,
University of New Hampshire

Visual Key by
Susan Rayfield

Alfred A. Knopf, Ne

CONTENTS

NATIONAL AUDUBON SOCIETY

The mission of the NATIONAL AUDUBON SOCIETY *is to conserve and restore natural ecosystems, focusing on birds and other wildlife, for the benefit of humanity and the earth's biological diversity.*

With more than 560,000 members and an extensive chapter network, our staff of scientists, educators, lobbyists, lawyers, and policy analysts works to save threatened ecosystems and restore the natural balance of life on our planet. Through our sanctuary system we manage 150,000 acres of critical habitat. *Audubon* magazine, sent to all members, carries outstanding articles and color photography on wildlife, nature, and the environment. We also publish *Field Notes,* a journal reporting bird sightings, and *Audubon Adventures,* a bimonthly children's newsletter reaching 600,000 students.

NATIONAL AUDUBON SOCIETY produces television documentaries and sponsors books, electronic programs, and natu travel to exotic places.

For membership information:

NATIONAL AUDUBON SOCIE
700 Broadway
New York, NY 10003-9
(212) 979-3000

THE AUTHORS

Lorus and Margery Milne, a husband-and-wife team of naturalists, have studied the insects and spiders of North America for many years. They have written 39 books, including *Insect Worlds, Invertebrates of North America, The Secret Life of Animals* (with Franklin Russell), *The Nature of Life, The World of Night,* and *Living Plants of the World.* The Milnes have studied ecology and animal behavior around the world, travelling to South America, Africa, Europe, and the Middle East, as well as Southeast Asia, Australia, New Zealand, and Hawaii. Both authors received their doctorates from Harvard University and are currently on the faculty of the University of New Hampshire.

ACKNOWLEDGMENTS

Many people have been involved in the preparation of this book. Our sincere appreciation goes to the following individuals for reviewing portions of the text, identifying photographs, and offering helpful criticisms and suggestions: Allen Brady, Hope College; Steve Bullington and Fred Lawson, University of Wyoming; Herbert W. Levi, Harvard University; Charles R. Meck, Pennsylvania State University; Robert W. Mitchell, Texas Tech University; and E. S. Ross and D. D. Wilder, California Academy of Sciences. We would also like to thank Wallace J. Morse and John F. Burger, University of New Hampshire, and the staff of the University of New Hampshire Library, who facilitated our research. We are grateful to many others for their contributions to this book: Robert C. Dalgleish, Edmund Niles Huyck Preserve, Inc.; G. B. Edwards, Florida State Collection of Arthropods; B. A. Foote, Kent State University; Oscar Francke, Texas Tech University; Donald E. Johnston, Ohio State University; Stanley E. Malcomb, University of Connecticut; Paul Miliotis, Harvard University; Edward Mockford, Illinois State University; William Muchmore, University of Rochester; Daniel Otte, Academy of

Natural Sciences; Jan Peters, Florida
A&M University; Roland M. Shelly,
North Carolina State Museum; Daniel
E. Sonenshine, Old Dominion
University; Lewis J. Stannard,
University of Illinois; and R. G.
Weber, University of Delaware.
It has been a pleasure working with the
staff of Chanticleer Press. Paul Steiner,
Milton Rugoff, and Gudrun Buettner
provided constant support and
encouragement. Our special thanks go
to Susan Rayfield, who developed the
idea for the book and the organization
for the visual key; to Jane Opper, who
edited the text and provided countless
valuable suggestions; to Mary Beth
Brewer, who coordinated all phases of
the project; to John Farrand, who
served as a scientific consultant
throughout; and to Carol Nehring, who
supervised the art and layout. We
would also like to thank Susan Linder
for locating many of the photographs;
Constance Mersel, Ann Whitman, and
Charles Ervin for their valuable
editorial assistance; Mary deButts for
the picture and text layout; and Dean
Gibson, Helga Lose, and Ray Patient
for their special efforts in the
production of this book.

INTRODUCTION

Insects are found everywhere—from the tropics to the tundra, in water, wood, plants, soil, and even inside the bodies of other animals. Over a million different species have been identified, more than all other animals combined. Nearly 100,000 species live in North America.

Many insects are beneficial, pollinating flowers, fertilizing the soil, and providing such commercially valuable products as honey, beeswax, and silk. Others are pests that cause considerable damage to crops and trees.

Observing insects as they go about their daily lives can be as fascinating as watching birds. This book will help you learn about their secret world. Familiarity with the 550 common and conspicuous insects in this guide will enable you to recognize their relatives elsewhere. Although spiders are not insects, many people group them with insects, and we have therefore included over 60 common spiders and their kin. This book is designed for everyone who wants to learn how to identify the insects and spiders found in woods, fields, and ponds, as well as gardens and homes. The striking color photographs and clear, nontechnical descriptions make this guide both a pleasure to look at and easy to use.

What is an insect? An insect is an invertebrate animal. It belongs to a large group known as the Phylum Arthropoda along with spiders, horseshoe crabs, sea spiders, crustaceans, millipedes, and centipedes. The Phylum Arthropoda is divided into classes, with insects forming the Class Insecta. All arthropods have in common segmented bodies with paired, many-jointed legs. Members of the Class Insecta are further distinguished from the others in having 3 major body sections and, as everyone can easily recognize, 6 legs and 2 antennae. Within their class, insects are subdivided into different orders. Beetles belong to the Order Coleoptera, and butterflies are part of the Order Lepidoptera. These orders are, in turn, divided into families; for example, ladybug beetles are grouped in the Family Coccinellidae. Families are divided into genera, and finally genera into species. When people speak of a specific kind of insect, what they usually have in mind is a species, such as the Five-spotted Ladybug or the Monarch Butterfly. Members of each species have common characteristics and are capable of interbreeding.

What is a spider? A spider is not an insect. But like an insect, it is an invertebrate animal and member of the Phylum Arthropoda. Spiders belong to the Class Arachnida, along with ticks, mites, scorpions, and their kin. Within the Class Arachnida, spiders comprise the Order Araneae. Unlike insects, arachnids have only 2 major body sections and do not bear antennae. A spider can be instantly recognized by its 8 legs. The order is subdivided into different families. For example, one of the largest spider groups—the orb weavers—constitutes the Family Araneidae. Then, in accordance with scientific classification, each family is divided into genera, and genera into species.

Classification of Insects and Spiders: Zoologists classify insects and spiders according to common structural characteristics. The sequence of orders, families, genera, and species is based on interpretations of evolutionary relationships. Not all scientists, however, agree on the same evolutionary sequence.

We have tried to present the latest information derived, wherever possible, from a consensus among specialists. For all insects except butterflies and moths we have used the phylogenetic sequence presented in *An Introduction to the Study of Insects* by D. J. Borror, D. M. DeLong, and C. A. Triplehorn (New York: Holt, Rinehart and Winston, 1976). The arrangement of butterflies and moths is based on the new sequence given in *Moths of North America,* edited by Ronald Hodges (London: E. W. Classey Ltd., 1980). The organization of spiders follows *How to Know the Spiders* by B. J. Kaston (3rd ed., Dubuque: Wm. C. Brown, 1978).

Parts of Insects: An insect's body is composed of many segments, grouped in 3 major parts—the head, thorax, and abdomen. The head bears the antennae, eyes, and mouthparts; the thorax bears legs and often wings; the abdomen may have a pair of sensory appendages, called cerci, at the tip. All parts are protected by a hard outer covering, called the exoskeleton.

Head: Insects have 1 pair of segmented antennae, usually located between or in front of the compound eyes. Their shape varies greatly among insects. Antennae are used mostly for smell and touch, but some insects also use these appendages for hearing.

In addition to antennae, most insects also have 2 kinds of eyes—simple eyes, called ocelli, which are merely sensitive to light, and compound eyes made up of many minute facets or lenses which

Types of Antennae

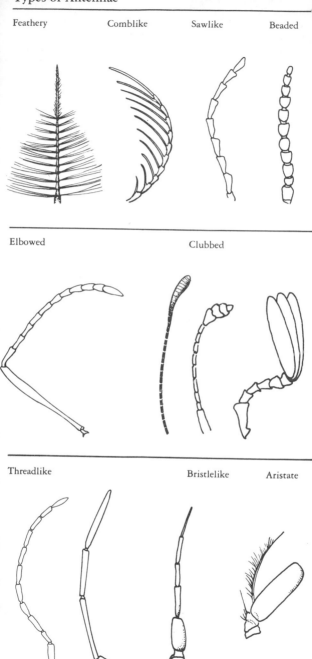

Feathery   Comblike   Sawlike   Beaded

Elbowed   Clubbed

Threadlike   Bristlelike   Aristate

record multiple images. A House Fly, for example, has about 4,000 facets in each compound eye, and a dragonfly has more than 50,000 facets.

Mouthparts: Mouthparts usually consist of the upper lip (labrum) and lower lip (labium), the jaws (mandibles), and two smaller jawlike appendages (maxillae). Some insects also have appendages, called palps, on the labium and maxillae. The mouthparts are generally used for either biting or sucking. Insects with biting mouthparts, such as beetles and dragonflies, work their mandibles from side to side. Those with sucking mouthparts generally possess a beak or beaklike tongue for sucking liquids. Mosquitoes, biting flies, true bugs, and some homopterans, such as leafhoppers, have piercing-sucking mouthparts with daggerlike structures, called stylets. Most flies have sponging mouthparts for lapping up liquids. Butterflies and moths have a coiled, tonguelike proboscis for sipping. Bees and wasps have biting and lapping mouthparts.

Thorax: The thorax is composed of 3 segments —the prothorax, mesothorax, and metathorax. Each segment bears a pair of legs. In winged insects the mesothorax and usually the metathorax bear wings. In many insects, such as cockroaches, grasshoppers, and mantids, the upper surface of the prothorax, the pronotum, is large, extending from the head to the base of the wings. Bugs and many beetles have a distinct, triangular shield, called the scutellum, located immediately behind the pronotum. Some flies, such as tachinids, have a large, conspicuous swelling underneath the scutellum, called the postscutellum.

Legs: Insects have 3 pairs of legs. Each leg is composed of 5 parts—coxa, trochanter, femur, tibia, and tarsus. The tarsus contains 2–5 segments and often bears a pair of claws at the tip and 1 or more pads. Leg size and shape differ greatly

Biting Mouthparts

Wasp

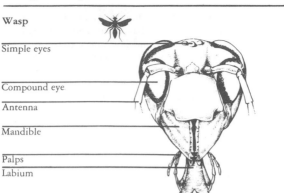

Simple eyes
Compound eye
Antenna
Mandible
Palps
Labium

Dragonfly

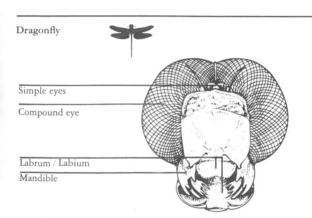

Simple eyes
Compound eye
Labrum / Labium
Mandible

Beetle

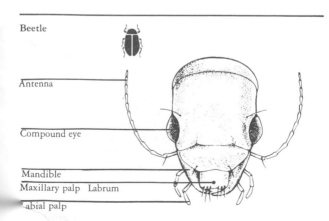

Antenna
Compound eye
Mandible
Maxillary palp   Labrum
Labial palp

Sucking Mouthparts

Leafhopper

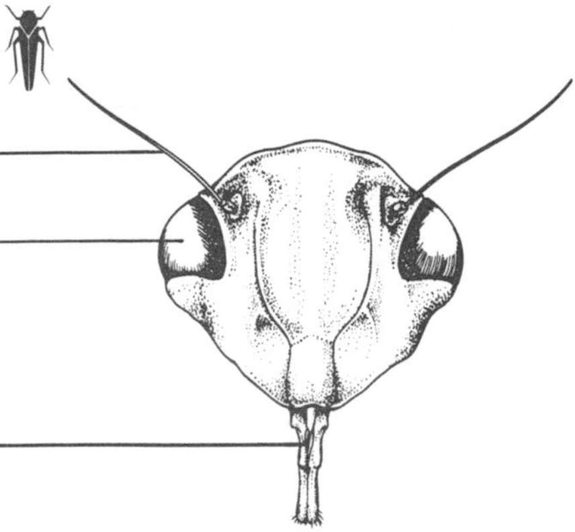

Antenna
Compound eye
Beak (Proboscis)

House Fly

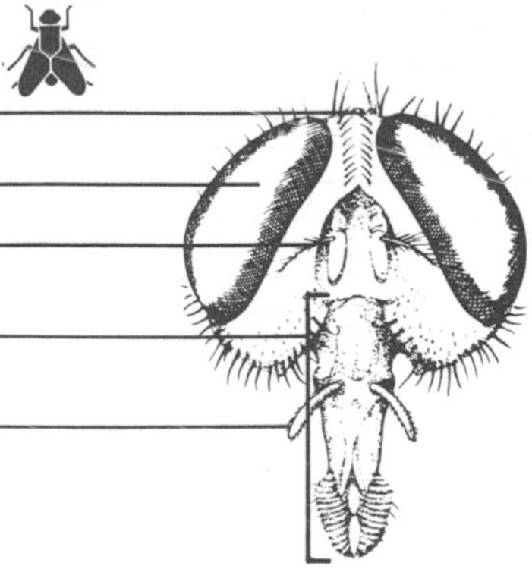

Simple eyes
Compound eye
Antenna
Proboscis
Maxillary palp

Mosquito

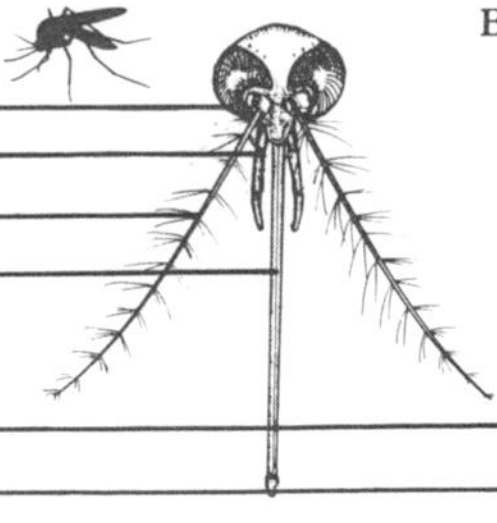

Compound eye
Maxillary palp
Antenna
Proboscis

Butterfly

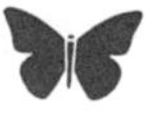

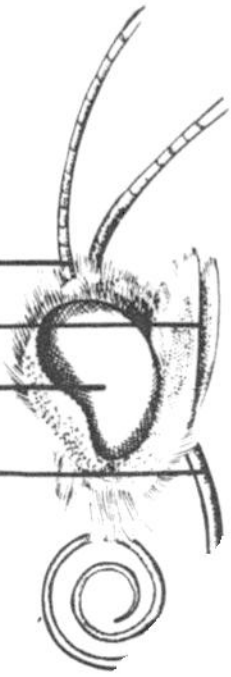

Antenna
Labial palp
Compound eye
Proboscis

Parts of Insects

Grasshopper

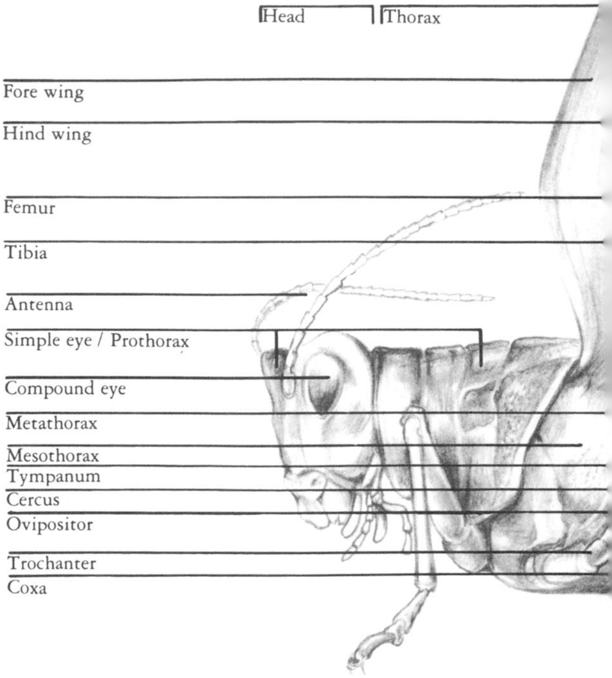

Head | Thorax

Fore wing

Hind wing

Femur

Tibia

Antenna

Simple eye / Prothorax

Compound eye

Metathorax

Mesothorax

Tympanum

Cercus

Ovipositor

Trochanter

Coxa

Fly

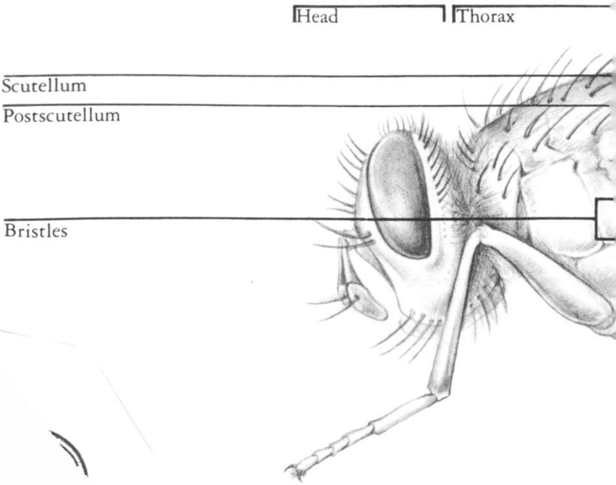

Head | Thorax

Scutellum

Postscutellum

Bristles

Abdomen

Abdomen

among insects. Mantids and mantidt.
have enlarged fore legs for grasping
prey; grasshoppers and crickets have
swollen hind femora for leaping. Most
female bees have pollen baskets or
brushes on the hind legs.

Wings: Most adult winged insects have 2 pairs
of membranous wings, but flies have
only 1 pair of flying wings and a pair of
clublike halteres. In beetles the fore
wings, or elytra, are hard and
armorlike; they cover the membranous
hind wings used for flying. The fore
wings, or hemelytra, of true bugs are
leathery and have membranous tips.
The arrangement of veins on the wings,
known as wing venation, is useful in
identifying many insects.
Although the number of veins varies
greatly among genera, most
entomologists use a standard
terminology. Starting at the front edge
of the wing, the main longitudinal
veins are the costa (C), which forms the
front or costal margin of the wing; the
subcosta (Sc); the radius (R); the media
(M); the cubitus (Cu); and the anal
veins (A). The branches which spring
from each vein are given subscript
numbers. For example, the five
branches of radial veins in the wings of
butterflies and moths are named R_1,
R_2, R_3, R_4, and R_5. The veins that
cross the wing, known as cross veins,
are named in relationship to the
longitudinal veins. For example, the
cross vein between Cu_1 and M_3 is
m-cu. The spaces enclosed by veins are
called cells. These are designated
according to their position in the wing;
thus, the cells next to the wing margin
are marginal cells, and the cells behind
them are submarginal cells. Important
cells are given a special name and are
used for identification. For example,
the large cell at the base of the wings of
butterflies and moths is called the discal
cell. In many insects wing venation is
clearly visible, but veins in the wings of

eralized Wing Venation

atterfly or Moth

ostal (C)
Subcostal (Sc)
Radius (R)
Radial sector (Rs)
Medial (M)
Cubitus (Cu)
Anal veins (A)
Humeral (h)
Discal cell (D)
Frenulum (f)

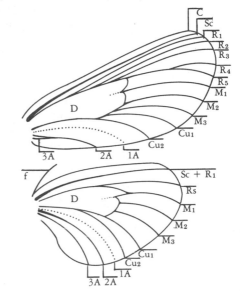

Wasp

Submedian cell (SMD)
Costal cell (C)
Basal vein (bv)
Submarginal cell (SM)
Marginal cell (MC)
Recurrent vein (rv)
Jugal lobe (jl)

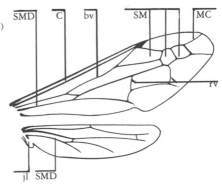

butterflies and moths are usually covered by scales. In addition to venation, insect wings may have distinctive structural features. Some bees and wasps have a lobe at the base of the hind wing called the jugal lobe. Many butterflies and moths have a bristle or group of bristles called the frenulum at the base of the front margin of the hind wing.

Abdomen: The abdomen usually consists of 11 segments, but some insects have fewer. Female genitalia are located between the 7th and 8th segments. In many insects these organs are associated with an ovipositor, which protrudes from the tip of the abdomen. The ovipositor enables a female to insert eggs into otherwise inaccessible places—the soil, plants, or bodies of other insects. Male reproductive organs are located on the 9th segment, and in some insects there is also a pair of claspers for grasping mates. By noting small differences in genitalia, entomologists can distinguish closely related species. Often the last segment bears a pair of tail-like or pincerlike sensory appendages, called cerci. All insects have pairs of breathing pores, or spiracles, along the sides of the abdomen in addition to 1 or 2 pairs of spiracles on the thorax.

Parts of Spiders: A spider's body has 2 parts—the cephalothorax, consisting of the head and thorax combined, and the abdomen. These 2 parts are joined by a short, thin stalk, or pedicel. The cephalothorax bears the eyes, mouthparts, and 4 pairs of legs; it is protected by a shieldlike covering, the carapace. The abdomen contains the respiratory, digestive, and reproductive organs as well as the spinnerets, which spin silk.

rax: Most spiders have 8 simple eyes at the front of the cephalothorax, but some have fewer, and a few families have

ts of a Spider

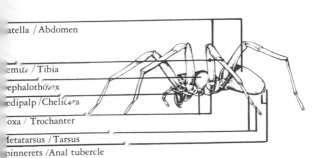

atella / Abdomen

emur / Tibia

ephalothorax

edipalp / Chelicera

oxa / Trochanter

Metatarsus / Tarsus

pinnerets / Anal tubercle

Pedipalps

♀ ♂

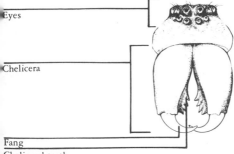

Eyes

Chelicera

Fang

Cheliceral teeth

none. The arrangement of the eyes differs greatly among spiders and is one means of recognizing different families. Below the eyes are the jaws, or chelicerae. The chelicerae end in fangs containing the openings for the poison glands. Small leglike appendages, called pedipalps, are located between the chelicerae and the first pair of legs. Each pedipalp has an enlarged base against which the spider crushes prey held between the chelicerae. The male's pedipalps end in a club and are used in mating; the female's pedipalps are not clubbed.

Legs: The spider's leg, unlike the insect's, has 7 segments—the coxa, trochanter, femur, patella, tibia, metatarsus, and tarsus. The tip of the tarsus usually bears 2 or 3 claws and many bristles and hairs.

Abdomen: On the rear underside of the abdomen, most spiders have 6 fingerlike structures, called spinnerets, which are used to spin silk. Members of a few spider families have only 2 or 4 spinnerets. In some families, there is also a sievelike spinning plate, called the cribellum, in front of the spinnerets. In all spiders the anal opening is situated on a small tubercle located behind and above the spinnerets.

How Insects and Spiders Grow: Like all arthropods, insects and spiders have a hard outer covering, the exoskeleton. Because this covering cannot stretch, insects and spiders can grow only by shedding or molting the exoskeleton several times. Many insects pass through 4 distinct growing stages —the egg, larva, pupa, and adult. The larva is the active feeding stage, while the pupa is an inactive resting stage during which the larva transforms into an adult. This type of development, termed complete metamorphosis, occurs in such insects as butterflies, moths, flies, beetles, wasps, and bees.

In other insects, the pupal stage is omitted, and the young often resemble small versions of adults, except for the absence of sexual organs and, in some orders, wings. This is called simple metamorphosis and is found in grasshoppers, true bugs, dragonflies, and in certain other orders. The young are called nymphs, except for the aquatic offspring of dragonflies, damselflies, mayflies, and stoneflies, which are termed naiads. Spiders also have simple metamorphosis with a series of stages and molts. The spiderlings resemble adults in form but are not sexually mature.

A New Approach to Identification: Because insects and spiders are so numerous, learning to recognize different species can seem overwhelming. For this reason, we have chosen a new approach to identification —one that we hope will be both easier and more enjoyable. This is the only field guide to use vivid color photographs of insects and spiders as they appear in their natural habitats, and the only guide to identify them to species wherever possible. You can quickly find the insects or spiders you have seen (or their near relatives) because the photographs are arranged by shape and color. All other field guides to insects or spiders are illustrated with black-and-white drawings and paintings, and are grouped by scientific order according to structural features which are unfamiliar to most people. In contrast, our simple organization of close-up color photographs allows for easy identification and at the same time lets you appreciate the natural beauty of each insect and spider.

Organization of the Color Plates: We have arranged the photographs of insects and spiders according to the features you see in the field—shape and color. Thus, the group called Aphidlik

Insects contains small, similar-shaped insects, such as aphids, mealybugs, and whiteflies. The group called Ants and Termites includes two spiders that mimic ants; similarly, the group Bees, Wasps, and Kin contains some beelike flies that could be easily mistaken for bees.

The color plates are arranged in the following order:

Caterpillars
Water Nymphs and Larvae
Aphidlike Insects
Flealike Insects
Earwigs and Silverfish
Diving Beetles and Water Bugs
Hopperlike Insects
Plant Bugs, Toad Bugs, and
Cockroaches
Weevils
Beetles
Grasshoppers, Crickets, and Cicadas
Mantids and Walkingsticks
Ants and Termites
Lacewings and Other Long-winged
Insects
Dragonflies and Damselflies
Flies
Bees, Wasps, and Kin
Moths
Butterflies
Ticks, Mites, and Scorpions
Spiders

Key to the Color Plates: The organization of the color plates is explained in a table preceding them. A silhouette of a typical member of each group appears on the left. Silhouettes of insects and spiders within that group are shown on the right. For example, the silhouette of a scarab beetle represents the group Beetles; a silhouette of a mantid represents Mantids and Walkingsticks. This representative silhouette is inset on a thumb tab at the left edge of each double page of color plates devoted to that group of insects and spiders.

Captions: The caption under each photograph
gives the plate number, common name
of the insect or spider, its size, and the
page number of the text description.
The measurement given is usually body
length; however, for butterflies and
moths, we indicate wingspan (w.).
Where easily determined the sex of the
insect or spider—male ($\male$) or female
($\female$)—as well as social caste and any
interesting behavior, such as mating,
laying eggs, and stalking prey are also
provided. The color plate number is
repeated in front of each text
description as a cross reference.

Organization of Insects are described first, followed by
the Text: spiders and their kin. The families
within each insect or spider order are
arranged in phylogenetic sequence,
beginning with the most primitive and
ending with the most advanced. The
arrangement of genera and species
within each family is alphabetical by
scientific name. Preceding the text
section, we include a chart showing
representative members of each order to
help you recognize different insects,
spiders, and spider kin. Each account
starts with the common name followed
by the scientific name. Scientific names
for genera and species are always
italicized, those for orders and families
are not. Each scientific name consists of
two words usually derived from Latin or
Greek. The first, always capitalized, is
the name of the genus; the second, in
lower case, indicates the species within
the genus. For example, the common
name of one species of bee is Honey
Bee; its generic name is *Apis,* its
specific name is *mellifera,* and the
scientific name is *Apis mellifera.* The
number of the color plate appears to the
left of the common name. Wherever
possible, photographs are identified as
to species or genus, but in a few cases
the identification is a family or order.
Common names of insects used in this

guide are based on the *Bulletin of the Entomological Society of America* and the consensus of entomologists. Spider names reflect those commonly used by arachnologists. Each species account includes the following information:

Description: Every description begins with measurements (in both inches and millimeters). They represent the approximate length of the adult insect or spider, from the head to the tip of the abdomen, excluding antennae and appendages. Wingspan (w.) is given for butterflies and moths. For dragonflies, both length and wingspan are indicated. Measurements for caterpillars, aquatic larvae, and naiads refer to their maximum size. For easy use, a measurement in inches to the nearest ruler fraction is given first with the scientific measurement in millimeters following in parentheses. Because most insects are small, the metric equivalent in inches must cover a wide range. For example, ⅜" is the equivalent for 8–11 mm, thus both the Meadow Spittlebug (9–10 mm) and the Dogbane Leaf Beetle (9–11 mm) are approximately ⅜" long. Next is a description of the adult, indicating its shape, color, and distinguishing marks. Prominent features are italicized for quick identification. Diagnostic details, which help entomologists distinguish between species, such as the number of spines or antennal segments, are only included if clearly visible. Because details of wing venation and the reproductive organs are unnecessary for the field identification of most insects, they are usually omitted. In moths, some flies, wasps, and bees, where wing venation is essential for identifying families, marginal drawings of wings or comparative wing venation charts are included. A brief description of the immature insect (if it is commonly seen) appears at the end of each description. Wherever possible, familiar terms are used; when technical terms are needed, they are defined in the glossary or illustrated with labeled drawings in the introduction. As an additional aid in distinguishing similar-looking insects, comparative wing venation charts appear at the beginning of the sections on Butterflies and Moths and Bees. For some flies and wasps, wing venation drawings are included in the text margin.

Habitat: Although insects and spiders move about freely, they generally are found in a particular environment, or habitat. Within the same species it is quite common for members to live in several different types of habitat in different geographical localities. Knowing the typical habitat helps in locating the species.

Range: This is the first field guide to cover the most common species of insects and spiders in North America from Alaska to Mexico. The geographical range of each species is given from north to south, starting in southeastern Canada and the northeastern United States and moving clockwise—south, west, north, and then east.

Food: The food an insect or spider eats is often a major aid in locating and identifying the species. In many groups both the adult and young eat the same food. But in some orders, such as beetles, flies, or butterflies and moths, the adult and its young feed on different foods; the food of each is indicated in the text. Occasionally the adult food is unknown, and only the food for the young is specified. Knowing the food for the young often is the first step in controlling economic pests.

Sound: Scraping and drumming activities of various insects produce audible sounds. Some, such as the buzzing of bees, wasps, and some flies, may be merely warning noises to drive off intruders.

Other sounds, among these the singing of cicadas, grasshoppers, and crickets, appear to be communication—between potential mates as a form of courtship, or between rivals as indicators of territory. Insects that live in social groups, such as ants and bees, make sounds to coordinate their activities. Recognizing all of these sounds can help you find the soundmaker and identify it.

Flight: This section appears in Butterflies and Moths to indicate the months or season during which the adult can be seen in the mid-temperate part of its range. This period covers all times the species is seen. Where there is more than 1 generation a year, the flight sequence represents a collective time for all generations and is not necessarily a continuous period for 1 generation.

Web: Spider species in web-spinning families often can be identified by the distinctive webs they construct. For each web-spinner this section indicates web type and weaving patterns as well as web locations.

Life Cycle: This paragraph covers different aspects of the insect's or spider's life—where and when eggs are laid, activities of the young, courtship and mating of adults, and the interactions of all stages with predators and parasites.

Comments: Each account concludes with comments on the species' behavior, its impact on the environment, and its folklore. Closely related species are also briefly described, including their size and geographical range.

HOW TO USE THIS GUIDE

Example 1 You see a slender, elongate bug with
Bug on a leaf in red and green diagonal stripes and
an eastern garden yellow legs.

1. Turn to the Thumb Tab Guide
   preceding the color plates and look for
   the silhouette that best resembles the
   insect you have seen—the leafhopper
   silhouette in the group Hopperlike
   Insects, color plates 112–114.
2. Check the color plates. Three insects
   have the same basic shape, but only one
   has the right colors—the Scarlet-and-
   green Leafhopper. The caption gives its
   size and the text page 497.
3. The text confirms your identification:
   the Scarlet-and-green Leafhopper is in
   your range, often found in gardens.

Example 2 You catch a yellowish-tan, long-winged
Grasshopper on a grasshopper that has a blackish
southwestern plain pattern.

1. In the Thumb Tab Guide you find the
   silhouette for grasshoppers and crickets,
   color plates 247–281.
2. Turn to the color plates. Your
   grasshopper resembles 3 photographs—
   the Spur-throat, Alutacea Bird, and
   Southeastern Lubber, color plates 265–
   267. You eliminate the Southeastern
   Lubber, because it has short wings.
3. Checking the descriptions on pages 422
   and 425, you learn that only the Spur-
   throat bears a projection below its fore
   legs, which your grasshopper also has.

Key to the Color Plates

The color plates on the following pages are divided into 21 groups:

Caterpillars
Water Nymphs and Larvae
Aphidlike Insects
Earwigs and Silverfish
Diving Beetles and Water Bugs
Hopperlike Insects
Plant Bugs, Toad Bugs, and
Cockroaches
Weevils
Beetles
Grasshoppers, Crickets, and Cicadas
Mantids and Walkingsticks
Ants and Termites
Lacewings and Other Long-winged
Insects
Dragonflies and Damselflies
Flies
Bees, Wasps, and Kin
Moths
Butterflies
Ticks, Mites, and Scorpions
Spiders

Thumb Tab Guide: To help you find the correct group, a table of insect and spider silhouettes precedes the color plates. Each group is represented by a silhouette of a typical member of that group on the left side of the table. On the right, you will find the silhouettes of insects and spiders found within that group.

The representative silhouette for each group is repeated as a thumb tab at the left edge of each double page of color plates, providing a quick and convenient index to the color section.

Part I
Color Plates

Thumb Tab	Group	Plate Numbers
	Caterpillars	1–30
	Water Nymphs and Larvae	31–48
	Aphidlike Insects	49–66

Typical Shapes		Plate Numbers
	caterpillars	1–30
	mosquito, midge, and black fly larvae	31–33, 48
	diving beetle larvae	34, 44
	dragonfly and damselfly naiads	35, 39, 42–43
	stonefly and mayfly naiads	36–38, 40–41
	dobsonfly larvae	45
	caddisfly larvae	46–47
	scales	49–50
	mealybugs, cochineal bugs, woolly aphids, and psyllids	51–54

Typical Shapes		Plate Numbers
	aphids	55, 57, 59–63
	whiteflies and the Oak Lace Bug	56, 58
	spittlebugs, thrips, and entomobryid springtails	64–66
	lice, bed bugs, and the Sheep Ked	67, 70–71, 73, 78
	fleas	68–69, 76
	globular springtails	72
	thrips, snow scorpionflies, and podurid springtails	74–75, 77, 79–81
	proturans	82
	earwigs, forcipate diplurans, and timemas	83–84, 88–90

Thumb Tab	Group	Plate Numbers
	Earwigs and Silverfish (continued)	82–90
	Diving Beetles and Water Bugs	91–102
	Hopperlike Insects	103–114

Thumb Tab	Group	Plate Numbers
	Plant Bugs, Toad Bugs, and Cockroaches	115–129
	Weevils	130–138
	Beetles	139–246

Typical Shapes		Plate Numbers
	plant bugs, conenoses, and bee assassins	115–122
	Long-necked Seed Bug	123
	cockroaches	124–126
	ambush bugs	127
	toad bugs and brochymenas	128–129
	weevils	130–138
	ladybugs; cucumber, asparagus, fungus, hister, and other beetles	139, 145–151, 153, 182, 183, 199–200, 214–215, 243
	fireflies; bark, click, checkered, milkweed, and other beetles	140–144, 169–174, 179–180, 186, 210, 222–223, 230, 237, 244–246

Typical Shapes		Plate Numbers
	locust, laurel, and root borers; leather-wings; pine sawyers; longhorned, tiger, and other beetles	157–159, 161–162, 164, 166–168, 176–178, 181, 184, 193–194, 202, 207, 220–221, 229, 231–232, 238–239, 241, 245
	cabbage and stink bugs	152, 156, 175, 235–236
	darkling, ground, stag, June, Japanese, and other beetles	154–155, 163, 185, 187–192, 195–198, 201, 203–206, 208–209, 211–213, 216–219, 224–228, 233–234, 240, 242
	net-wings and the Lichen Moth	160, 163, 165
	grasshoppers, crickets, locusts, and shield-backed katydids	247–276, 281
	tree crickets, cone-heads, and katydids	277–280, 282–288
	cicadas	289–291

Typical Shapes		Plate Numbers
	lacewings, dobson-flies, alderflies, and snakeflies	328–334, 336
	caddisflies	335, 337–338
	stoneflies	339–342
	dragonflies	343–349, 358–365, 367–377
	damselflies	350–357, 366, 378
	mosquitoes, midges, gnats, and dance flies	379–381, 384–390, 410–411, 415, 419
	mayflies	391–396
	crane flies and the Stilt Bug	382–383, 404–408
	robber and march flies; bee killers, owlflies, and antlions	397–403, 412

Thumb Tab	Group	Plate Numbers
	Flies (continued)	379–441
	Bees, Wasps, and Kin	442–519

Typical Shapes		Plate Numbers
	house, fruit, deer, horse, black, bottle, and other flies	409, 413–414, 416–418, 420–441
	paper, digger, potter, mud dauber, and other wasps; wasplike flies, moths, and ants	442–443, 446–447, 451–452, 454–462, 466–467, 470, 473, 475–476, 479–480
	gall wasps, chalcids and braconids	448–449, 463 469, 484
	ichneumons and pelecinids	445, 453, 464–465
	hornets, yellow jackets, and hornetlike bees	444, 468, 471, 478, 483, 486
	horntails and sawflies	450, 472, 474, 477, 481
	bees and beelike flies	482, 485, 487–514, 517–519
	beeflies	515–516

Thumb Tab	Group	Plate Numbers
	Moths	520–573
	Butterflies	574–627

Typical Shapes		Plate Numbers
	underwings; tiger, gypsy, and other moths	520–521, 526–529, 531, 535–537, 540–546, 551, 553, 555–557, 559
	flour, meal, tuber, clothes, and yucca moths	522, 524–525, 530, 532–534, 552
	sphinx and plume moths	523, 538, 547–549, 554, 558, 565, 569–570
	giant silkworm and measuringworm moths	539, 550, 560–564, 566–568, 571–573
	skippers and the Snout Butterfly	574–575, 597, 606
	swallowtails	576, 578, 580, 583–584, 593–594
	hairstreaks and blues	577, 586–588, 595, 601, 609, 627
	Zebra, Monarch, Mourning Cloak, and other large butterflies	585, 591, 611, 614–615, 620–622, 624–625

Typical Shapes		Plate Numbers
	whites, sulphurs, coppers, metalmarks, crescents, satyrs, and other small butterflies	579, 581–582, 589–590, 592, 596, 598–600, 602–605, 607–608, 610, 612–613, 616–619, 623, 626
	ticks and mites	628–633
	scorpions and kin	634–639
	orb weavers; wolf, orchard, house, fishing, and jumping spiders; tarantulas and the Black Widow	640–656, 659–660, 663, 665–672, 675–677, 680–681, 686–693, 694–699
	crab spiders and goldenrod spiders	657–658, 661–662, 664, 678–679, 682–683
	spider beetles	684–685
	daddy-long-legs	673–674
	micrathenas and crablike orb weavers	700–702

The color plates on the following pages are numbered to correspond with the numbers preceding the text descriptions. The caption under each photograph gives the plate number, common name of the insect or spider, its size, and the page number of the text description. The measurement given is usually body length; however, for butterflies and moths, we indicate wingspan (w.). Where easily determined, the sex of the insect or spider—male (♂) or female (♀)—as well as social caste and any interesting behavior, such as mating, laying eggs, and stalking prey, are also given. The color plate number is repeated in front of each text description as a cross reference.

Caterpillars

Caterpillars are the young of butterflies and moths. Most are cylindrical with distinctive colors and patterns, often with many spines, horns, and bristly hairs. All caterpillars have 3 pairs of walking legs on the thorax and up to 5 pairs of leglike appendages on the abdomen.

1 Buckeye, *to* 2″, *p.* 749

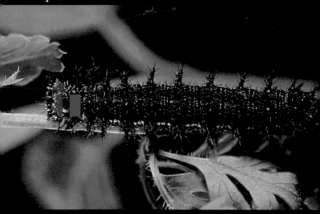

2 Mourning Cloak, *to* 2⅜″, *p.* 746

3 Gulf Fritillary, *to* 2″, *p.* 739

4 Acraea Moth, *to* 2½", *p.* 787

5 Woolly Bear, *to* 2⅛", *p.* 790

6 Milkweed Tiger Moth, *to* 1⅝", *p.* 788

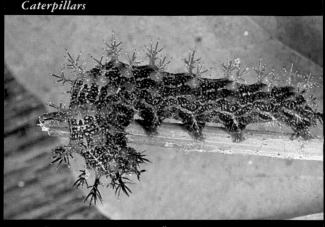

7 Question Mark, *to* 2½″, *p. 748*

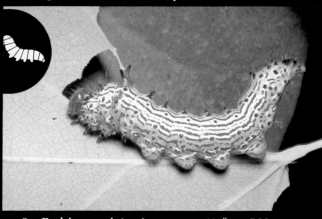

8 Red-humped Appleworm, *to* 1½″, *p. 785*

9 Silver-spotted Skipper, *to* 2″, *p. 716*

10 Bagworm, *to* 1⅝", *p.* 707

11 Painted Lady, *to* 2", *p.* 741

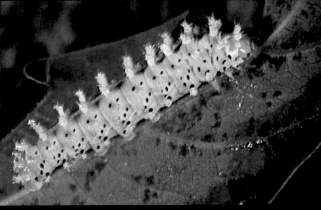

12 Cynthia Moth, *to* 4", *p.* 775

13   White-marked Tussock Moth, *to* 1¼″, *p.* 794

14   Large California Spanworm, *to* 2⅛″, *p.* 767

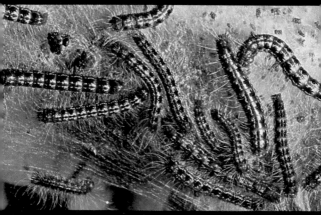

15   Tent Caterpillar, *to* 3″, *p.* 767

16 Monarch, *to* 2¾", *p.* 758

17 Queen, *to* 2½", *p.* 757

18 Eight-spotted Forester, *to* 1½", *p.* 796

19   Imperial Moth, *to* 5⅛", *p.* 773

20   Cecropia Moth, *to* 4⅜", *p.* 775

21   European Cabbage Butterfly, *to* 1⅜", *p.* 728

22 Eastern Tiger Swallowtail, *to* 2½″, *p. 721*

23 Polyphemus Moth, *to* 3½″, *p. 770*

24 Luna Moth, *to* 3⅛″, *p. 769*

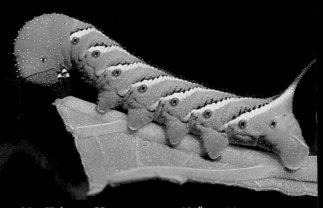

25   Tobacco Hornworm, to 3¾", *p. 781*

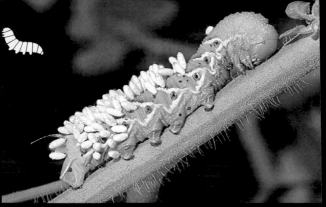

26   Tomato Hornworm, *with parasites, to 4", p. 780*

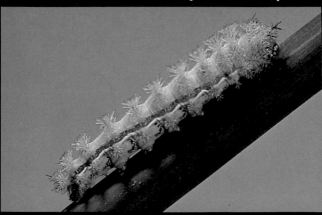

27   Io Moth, *to 3", p. 771*

28 Regal Moth, *to 5⅞"*, *p. 772*

29 Tentacled Prominent, *to 1¼"*, *p. 784*

30 Saddleback, *to 1"*, *p. 760*

These common, immature insects are aquatic. They do not resemble their adult forms and may be mistaken for other insect species. Some go through simple metamorphosis before changing to adults and are called naiads, such as the young of dragonflies, damselflies, mayflies, and stoneflies. Others, which have complete metamorphosis, are known as larvae; these include the young of mosquitoes, midges, black flies, diving beetles, dobsonflies, alderflies, and caddisflies.

31   Mosquito, *pupa, to ¼", p. 639*

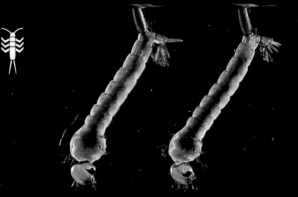

32   Mosquito, *larvae, to ¾", p. 639*

33   Phantom Midge, *larva, to ¾", p. 638*

34 Predacious Diving Beetle, *larva, to 2¾",* *p. 540*

35 Narrow-winged Damselfly, *naiad, to 1",* *p. 385*

36 Small Mayfly, *naiad, to ⅜",* *p. 356*

37   Small Mayfly, *naiad, to* ⅜″, *p. 356*

38   Stream Mayfly, *naiad, to* ⅝″, *p. 357*

39   Narrow-winged Damselfly, *naiad, to* 1″, *p. 385*

40 Common Stonefly, *naiad, to 1″, p. 413*

41 Common Stonefly, *naiad, to 1″, p. 413*

42 Common Skimmer, *naiad, to 1¼″, p. 369*

43   Darner, *naiad, to* 2½″, *p. 364*

44   Predacious Diving Beetle, *larva, to* 2¾″, *p. 540*

45   Dobsonfly, *larva, to* 3½″, *p. 521*

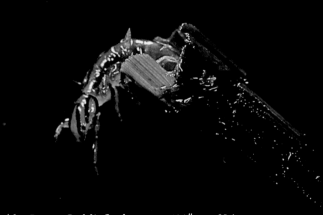

46 Large Caddisfly, *larva, to* 1⅛″, *p. 694*

47 Long-horned Caddisfly, *larvae, to* ½″, *p. 696*

48 Black Fly, *larvae, to* ¼″, *p. 646*

All of these small insects are pests on plants. Most species, such as aphids, whiteflies, psyllids, mealybugs, and cochineal bugs, produce a waxy, woolly, or powdery material that often covers part of their bodies. The tiny bodies of scale insects are completely hidden under crusty scales; nymphs of spittlebugs are surrounded by frothy bubbles. Other similar-shaped insects included in this group are thrips, springtails, and the Oak Lace Bug.

49 Oyster Shell Scale, ¹⁄₁₆–¹⁄₈″, *p. 510*

50 Cottony Cushion Scale, ¹⁄₈–¹⁄₄″, *p. 508*

51 Woolly Apple Aphid, ¹⁄₁₆″, *p. 507*

52   Cochineal Bug, ¹⁄₁₆–¹⁄₈″, *p. 512*

53   American Alder Psyllid, ¹⁄₁₆–¹⁄₈″, *p. 500*

54   Long-tailed Mealybug, ¹⁄₁₆–¹⁄₈″, *p. 511*

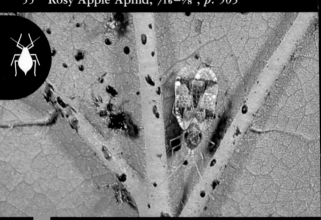

55 Rosy Apple Aphid, ¹⁄₁₆–¹⁄₈″, *p. 503*

56 Oak Lace Bug, ¹⁄₈″, *p. 477*

57 Cloudy-winged Cottonwood Aphid, ¹⁄₈″, *p. 505*

58　Greenhouse Whitefly, ¹⁄₁₆″, *p. 501*

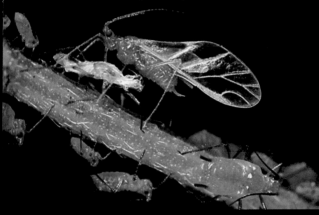

59　Green Apple Aphid, ¹⁄₁₆–¹⁄₈″, *p. 502*

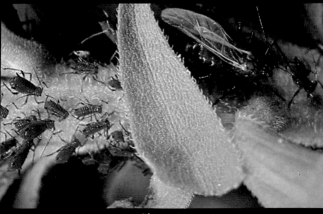

60　Rosy Apple Aphid, ¹⁄₁₆–¹⁄₈″, *p. 503*

61   Root Aphid, *with ants,* 1/8″, *p. 505*

62   Giant Willow Aphid, 1/8–1/4″, *p. 506*

63   Aphid, *giving birth,* 1/8″, *p. 504*

64　Ainsley's Springtail, ⅛", *p. 345*

65　Meadow Spittlebug, *nymph*, ⅜", *p. 495*

66　Common Thrips, ¹⁄₁₆", *p. 517*

Flealike Insects

This group includes several parasites of animals and people, such as fleas, lice, and bed bugs, as well as a few insects that look like fleas—thrips, snow scorpionflies, and snow fleas and other springtails.

67 Bed Bug, ⅛–¼″, *p. 469*

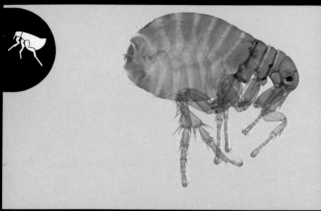

68 Human Flea, ¹⁄₁₆″, *p. 630*

69 Oriental Rat Flea, ¹⁄₁₆–⅛″, *p. 631*

70   Human Body Louse, ⅛″, *p. 457*

71   Hog Louse, ¼″, *p. 456*

72   Globular Springtail, ¹⁄₁₆″, *p. 346*

73 Sheep Ked, ¼″, *p. 683*

74 Snow Scorpionfly, ⅛″, *p. 626*

75 Banded-wing Thrips, ¹⁄₁₆″, *p. 516*

76 Cat Flea, 1/16″, *p. 630*

77 Snow Flea, 1/16″, *p. 344*

78 Bird Louse, 1/16″, *p. 454*

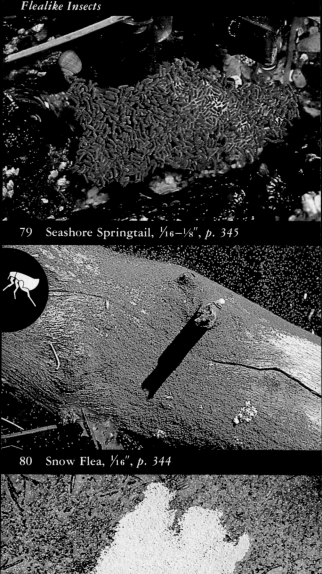

79 Seashore Springtail, ¹/₁₆–¹/₈″, *p. 345*

80 Snow Flea, ¹/₁₆″, *p. 344*

81 Springtail, ¹/₁₆–¹/₄″, *p. 343*

Earwigs and Silverfish

Several of these small insects are familiar pests. Silverfish, firebrats, and jumping bristletails have 2 or more tail-like appendages at the tip of the abdomen. Earwigs, timemas, and forcipate diplurans have pincers on the abdomen instead of tails. All have elongate bodies composed of many visible segments. This group also includes the most primitive insects—proturans.

82 Proturan, ¹⁄₁₆″, *p. 341*

83 Forcipate Dipluran, ⅛–¼″, *p. 348*

84 California Timema, ♀ *carrying* ♂, ½–⅞″, *p. 448*

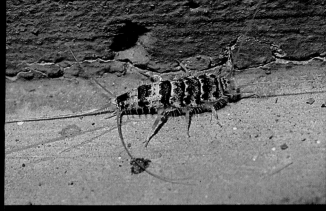

85 Firebrat, 3⁄8–1⁄2″, *p. 353*

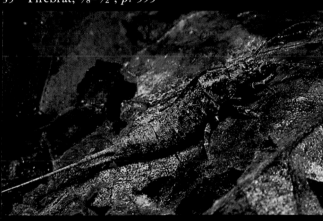

86 Jumping Bristletail, 5⁄8″, *p. 350*

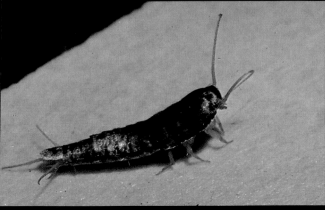

87 Silverfish, 3⁄8–1⁄2″, *p. 352*

88   European Earwig, ⅜–⅝″, *p. 409*

89   Ring-legged Earwig, ⅜–1″, *p. 408*

90   Riparian Earwig, ¾–1″, *p. 409*

Diving Beetles and Water Bugs

All of these insects live in water—ponds, rivers, lakes, streams, puddles, and sometimes even birdbaths. When swimming, all use their flattened, paddle-shaped hind legs like oars. Some are predators, others feed on plants. This group includes predacious diving beetles, whirligig beetles, water scavenger beetles, water bugs, water boatmen, and backswimmers.

91   Small Whirligig Beetle, ⅛–¼″, *p. 544*

92   Large Whirligig Beetle, ⅜–⅝″, *p. 543*

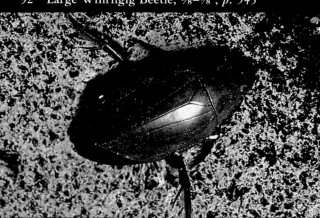

93   Giant Water Scavenger Beetle, 1⅜–1½″, *p. 545*

94　Small Flat Diving Beetle, ½", *p. 541*

95　Large Diving Beetle, 1–1⅝", *p. 541*

96　Marbled Diving Beetle, ⅜–⅝", *p. 542*

97 Eastern Toe-biter, 1¾–2⅜″, *p. 464*

98 Common Backswimmer, ⅜–½″, *p. 462*

99 Kirby's Backswimmer, ⅜–½″, *p. 461*

100 Water Boatman, ¼–½″, *p. 460*

101 Giant Water Bug, 1¾–2⅜″, *p. 464*

102 Ferocious Water Bug ♂, *with eggs,* 1¼″, *p. 463*

Hopperlike Insects

This group includes treehoppers, leafhoppers, planthoppers, stink bugs, some plant bugs, and the Wheel Bug. Many of these insects hop from plant to plant. Several have a conspicuous projection on their bodies—a hump, horn, or wheel-like crest. All are relatively small.

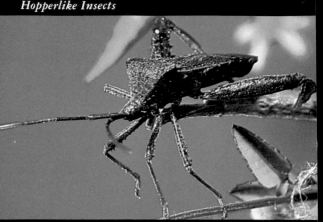

103   Florida Leaf-footed Bug, ¾–⅞″, *p. 481*

104   Squash Bug, ⅝″, *p. 482*

105   Wheel Bug, 1⅛–1⅜″, *p. 474*

06 Locust Treehopper, 3/8–1/2", *p. 494*

07 Thorn-mimic Treehopper, 1/4", *p. 493*

08 Partridge Scolops, 1/4–3/8", *p. 499*

109   Oak Treehopper, ⅜", *p. 493*

110   Buffalo Treehopper, ⅜", *p. 494*

111   Green Stink Bug, ½–¾", *p. 484*

112 **Scarlet-and-green Leafhopper,** ⅜″, *p. 497*

113 **Grape Leafhopper,** ¹⁄₁₆–⅛″, *p. 497*

114 **Sharpshooter,** ⅜″, *p. 498*

Plant Bugs, Toad Bugs, and Cockroaches

Most of these insects are found on plants, although most cockroaches are usually seen indoors. Some are brightly colored with patterns that often emphasize the way the fore wings overlap flat over the body at rest. Included here are a few strange-looking creatures, such as ambush bugs, toad bugs, and brochymenas, which blend so well with their environment that they are hard to see.

115   Large Milkweed Bug, ⅜–⅝″, *p. 479*

116   Small Eastern Milkweed Bug, ⅜–½″, *p. 478*

117   Eastern Boxelder Bug, ⅜–⅝″, *p. 483*

118 Bee Assassin, ½–⅝″, p. 473

119 Bee Assassin, ½–⅝″, p. 473

120 Eastern Blood-sucking Conenose, ¾″, p. 474

121 Scarlet Plant Bug, ¼–⅜″, *p. 471*

122 Adelphocoris Plant Bug, ¼–⅜″, *p. 470*

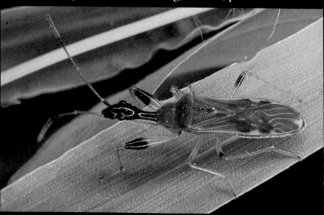

123 Long-necked Seed Bug, ⅜″, *p. 478*

124　German Cockroach, ½–⅝″, *p. 394*

125　American Cockroach, 1½–2″, *p. 393*

126　Oriental Cockroach, 1–1⅜″, *p. 392*

127   Jagged Ambush Bug, ⅜–½″, *p. 476*

128   Brochymena, ½–⅝″, *p. 484*

129   Toad Bug, *with prey*, ⅜″, *p. 467*

Weevils

 Weevils are easily recognized by their downcurved snouts, which they use to bore into plants, seeds, and fruit. Some are serious pests, such as the Boll Weevil, which once decimated cotton crops in the South. Several of these small beetles are dark, others range from tan to green or red, and some are coated with hair or hairlike scales. All have hard bodies, usually with grooves and pit marks.

130 Agave Billbug, ⅜–¾″, *p. 617*

131 Boll Weevil, ⅛–¼″, *p. 613*

132 Pine Weevil, ⅛–¼″, *p. 615*

133   Alfalfa Weevil, ⅛–¼″, *p. 615*

134   Black Oak Acorn Weevil, ⅜″, *p. 614*

135   Bean Weevil, ¹⁄₁₆–⅛″, *p. 602*

136   Lesser Cloverleaf Weevil, ⅛″, *p. 614*

137   Stored-grain Billbug, ⅛″, *p. 617*

138   Rose Weevil, ¼″, *p. 616*

Beetles

Beetles, unlike plant bugs, have hard, armorlike fore wings that usually meet in a straight line down the middle of the back. This large and amazingly varied group contains many colorful insects with striking iridescent hues or bold dots, stripes, and patterns. The vast majority, however, are somber shades of brown and black. They range from round to elongate, and several have unique shapes, such as the antlered stag beetles or the bumpy Horned Fungus Beetle. A few beetlelike plant bugs and a beetle mimic, the Lichen Moth, are also included here.

139 Rough Fungus Beetle, ½–⅝″, *p. 579*

140 Red Flat Bark Beetle, ⅜–½″, *p. 578*

141 Elegant Checkered Beetle, ⅜″, *p. 575*

42 Red-blue Checkered Beetle, ⅜″, *p. 577*

143 Red Milkweed Beetle, *mating*, ⅜–½″, *p. 600*

44 California Checkered Beetle, *mating*, ⅜″, *p. 575*

145 Spotless "Nine-spotted" Ladybug, ¼", *p. 581*

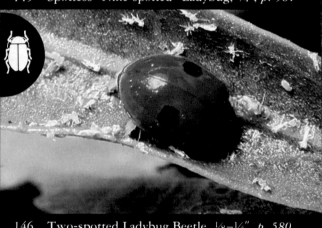

146 Two-spotted Ladybug Beetle, ⅛–¼", *p. 580*

147 Nine-spotted Ladybug Beetle, ¼", *p. 581*

148   Convergent Ladybug Beetle, ¼–⅜″, *p. 582*

149   Spotted Asparagus Beetle, ¼–⅜″, *p. 606*

150   Willow Leaf Beetle, ¼–⅜″, *p. 606*

151 Swamp Milkweed Leaf Beetle, ⅜–½″, *p. 609*

152 Harlequin Cabbage Bug, ⅜″, *p. 486*

153 Spotted Cucumber Beetle, ¼″, *p. 607*

54 Three-lined Potato Beetle, ¼″, *p. 609*

55 Girdled Leaf Beetle, ¹⁄₁₆–¼″, *p. 611*

56 Four-lined Plant Bug, ⅛″, *p. 472*

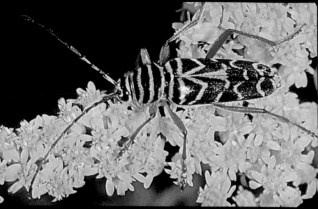

157 Locust Borer, ½–¾″, *p. 594*

158 Long-jawed Longhorn, ¾–1¼″, *p. 591*

159 Notch-tipped Flower Longhorn, ⅜–½″, *p. 601*

50   Golden Net-wing, ¼–⅜″, *p. 571*

61   Striped Blister Beetle, ½–⅝″, *p. 588*

62   Pennsylvania Leather-wing, ⅜–½″, *p. 569*

163   Banded Net-wing, *mating,* ⅜–¾", *p. 570*

164   Elder Borer, ⅝–1", *p. 592*

165   Lichen Moth, ⅜", *p. 790*

166 Bombardier Beetle, ¼–⅝″, *p. 537*

167 Arizona Blister Beetle, ⅝–1⅛″, *p. 588*

168 Downy Leather-wing, ⅜–½″, *p. 569*

169   Cottonwood Twig Borer, ⅜–½″, *p. 596*

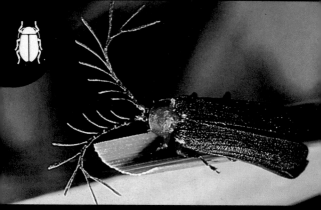

170   Fire Beetle, ⅜–½″, *p. 587*

172   Pennsylvania Firefly, ⅜–⅝″, *p. 568*

173   Pyralis Firefly, ⅜–½″, *p. 567*

174   American Carrion Beetle, ⅝–⅞″, *p. 550*

175 Two-spotted Stink Bug, *mating*, ¼–⅜″, *p. 485*

176 Tomentose Burying Beetle, ½–⅝″, *p. 550*

177 Margined Burying Beetle, ¾–1⅛″, *p. 549*

78   Ladybug Beetle, *larva*, ¹⁄₁₆–³⁄₈″, *p. 579*

79   Western Pine Borer, 1–1¹⁄₈″, *p. 563*

80   Ribbed Pine Borer, ¹⁄₂–³⁄₄″, *p. 598*

181   Cottonwood Borer, ⅞–1⅝″, *p. 597*

182   Dogwood Calligrapha, *mating,* ⅜″, *p. 603*

183   Ashy Gray Ladybug Beetle, ⅛–¼″, *p. 582*

184 California Laurel Borer, 1–1½″, *p. 599*

185 Ironclad Beetle, ¾–1″, *p. 586*

186 Eastern Eyed Click Beetle, 1–1¾″, *p. 565*

187 Agassiz's Flat-horned Stag Beetle, ⅜″, *p. 553*

188 Black-horned Pine Borer, ⅜–½″, *p. 590*

189 Rugose Stag Beetle ♂, ⅜–¾″, *p. 554*

190 Yellow Mealworm Beetle, ½–⅝″, p. 585

191 Common Black Ground Beetle, ½–⅝″, p. 539

192 Patent-leather Beetle, 1¼–1⅜″, p. 555

193   Black Pine Sawyer, ⅝–1″, *p. 595*

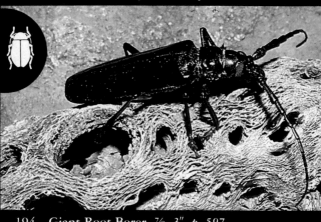

194   Giant Root Borer, ⅞–3″, *p. 597*

195   Dejean's Flightless Tiger Beetle, ⅝–⅞″, *p. 536*

196 European Ground Beetle, ¾–1″, *p. 538*

197 Broad-necked Darkling Beetle, ⅞–1″, *p. 584*

198 Waterlily Leaf Beetle, ¼–½″, *p. 608*

199   Hister Beetle, ⅛–¼″, *p. 546*

200   Tumblebug, ⅜–¾″, *p. 556*

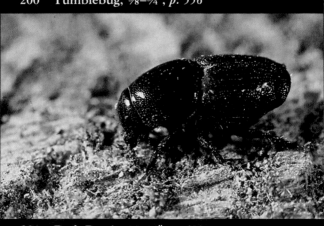

201   Bark Beetle, ⅛–¼″, *p. 620*

202 Short-winged Blister Beetle, ½–⅝", *p. 589*

203 Boat-backed Ground Beetle, ⅝–⅞", *p. 540*

204 Glossy Pillbug, ½–¾", *p. 559*

205 Milkweed Leaf Beetle, ⅜″, *p. 605*

206 Green Pubescent Ground Beetle, ⅜–⅝″, *p. 539*

207 Six-spotted Green Tiger Beetle, ½–⅝″, *p. 535*

208 Green June Beetle, ¾–⅞", *p. 557*

209 Fiery Searcher, 1–1⅜", *p. 537*

211 Dogbane Leaf Beetle, ⅜″, *p. 605*

212 Japanese Beetle, ⅜–½″, *p. 561*

213 Red Turpentine Beetle, ¼–⅜″, *p. 619*

214 Clavate Tortoise Beetle, ¼″, *p. 607*

215 Milkweed Tortoise Beetle, ⅜–½″, *p. 604*

216 Engraver Beetle, ⅛–¼″, *p. 619*

217 Elephant Stag Beetle ♂, 1¾″–2⅜″, *p. 552*

218 Reddish-brown Stag Beetle ♂, ⅞–1⅜″, *p. 553*

219 May Beetle, ¾–1⅜″, *p. 560*

220   Giant Root Borer, ⅞–3″, *p. 597*

221   Pine Sawyer, 1⅝–2¼″, *p. 593*

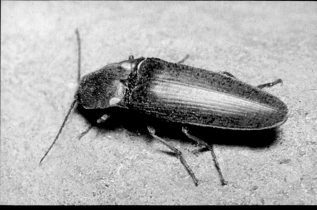

222   Fire Beetle, ¾–1″, *p. 566*

223 Northern Carrion Beetle, ½″, *p. 551*

224 Plicate Beetle, ½″, *p. 585*

225 Horned Fungus Beetle, ⅜–½″, *p. 583*

226 Apple Twig Borer, ¼–⅜″, *p. 573*

227 Gold-and-brown Rove Beetle, ½–¾″, *p. 547*

228 Brown Fruit Chafer, ½–⅝″, *p. 558*

229 Twig Pruner, ⅜–⅝″, *p. 592*

230 Slender Checkered Beetle, ¼–⅜″, *p. 576*

231 Golden-haired Flower Longhorn, ⅜–⅝″, *p. 593*

232 Yellow Douglas Fir Borer, ¾–1″, *p. 591*

233 Grapevine Beetle, ¾–1″, *p. 559*

234 Goldsmith Beetle, ¾–1″, *p. 557*

235 Tarnished Plant Bug, ¼″, *p. 472*

236 Spined Soldier Bug, ⅜–½″, *p. 486*

237 Divergent Metallic Wood Borer, ⅝–⅞″, *p. 564*

238 Beautiful Tiger Beetle, ⅝–¾″, *p. 534*

239 Dainty Tiger Beetle, *mating*, ⅜–½″, *p. 535*

240 Pictured Rove Beetle, ⅝–¾″, *p. 548*

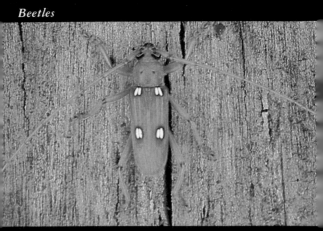

241 Ivory-marked Beetle, ½–¾″, *p. 599*

242 Ten-lined June Beetle, 1–1⅜″, *p. 561*

243 Colorado Potato Beetle, ¼–⅜″, *p. 610*

244 Willow Borer, ½–⅝″, *p. 601*

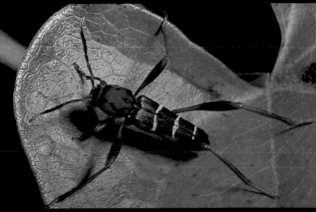

245 Cylindrical Hardwood Borer, ¼–¾″, *p. 596*

246 Oak Timberworm Beetle, ¼–1⅛″, *p. 612*

Grasshoppers, Crickets, and Cicadas

Most of these familiar summer insects are known for the musical trilling of the males. Grasshoppers, locusts, crickets, and katydids have conspicuous, large hind legs that are used for jumping. Their big, flat heads have huge eyes and large chewing mouthparts. Although cicadas are not related to grasshoppers and do not have jumping legs, they too are known for their loud buzzing sounds. Many of these insects are green, brown, or black but some have bright patterns. They live in a wide variety of habitats, ranging from meadows, prairies, and deserts to trees, caves, and houses.

247 Jerusalem Cricket, 1⅛–2″, *p. 437*

248 Spotted Camel Cricket, ⅜–¾″, *p. 436*

249 Secret Cave Cricket ♀, ⅜–1″, *p. 437*

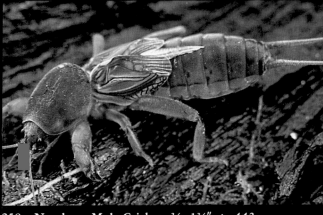

250 Northern Mole Cricket, ¾–1⅜″, *p. 443*

251 House Cricket ♀, ⅝–⅞″, *p. 439*

252 Field Cricket, ⅝–1″, *p. 439*

253 Aztec Pygmy Grasshopper, ¼–½″, *p. 416*

254 Keeled Shield-back Katydid ♀, ¾–1½″, *p. 432*

255 Short-legged Shield-back Katydid ♀, 2″, *p. 430*

256 Dragon Lubber Grasshopper, ¾–1¾″, *p. 420*

257 Mormon Cricket ♀, 1–2⅜″, *p. 430*

258 Mescalero Shield-back Katydid ♀, ¾–2″, *p. 43*

259 Painted Grasshopper, ¾–1⅜″, *p. 418*

260 Three-banded Grasshopper, 1–1¾″, *p. 420*

261 Pallid-winged Grasshopper, 1¼–1⅝″, *p. 427*

262 Lubber Grasshopper, *mating,* 1½–3⅛″, *p. 418*

263 American Bird Grasshopper, 1⅝–2⅛″, *p. 425*

264 Differential Grasshopper, 1⅛–1¾″, *p. 421*

265 Spur-throated Grasshopper, 1–1⅜″, *p. 422*

266 Alutacea Bird Grasshopper, 1⅛–1¾″, *p. 425*

267 Southeastern Lubber Grasshopper, 2–2¾″, *p. 42*

268 Carolina Locust, 1⅜–2″, *p. 419*

269 Alutacea Bird Grasshopper, 1⅛–1¾″, *p. 425*

270 Green Valley Grasshopper, 1½–2¾″, *p. 426*

271   Red-legged Locust, ¾–1″, *p. 421*

272   Gladiator Katydid, 1⅜″, *p. 433*

273   Creosote Bush Grasshopper, ¾–1″, *p. 417*

274   Great Crested Grasshopper ♀, 1½–1¾″, p. 428

275   Horse Lubber Grasshopper, 1½–2½″, p. 427

276   Panther-spotted Grasshopper, 1–1⅜″, p. 423

277 California Tree Cricket, ½–⅝″, *p. 440*

278 Two-striped Grasshopper, 1⅛–2⅛″, *p. 423*

279 Black-horned Tree Cricket, ½″, *p. 441*

280 Snowy Tree Cricket, ½–⅝″, *p. 441*

281 Toothpick Grasshopper, *mating,* 1–1⅛″, *p. 426*

282 Nebraska Cone-head, ♀ 1⅛–1⅜″, *p. 433*

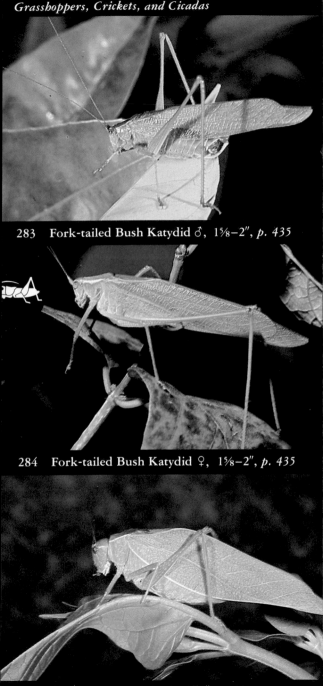

283 Fork-tailed Bush Katydid ♂, 1⅝–2″, *p. 435*

284 Fork-tailed Bush Katydid ♀, 1⅝–2″, *p. 435*

285 California Katydid, 1¾–2⅜″, *p. 431*

286   Angular-winged Katydid, 2–2½″, *p. 431*

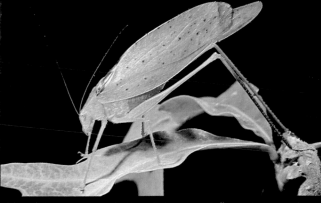

287   Oblong-winged Katydid, 1⅝–1¾″, *p. 429*

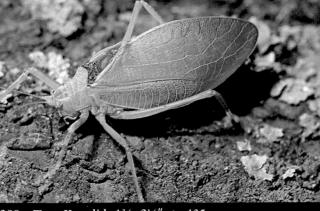

288   True Katydid, 1¾–2⅛″, *p. 435*

289 Grand Western Cicada, 1⅜–1⅝″, *p. 492*

290 Dogday Harvestfly, 1⅛–1¼″, *p. 491*

Mantids and Walkingsticks

These insects have long, slender bodies and long legs. The green or brown mantids and their look-alikes, mantidflies, have enlarged fore legs for grasping their prey. All are masters of camouflage. Walkingsticks resemble twigs, while water striders and waterscorpions resemble floating sticks as they drift across the water.

292 Common Water Strider, ½–⅝″, *p. 468*

293 Western Waterscorpion, 1″, *p. 466*

294 Brown Waterscorpion, 1⅜–1¾″, *p. 466*

295 Northern Walkingstick, 3–3¾″, p. 446

296 Giant Walkingstick, 3–5⅞″, p. 447

297 Northern Walkingstick, 3–3¾″, p. 446

298   Chinese Mantid, 2½–3⅜", *p. 398*

299   Praying Mantis ♀, 2–2½", *p. 397*

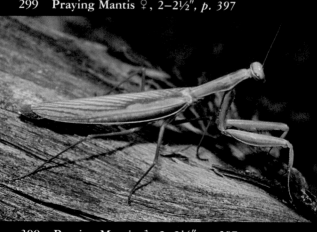

300   Praying Mantis ♂, 2–2½", *p. 397*

01 Obscure Ground Mantid, ⅝–1⅛″, *p. 396*

02 Carolina Mantid, 2⅜″, *p. 397*

03 Mantidfly, *with prey,* ¾–1″, *p. 527*

304 Mantidfly, ¾–1″, *p. 527*

305 Brown Mantidfly, ⅞″, *p. 526*

306 Brown Mantidfly, *mating*, ⅞″, *p. 526*

Ants and Termites

At first glance, ants and termites strongly resemble each other. Both are small, soft-bodied insects that live in colonies composed of queens, workers, soldiers, and mating males and females. Some have wings, while others are wingless. Ants are black, brown, or reddish and have a waistlike constriction between the base of the abdomen and the thorax. Termites are usually paler and do not have a "waist." Velvet ants, actually wasps, are densely covered with red, orange, yellow, or white hair. Two spiders that mimic ants are included in this group, along with a minute zorapteran that lives in rotting wood like termites but does not form colonies.

307   Zorapteran, ⅛″, *p. 404*

308   Pacific Coast Termite, *soldier,* ¾–1″, *p. 400*

309   Honey Ant, *repletes and workers,* ¹⁄₁₆–⅛″, *p. 830*

310 Texas Carpenter Ant, *workers,* ¾–1″, *p. 823*

311 Big-headed Ant, *worker,* ⅛″, *p. 829*

312 Leafcutting Ant, *worker,* ¹⁄₁₆–½″, *p. 822*

313 Fire Ant, *worker*, 1/16–1/4", p. 831

314 Crater-nest Ant, *workers*, 3/8–1/2", p. 825

315 Texas Shed-builder Ant, *workers*, 1/8", p. 825

316   Red Ant, *worker*, ⅛–½", *p. 826*

317   Rough Harvester Ant, *workers*, ¼–½", *p. 830*

318   Black Carpenter Ant, *worker*, ¼–½", *p. 824*

319 Ant-mimic Spider, ¼–⅜″, *p. 902*

320 Ant-mimic Jumping Spider, ⅛–¼″, *p. 912*

321 Little Black Ant, *workers,* ¹⁄₁₆″, *p. 827*

322 Spine-waisted Ant, *workers*, 3/8", *p. 822*

323 Arid Lands Honey Ant, *reproductives*, 5/8", *p. 828*

324 Subterranean Termite, *reproductives*, 1/4", *p. 401*

325 Cow Killer ♀, ⅝–1″, *p. 818*

326 Red Velvet-ant ♀, ⅞″, *p. 818*

327 Thistledown Velvet-ant ♀, ½–⅝″, *p. 817*

Lacewings and Other Long-winged Insects

These medium to large, green or brown insects have long, slender antennae and elongate wings, often with many visible veins and cross-veins. All are clumsy fliers that flutter around lights at night and rest inconspicuously on logs or the ground during the day. Dobsonflies, alderflies, snakeflies, lacewings, and caddisflies fold their wings rooflike over the body at rest. Stoneflies fold their wings flat, one on top of the other.

328 Brown Lacewing, 3/8–5/8″, *p. 528*

329 Fishfly, 3/4–1″, *p. 521*

330 Giant Lacewing, 1 1/8″, *p. 530*

331 **Alderfly**, ⅜″, *p. 520*

332 **Eastern Dobsonfly** ♂, 2″, *p. 522*

333 **Eastern Dobsonfly** ♀, 2″, *p. 522*

334   Texan Snakefly ♀, ⅝–⅞″, *p. 524*

335   Ash-winged Large Caddisfly, ¾″, *p. 694*

336   Green Lacewing, ⅜–⅝″, *p. 528*

37 Betten's Silverstreak Caddisfly, ¼–⅞", *p. 695*

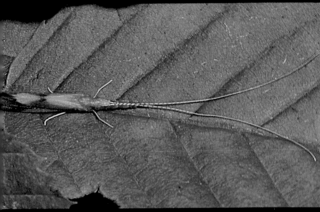

38 Tawny Brown-marked Longhorn, ¼–¾", *p. 696*

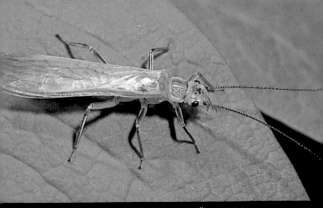

339 Green-winged Stonefly, ¼–⅝", *p. 413*

340 Giant Stonefly, 2½″, *p. 412*

341 Californian Pteronarcys, 1¼–1¾″, *p. 412*

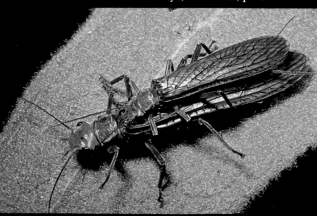

342 Californian Acroneuria, *mating*, ⅞–1¼″, *p. 413*

Dragonflies and Damselflies

Found near water, dragonflies and damselflies have 2 sets of long transparent wings, often iridescent bodies, and huge compound eyes. Unlike other insects, they can move their wings independently and fly both forward and backward with amazing speed. At rest dragonflies hold their wings outstretched whereas damselflies extend theirs vertically to the rear.

343   Green Darner, 2¾–3⅛″, *p. 364*

344   Western Mountain Gomphid, 2″, *p. 367*

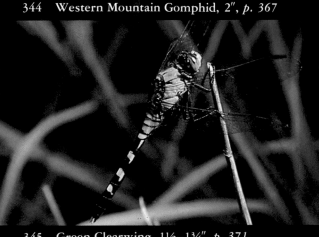

345   Green Clearwing, 1½–1¾″, *p. 371*

346 Green Darner, 2¾–3⅛", *p. 364*

347 White Tail, 1⅝–1⅞", *p. 378*

348 Twelve-spot Skimmer, 1¾–2¼", *p. 373*

349   Swift Long-winged Skimmer, 1⅛–1¾″, *p. 375*

350   Stocky Lestes, 1¼–1½″, *p. 385*

351   Violet Tail, 1⅛–1¼″, *p. 387*

52 Short-stalked Damselfly, *laying eggs,* 1¾", *p. 386*

53 Doubleday's Bluet, 1⅛–1¼", *p. 388*

54 Circumpolar Bluet, *with prey,* 1⅛–1½", *p. 388*

355   Red Bluet, 1–1¼″, *p. 389*

356   Southwestern Short Damselfly, 1–1⅛″, *p. 386*

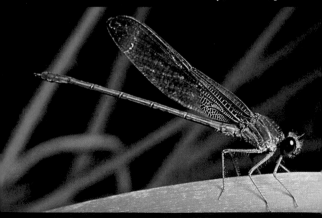

357   Ruby Spot, 1½–1¾″, *p. 383*

358 Half-banded Toper, 1⅝–1¾″, *p. 380*

359 Streak-winged Red Skimmer, 1⅝–1¾″, *p. 379*

360 Red Skimmer, 1⅛–2⅜″, *p. 374*

361 Elisa Skimmer, 1⅛–1⅜″, *p. 370*

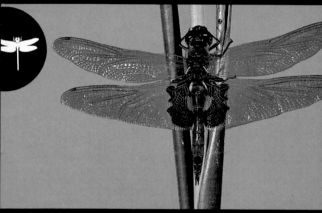

362 Red Saddlebag, 1¾–2¼″, *p. 381*

363 Jagged-edged Saddlebag, 1¾″–2¼″, *p. 380*

364 Brown-spotted Yellow-wing, 1⅜–1⅝″, *p. 370*

365 Widow, 1⅝–2″, *p. 372*

366 Black-winged Damselfly ♂, 1⅝–1¾″, *p. 382*

367 Small Western Gomphid, 2–2⅛″, *p. 366*

368 Western Flying Adder, 2¾–3⅜″, *p. 368*

369 Bluebell, ¾″, *p. 375*

370 Robust Pink Skimmer, 1⅝–1¾", *p. 378*

371 Black-faced Skimmer, 1⅝–1¾", *p. 372*

372 Climber Dragonfly, 1⅜–1⅞", *p. 369*

373 Four-spot Skimmer, 1⅝–1¾", *p.* 373

374 Low-flying Amber-wing, ¾–1", *p.* 377

375 Rusty Skimmer, 1⅞–2⅛", *p.* 376

376 Red Skimmer, 1⅛–2⅜″, *p. 374*

377 Brown Darner, 2¾–3⅛″, *p. 365*

378 Dark Lestes, 1¼–1⅝″, *p. 384*

Flies

Flies are found almost everywhere. They are easily distinguished from other flying insects by their single pair of functional wings. Most are small. Some, such as mosquitoes, gnats, and midges, have piercing mouthparts for drawing blood; others, such as the House Fly, have mouthparts adapted for sponging or lapping up food. Included in this group are some insects that resemble flies but belong to other orders, such as mayflies, owlflies, antlions, scorpionflies, and hangingflies. Unlike the true flies, these insects have 2 pairs of flying wings.

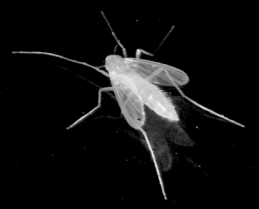

379   Green Midge, ¼–⅜″, *p. 646*

380   Golden Saltmarsh Mosquito ♀, ⅛–¼″, *p. 640*

381   Golden Saltmarsh Mosquito ♂, ⅛–¼″, *p. 640*

382 Giant Western Crane Fly, 1–1⅜″, p. 635

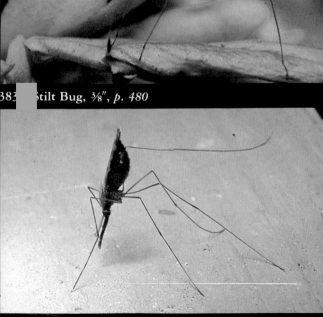

383 Stilt Bug, ⅜″, p. 480

384 Malaria-carrying Mosquito, ⅛″, p. 642

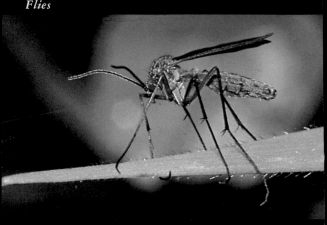

385 Fungus Gnat, ⅛–¼″, *p. 649*

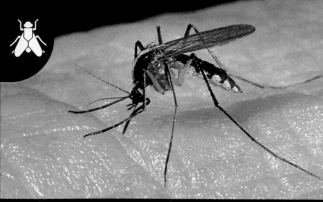

386 Summer Mosquito ♀, ⅛–¼″, *p. 639*

387 Tree-hole Mosquito ♀, ¼–⅜″, *p. 641*

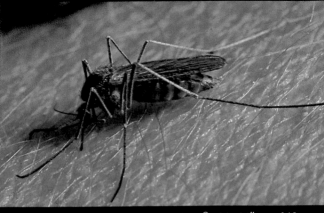

388 Unmarked Slender Mosquito ♀, ¼–½″, *p. 643*

389 Snow Mosquito ♀, ¼″, *p. 640*

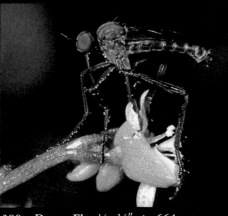

390 Dance Fly, ⅛–¼″, *p. 664*

391   Brown Stream Mayfly, *subimago*, ¼–⅝″, *p. 357*

392   Golden Mayfly,   ¾–1⅛″, *p. 361*

393   Coffin Fly,   ¾–⅞″, *p. 360*

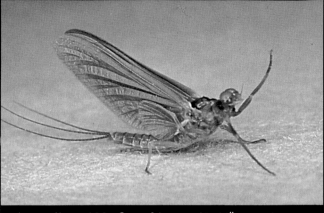

394 Midboreal Mayfly, *subimago*, ⅜–½″, p. 359

395 Early Brown Spinner, ⅜–½″, p. 358

396 Small Mayfly, ⅛–⅜″, p. 356

397　Flower-loving Fly, ⅛–¼″, *p. 656*

398　Bearded Robber Fly, ½–¾″, *p. 658*

399　Robber Fly, *mating*, ⅝–¾″, *p. 661*

400 Giant Robber Fly, *with prey,* 1–1⅛", *p. 660*

401 Bee Killer, *with prey,* ¾–1⅛", *p. 660*

402 Owlfly, 1⅜–1⅝", *p. 531*

403   Antlion, 1¾″, *p. 530*

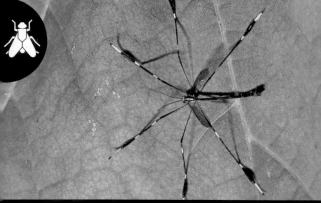

404   Phantom Crane Fly, ⅜–½″, *p. 637*

405   Green Stigma Hangingfly, 1″, *p. 628*

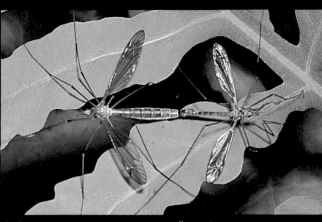

406   Crane Fly, *mating,* ⅜–2½″, *p. 635*

407   Crane Fly, ⅜–2½″, *p. 635*

408   Wood-boring Tipulid, ⅝–1″, *p. 634*

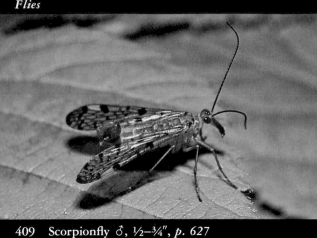

409 Scorpionfly ♂, ½–¾″, *p. 627*

410 Comstock's Net-winged Midge, ⅜″, *p. 638*

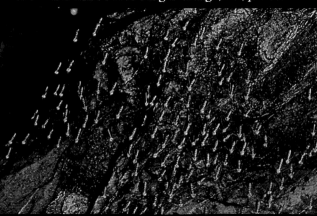

411 Comstock's Net-winged Midge, ⅜″, *p. 638*

412 March Fly, ⅜–½″, *p. 648*

413 Texan Long-legged Fly, ⅛–¼″, *p. 666*

414 Condylostylid Long-legged Fly, ⅙–¼″, *p. 666*

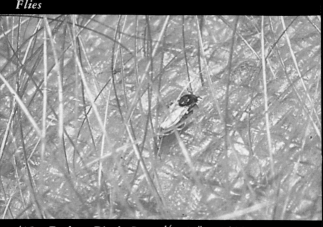

415 Bodega Black Gnat, ¹⁄₁₆–¹⁄₈″, *p. 644*

416 Black Fly, ¹⁄₁₆–¹⁄₈″, *p. 647*

417 Yellowstone Brine Fly, ¹⁄₈–¹⁄₄″, *p. 678*

418 Louse Fly, 1/16–3/8", *p. 683*

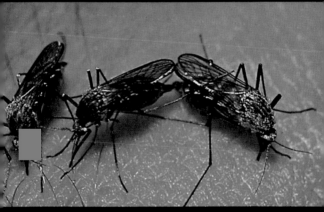

419 House Mosquito ♀, 1/8–1/4", *p. 643*

420 Californian Seaweed Fly, 1/8–1/4", *p. 676*

421   House Fly, ⅛–¼″, *p. 681*

422   Tachinid Fly, ⅛–½″, *p. 688*

423   Tachinid Fly ♀, *laying egg*, ⅛–½″, *p. 688*

424 Deer Fly, ⅜–⅝″, *p. 651*

425 Flesh Fly, ¼–½″, *p. 687*

426 Biting Stable Fly, ¼–⅜″, *p. 682*

427 Black Horse Fly, ¾–1⅛″, *p. 653*

428 Three-spot Horse Fly, ½–⅝″, *p. 653*

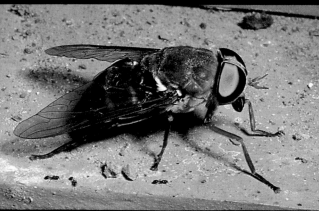

429 American Horse Fly, ¾–1⅛″, *p. 652*

430 Blue Bottle Fly, ½″, *p. 684*

431 Green Bottle Fly, ⅜–½″, *p. 686*

432 Screw-worm Fly, ½–⅝″, *p. 685*

433 Vinegar Fly, 1/16″, *p. 679*

434 Marsh Fly, 1/8–1/4″, *p. 677*

435 Apple Maggot Fly, 1/4″, *p. 675*

436 Walnut Husk Fly, ¼″, *p. 674*

437 Repetitive Tachinid Fly, ⅜–½″, *p. 690*

438 Tachina Fly, ¼–½″, *p. 691*

439   Beelike Tachinid Fly, ⅜–½″, *p. 689*

440   Dung Fly, ⅜″, *p. 680*

441   Pyrgotid Fly, ⅜–⅝″, *p. 673*

Bees, Wasps, and Kin

Many of these insects are brightly colored. Bees, hornets, and horntails have yellow bands. Most wasps and wasplike ichneumons are orange or orange and black. Others in this group have a solid metallic sheen. Bees and wasps have narrow waists, although some bees appear stocky because they are covered with fuzzy hair. Sawflies and horntails do not have waists. Many insects in this group can inflict a painful sting. Several beelike flies and wasplike moths are also included here.

442   Paper Wasp ♀, *workers*, ½–1″, *p. 834*

443   Doll's Clearwing Moth, ⅞″, *p. 711*

444   Giant Hornet, ¾–1⅛″, *p. 835*

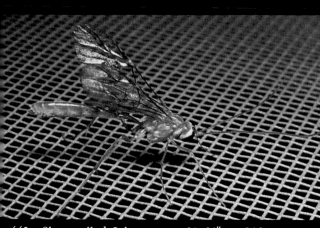

445 Short-tailed Ichneumon, ⅜–¾″, *p. 810*

446 Legionary Ant ♂, ¼–¾″, *p. 827*

447 Squash Vine Borer, ⅝″, *p. 710*

448 California Oak Gall Wasp, ⅛–¼″, *p. 813*

449 Live Oak Gall Wasp, ⅛–¼″, *p. 814*

450 Rusty Willow Sawfly, ¾″, *p. 802*

451 **Thick-headed Fly,** ⅜″, *p. 672*

452 **Florida Hunting Wasp,** ¾–⅞″, *p. 843*

453 **Red-tailed Ichneumon,** ⅝–¾″, *p. 810*

454 Elongate Aphid Fly, ⅜″, *p. 668*

455 Thread-waisted Wasp, ⅝–2⅛″, *p. 841*

456 Great Golden Digger Wasp, ⅝–⅞″, *p. 845*

457 **Digger Wasp,** ½–¾″, *p. 820*

458 **Tarantula Hawk** ♀, *stalking prey,* ½–¾″, *p. 839*

459 **Mydas Fly,** 1–1⅛″, *p. 657*

460   Blue-black Spider Wasp ♀, *with prey*, ¾″ p. 83

461   Steel-blue Cricket Hunter ♀, *with prey*, ⅝″, p. 8

462   Purplish-blue Cricket Hunter, 1–1⅛″, *p. 843*

463 California Torymus ♀, 1/8", *p. 811*

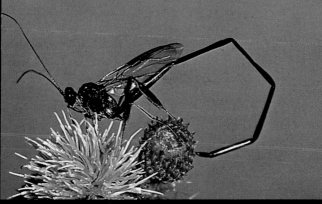

464 American Pelecinid ♀, 1¾–2", *p. 815*

465 Giant Ichneumon ♀, 1⅜–3", *p. 809*

466 Black-and-yellow Mud Dauber, 1–1⅛″, *p. 844*

467 Potter Wasp, ⅝–¾″, *p. 833*

468 Yellow-faced Bee, ¼″, *p. 851*

469 Golden-yellow Chalcid, ⅛–⅜″, *p. 812*

470 Polybiine Paper Wasp, ⅝″, *p. 834*

471 Western Cuckoo Bee, ⅜–½″, *p. 860*

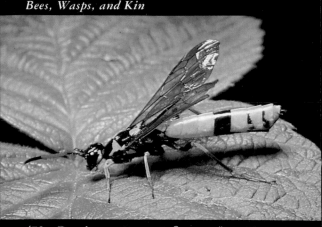

472   Raspberry Horntail ♀, ½–⅝″, *p. 806*

473   California Sycamore Borer, ¾″, *p. 711*

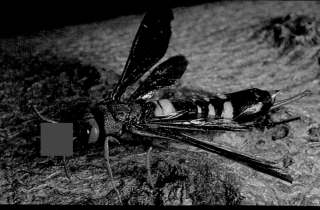

474   Pigeon Horntail ♀, 1–1½″, *p. 804*

475 Five-banded Tiphiid Wasp ♂, 1⅛–1⅜″, p. 816

476 Eastern Sand Wasp ♀, *with prey*, ½–⅝″, p. 841

477 Smoky Horntail ♀, 1–1⅜″, p. 805

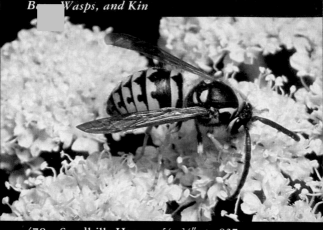

478   Sandhills Hornet, ⅝–¾″, *p. 837*

479   Cicada Killer ♀, *with prey,* 1⅛–1⅝″, *p. 845*

480   Scarab Hunter Wasp, 1–1⅛″, *p. 820*

481 Northeastern Sawfly, ⅜″, *p. 803*

482 Faithful Leafcutting Bee, ⅜–½″, *p. 857*

483 Bald-faced Hornet, ⅝–¾″, *p. 837*

484 Braconid Wasp, *emerging*, 1/16–1/8", *p. 807*

485 Fringe-legged Tachinid Fly, 1/4", *p. 691*

486 Yellow Jacket, *worker*, 1/2–5/8", *p. 836*

487 American Hover Fly, ⅜", *p. 670*

488 Toxomerus Hover Fly, ¼–½", *p. 671*

489 Gold-backed Snipe Fly, ⅜", *p. 654*

490   Western Leafcutting Bee, ⅜–½″, *p. 857*

491   Willow Mining Bee, ½″, *p. 855*

492   Plasterer Bee, ⅜–½″, *p. 851*

493 California Leafcutting Bee, ⅜–½″, *p. 859*

494 Clarkia Bee, ¼–⅜″, *p. 856*

495 Progressive Bee Fly, ¼–½″, *p. 663*

496 Bee Fly, ¼–⅜″, *p. 662*

497 California Carpenter Bee, ¾–1″, *p. 861*

498 Mason Bee, ⅜–½″, *p. 858*

499 Nevada Mining Bee, 3/8–1/2", *p. 855*

500 Early Tachinid Fly, 1/4–3/8", *p. 689*

501 Sacken's Bee Hunter, 5/8–7/8", *p. 659*

502　Digger Bee, ⅝″, *p. 859*

503　Mining Bee, ⅝″, *p. 854*

504　Woolly Bear Hover Fly, ½–⅝″, *p. 667*

505 Golden Northern Bumble Bee, ½–¾″, *p. 863*

506 Red-tailed Bumble Bee, ⅜–½″, *p. 864*

507 Bulb Fly, ⅜–½″, *p. 670*

508 Drone Fly, ⅝″, *p. 669*

509 Walnut Husk Fly, ¼″, *p. 674*

510 Alkali Bee, ⅜″, *p. 853*

511 Honey Bee, 3/8–5/8″, *p. 862*

512 Progressive Bee Fly, 1/4–1/2″, *p. 663*

513 Stiletto Fly, 3/8″, *p. 655*

514   Progressive Bee Fly, ¼–½″, *p. 663*

515   Large Bee Fly, ¼–½″, *p. 663*

516   Large Bee Fly, ¼–½″, *p. 663*

517 Virescent Green Metallic Bee ♀, ⅜–½″, *p. 852*

518 Augochlora Green Metallic Bee, ⅜″, *p. 853*

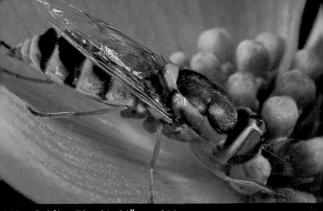

519 Soldier Fly, ⅜–½″, *p. 650*

Moths

Most moths are known for their destructive caterpillars, which often cause considerable damage to plants and stored foods. Like butterflies, moths have 4 delicate wings that are covered with scales. Most have somber colors and fly at night but some have bright patterns and are active during the day. At rest moths fold the wings rooflike over the body, curled around the body, or flat against a support.

520   Grape Leaf Skeletonizer, *w.* ¾–1″, *p.* 759

521   Eight-spotted Forester ♀, *w.* 1⅛–1¼″, *p.* 796

522   Fairy Moth, *w.* ½–⅝″, *p.* 705

523 Wild-cherry Sphinx ♂, *w.* 3–4½″, *p.* 783

524 Indian Meal Moth, *w.* ⅝″, *p.* 763

525 Mediterranean Flour Moth, *w.* ⅞″, *p.* 761

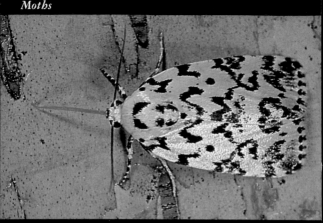

526 Hebrew, *w.* ⅞–1″, *p.* 799

527 Acraea Moth ♂, *w.* 1⅛–2″, *p.* 787

528 Gypsy Moth, *laying eggs, w.* 1⅛–2¾″, *p.* 793

529 Milkweed Tiger Moth ♀, *w.* 1–1¾", *p.* 788

530 Yucca Moth, *mating,* *w.* ¾–1", *p.* 705

531 Yellow Woolly Bear Moth ♂, *w.* 1½–2", *p.* 787

532   Sod Webworm Moth, *w.* ½–1½", *p. 762*

533   Potato Tuber Moth, *w.* ⅜–⅝", *p. 708*

534   Clothes Moth, *w.* ⅜–1⅜", *p. 706*

535 Orange Tortrix, *w.* ½–⅝″, *p. 713*

536 Fruit-tree Leaf Roller Moth, *w.* ¾″, *p. 712*

537 Meal Moth, *w.* ⅝–1″, *p. 763*

538 Artichoke Plume Moth, *w.* 3/4–1 1/8", *p.* 764

539 Large California Spanworm Moth, *w.* 1 3/4", *p.* 767

540 Spotted Tiger Moth ♂, *w.* 1 3/8–2", *p.* 789

541 Woolly Bear Caterpillar Moth ♀, *w. 2″, p. 790*

542 Codling Moth, *w. ⅝–¾″, p. 714*

543 California Cankerworm Moth ♂, *w. 1⅜″, p. 766*

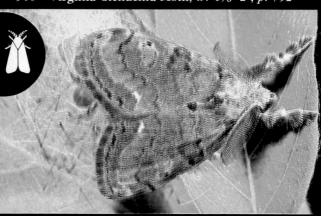

544 Virginia Ctenuchid Moth, *w.* 1⅜–2″, *p.* 792

545 White-marked Tussock Moth ♂, *w.* 1¼″, *p.* 794

546 Western Tent Caterpillar Moth ♀, *w.* 1⅝″, *p.* 768

547 Pandora Sphinx ♀, *w.* 3⅜–4″, *p.* 778

548 Big Poplar Sphinx ♂, *w.* 3½–5½″, *p.* 781

549 Virginia-creeper Sphinx ♀, *w.* 2–2¾″, *p.* 777

550   Imperial Moth ♂, *w.* 4–5⅞", *p.* 773

551   Rattlebox Moth ♀, *w.* 1¼–1¾", *p.* 791

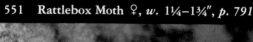

552   Ailanthus Webworm Moth, *w.* 1⅛", *p.* 709

553 Colona ♀, *w. 2–2¼", p. 789*

554 White-lined Sphinx ♂, *w. 2½–3½", p. 779*

555 Ornate Tiger Moth ♂, *w. 1⅛–1⅝", p. 786*

556   Locust Underwing ♂, *w.* 2½″, *p. 798*

557   Sweetheart Underwing ♂, *w.* 3–3⅜″, *p. 798*

558   Tobacco Hornworm Moth ♂, *w.* 3½–4½″, *p. 781*

559 **Alfalfa Looper** ♂, *w.* 1⅛–1⅝″, *p.* 797

560 **Tulip Tree Beauty** ♂, *w.* 1½–2″, *p.* 765

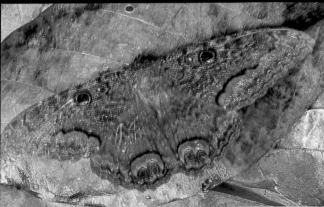

561 **Black Witch** ♂, *w.* 3½–5⅞″, *p.* 796

562 Cynthia Moth ♂, *w.* 3–5⅜", *p.* 775

563 Promethea Moth ♀, *w.* 2¾–4", *p.* 771

564 Cecropia Moth ♀, *w.* 4¾–5⅛", *p.* 775

565 Cerisy's Sphinx ♀, *w.* 2⅜–3⅜″, *p.* 782

566 Io Moth ♀, *w.* 2⅜–2¾″, *p.* 771

567 Polyphemus Moth *w.* 3½–5½″, *p.* 770

568   Regal Moth ♀, *w.* 4¾–5⅞″, *p.* 772

569   Hummingbird Moth ♂, *w.* 1½–2″, *p.* 778

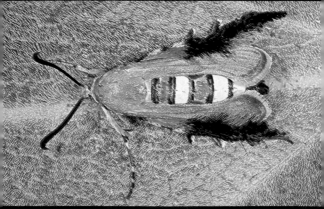

570   Manroot Borer, *w.* 1⅝–2⅜″, *p.* 710

571 Sheep Moth ♂, *w.* 2⅜–2¾″, *p. 774*

572 Rosy Maple Moth ♀, *w.* 1⅛–2″, *p. 773*

573 Luna Moth ♀, *w.* 3⅛–4½″, *p. 769*

Butterflies

Butterflies include many of our most beautiful and familiar insects. They fly only during the day and are often seen feeding and resting on flowers. Their wings are covered with scales like those of moths. But unlike moths, butterflies usually hold the wings together vertically over the body at rest.

574   Silver-spotted Skipper, *w.* 1¾–2″, *p. 716*

575   Long-tailed Skipper, *w.* 1⅝–2″, *p. 717*

576   Spicebush Swallowtail ♀, *w.* 4–4⅞″, *p. 722*

577 Spring Azure ♂, *w.* ¾–1¼″, *p. 730*

578 Green Swallowtail ♂, *w.* 3–4½″, *p. 718*

579 California Dog-Face ♂, *w.* 1¾–2″, *p. 726*

580 Eastern Tiger Swallowtail ♂, *w.* 4–5⅞", *p.* 721

581 Cloudless Sulphur, *w.* 2⅛–2¾", *p.* 726

582 European Cabbage Butterfly, *w.* 1⅛–2", *p.* 728

583　Eastern Black Swallowtail ♂, *w.* 2¾–3½″, *p.* 721

584　Giant Swallowtail ♀, *w.* 4–5½″, *p.* 719

585　Zebra, *w.* 3–3⅜″, *p.* 744

586   Eastern Tailed Blue ♂, *w.* ⅞–1⅛″, *p. 731*

587   Gray Hairstreak, *w.* 1⅛″, *p. 736*

588   Early Hairstreak, *w.* ¾–⅞″, *p. 731*

389 Great Southern White, *w.* 1¾–2⅜″, *p. 725*

390 Checkered White ♀, *w.* 1⅜–1⅝″, *p. 727*

391 Phoebus, *w.* 2⅜–3″, *p. 723*

592 Sara Orange Tip, *w.* 1–1¼″, *p. 724*

593 Zebra Swallowtail, *w.* 2½–4½″, *p. 719*

594 Pale Swallowtail ♂, *w.* 4–5⅜″, *p. 720*

595 Marine Blue, *mating, w.* ⅝–1⅛", *p.* 734

596 Painted Lady, *w.* 2–2¼", *p.* 741

597 Common Snout Butterfly, *w.* 1¾–2", *p.* 738

598   Eyed Brown, *w.* 1⅝–2″, *p. 755*

599   Little Wood Satyr, *w.* 1¾–2″, *p. 754*

600   Wood Nymph, *w.* 2–2¾″, *p. 753*

601 Silvery Blue ♀, *w. ⅞–1⅛", p. 733*

602 Pearly Eye, *w. 1⅝–2", p. 755*

603 Chryxus Arctic, *w. 1⅜–2½", p. 756*

604 Hackberry Butterfly, *w.* 1¾–2⅛″, *p.* 752

605 Brown Elfin, *w.* ⅞–1″, *p.* 734

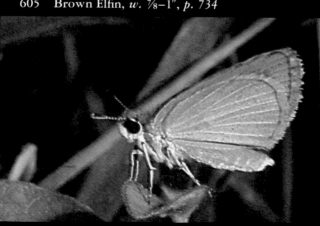

606 Least Skipper, *w.* ¾–⅞″, *p.* 716

607 Buckeye, *w.* 2⅛–2½", *p.* 749

608 White Peacock, *w.* 1¾–2¼", *p.* 740

609 Harvester, *w.* 1⅛–1¼", *p.* 732

610 Harris' Checkerspot, *w.* 1⅜–1¾″, *p. 741*

611 Great Spangled Fritillary, *w.* 3–3¾″, *p. 750*

612 Pearl Crescent, *w.* 1¼–1¾″, *p. 748*

613 Northern Metalmark, *w.* 1–1⅛″, *p. 737*

614 Gulf Fritillary, *w.* 2½–3⅛″, *p. 739*

615 Question Mark, *w.* 2½–2¾″, *p. 748*

616 American Copper ♂, *w.* 1″, *p.* 735

617 Milbert's Tortoise Shell, *w.* 1¾″, *p.* 747

Admiral, *w.* 2–2⅜″, *p.* 751

619 American Painted Lady, *w.* 1¾–2″, *p.* 742

620 Monarch, *w.* 3½–4″, *p* 758

621 Viceroy, *w.* 2½–2¾″, *p.* 745

622 Queen, *w.* 3⅛–3⅜″, *p.* 757

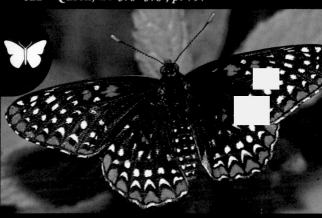

623 Baltimore, *w.* 1¾–2½″, *p.* 743

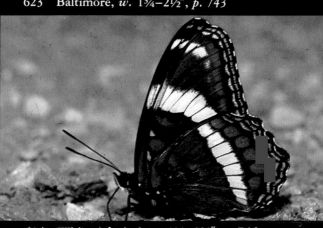

624 White Admiral, *w.* 3⅛–3⅜″, *p.* 746

625 Mourning Cloak, *w.* 2¾–3⅜", *p. 746*

626 Northern Metalmark, *w.* 1–1⅛", *p. 737*

627 Great Purple Hairstreak, *w.* 1¾", *p. 730*

Ticks, Mites, and Scorpions

This group includes a few examples of the most common spider relatives, some of which could be mistaken for insects. Ticks are external parasites of animals, reptiles, birds, and sometimes people. They are usually common in woods and farm areas. Most are brown. Mites are often more colorful, with red, yellow, green, or mottled bodies. They are found on land or in water, frequently on plants. Both ticks and mites are small. Scorpions and their kin are larger, up to 5″ long. Most have segmented shell-like bodies and large pincers. They are widespread in the South and Southwest.

628 Red Freshwater Mite, ⅛″, *p. 925*

629 Velvet Mite, ⅛″, *p. 926*

630 California Black-eyed Tick, ⅛″, *p. 929*

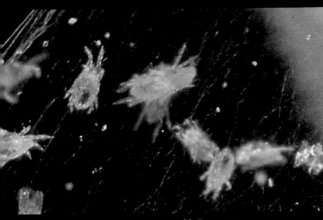

631 Two-spotted Spider Mite, ¹⁄₆₄″, *p. 924*

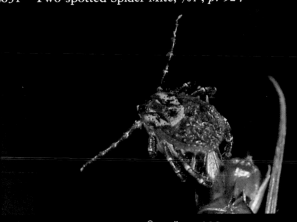

632 Eastern Wood Tick ♀, ⅛″, *p. 928*

633 Mammal Soft Tick ♀ and ♂, ¼″, *p. 927*

634 Giant Vinegarone ♀, *with young,* **3–3⅛″, p. 932**

635 Centruroides Scorpion ♀, *with young,* **2¾″, p. 914**

636 Giant Desert Hairy Scorpion, 5½″, p. 915

537 Chernetid, 1/16–1/8″, p. 918

538 Pale Windscorpion, 5/8–1¼″, p. 936

539 Side-spotted Tailless Whipscorpion, ¾″, p. 934

Spiders

Spiders are arachnids, not insects, and have 8 rather than 6 legs. Many spin intricate webs between plants, branches, or buildings but others do not spin webs. Daddy-long-legs are not spiders but because many people think of them as spiders, they are included here. All are predators. They live in a wide variety of habitats ranging from meadows, gardens, and woods to swamps, caves, and homes. Included with this group are 2 beetles that mimic spiders.

640   Ant-mimic Spider ♀, ¼–⅜″, *p. 902*

641   Lynx Spider ♀, ⅛–⅜″, *p. 900*

642   Nursery-web Spider ♀, *with young,* ⅝″, *p. 895*

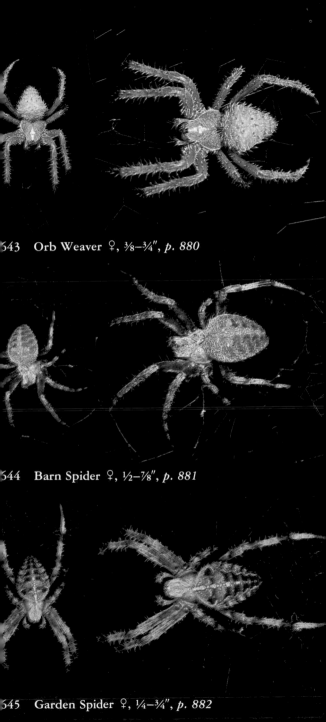

643 Orb Weaver ♀, ⅜–¾″, *p. 880*

644 Barn Spider ♀, ½–⅞″, *p. 881*

645 Garden Spider ♀, ¼–¾″, *p. 882*

646   Burrowing Wolf Spider, ½–⅞″, *p. 896*

647   Desert Tarantula, 2–2¾″, *p. 870*

648   Carolina Wolf Spider, ¾–1⅜″, *p. 897*

649 Turret Spider ♂, ½–⅝″, *p. 868*

650 California Trapdoor Spider ♂, ¾–1″, *p. 871*

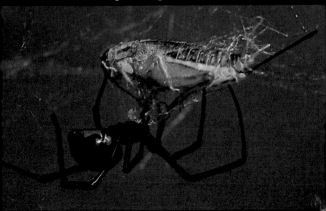

651 Black Widow Spider ♀, *with prey,* ⅜″, *p. 877*

652 Daring Jumping Spider, ¼–⅝″, *p. 912*

653 Dimorphic Jumping Spider, ¼–⅜″, *p. 911*

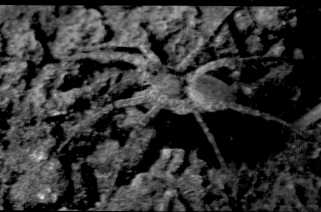

654 Thin-legged Wolf Spider, ⅛–⅜″, *p. 899*

555 Dimorphic Jumping Spider ♂, ¼″, p. 911

556 Metaphid Jumping Spider, ⅛–¼″, p. 911

658   Golden Huntsman Spider, ½–⅞″, *p. 905*

659   Desert Loxosceles, ¼–⅜″, *p. 875*

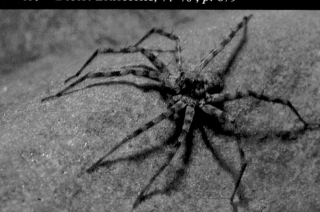

660   Thin-legged Wolf Spider, ⅛–⅜″, *p. 899*

661 Selenopid Crab Spider ♀, *with egg sac*, ⅝", *p. 905*

662 Selenopid Crab Spider, ⅜–⅝", *p. 905*

663 Brownish-gray Fishing Spider, ¼–1", *p. 894*

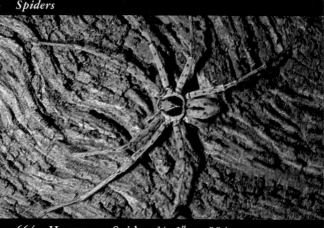

664   Huntsman Spider, ¾–1″, *p. 904*

665   Forest Wolf Spider ♂, ⅜″, *p. 898*

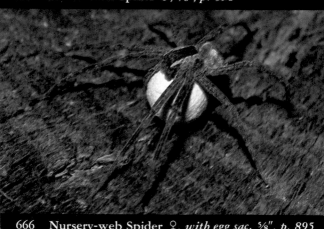

666   Nursery-web Spider ♀, *with egg sac,* ⅝″, *p. 895*

667 Rabid Wolf Spider, ½–⅞", *p. 898*

668 Six-spotted Fishing Spider, ⅜–¾", *p. 895*

669 Hammock Spider, ¼", *p. 879*

670 Grass Spider, ⅝–¾″, *p. 893*

671 Violin Spider, ¼–⅜″, *p. 875*

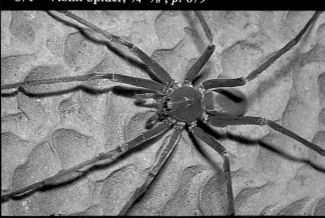

672 Wandering Spider, ¼–1″, *p. 903*

673 **Brown Daddy-long-legs,** ⅛–¼″, *p. 921*

674 **Eastern Daddy-long-legs,** ¼–⅜″, *p. 920*

675 **Spitting Spider,** ⅛–⅜″, *p. 874*

676 **Green Lyssomanes**, ¼–⅜″, *p. 910*

677 **Green Lynx Spider**, ½–⅝″, *p. 901*

678 **Goldenrod Spider** ♀, ¼–⅜″, *p. 907*

679 Goldenrod Spider ♀, ¼–⅜″, *p. 907*

680 Six-spotted Orb Weaver, ⅛–⅜″, *p. 884*

681 Green Lynx Spider, ½–⅝″, *p. 901*

682 Elegant Crab Spider ♀, ⅜″, *p. 907*

683 Thrice-banded Crab Spider ♀, ⅛–¼″, *p. 908*

684 Spider Beetle, 1/16″, *p. 572*

85 Texan Spider Beetle, ⅛″, *p. 573*

86 Bola Spider ♀, ½″, *p. 887*

87 Branch-tip Spider ♀ and ♂, ⅟₁₆–¼″, *p. 873*

688   Black Widow Spider, *immature*, ⅛–⅜″, *p. 877*

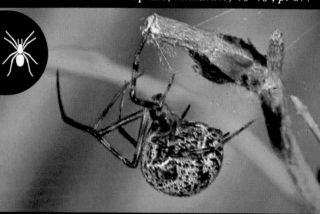

689   American House Spider ♀, ¼″, *p. 876*

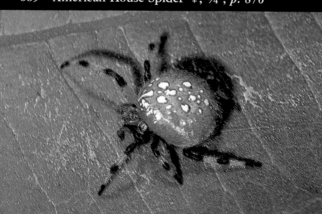

690   Shamrock Spider ♀, ⅜–¾″, *p. 883*

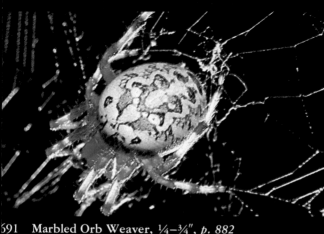

691 Marbled Orb Weaver, ¼–¾", *p. 882*

692 Venusta Orchard Spider, ⅛–⅜", *p. 890*

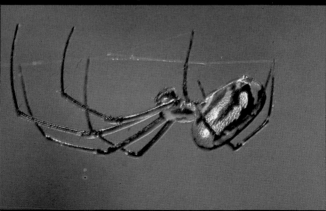

693 Mabel Orchard Spider, ♀, ¼", *p. 889*

694 Long-jawed Orb Weaver ♀, ¼″, *p. 891*

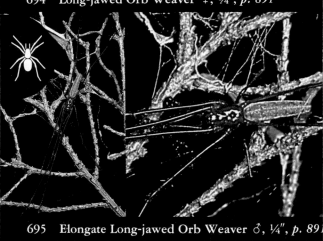

695 Elongate Long-jawed Orb Weaver ♂, ¼″, *p. 891*

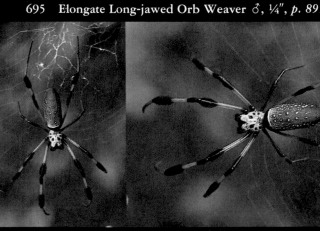

696 Golden-silk Spider ♀, ⅞–1″, *p. 888*

697 Black-and-yellow Argiope ♀, ¾–1⅛″, *p. 885*

698 Black-and-yellow Argiope ♀, ¾–1⅛″, *p. 885*

699 Silver Argiope ♀, ½–⅝″, *p. 885*

700 Arrow-shaped Micrathena ♀, ⅜″, *p. 888*

701 Arrow-shaped Micrathena ♀, ⅜″, *p. 888*

702 Crablike Spiny Orb Weaver ♀, ⅜″, *p. 886*

The chart on the following pages shows representative members of the 34 insect orders and the major arachnid orders, helping you recognize different insects, spiders, and kin.

The numbers preceding the species descriptions on the following pages correspond to the plate numbers in the color section.

Insect Orders

Order Protura   Proturan

Order Collembola   Springtail

Order Diplura   Dipluran

Order Microcoryphia   Jumping Bristletail

Order Thysanura   Silverfish

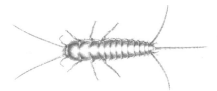

Order Ephemeroptera   Mayfly

Order Odonata   Dragonfly

Order Odonata   Damselfly

Order Blattodea   Cockroach

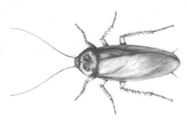

Order Mantodea   Mantid

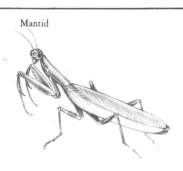

Order Isoptera   Reproductive Termite

Order Isoptera   Worker Termite

Order Isoptera   Soldier Termite

Order Zoraptera   Zorapteran

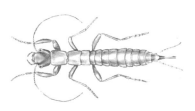

**Order
Grylloblattodea**   Ice Insect

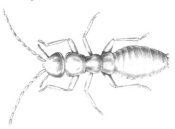

Order Dermaptera   Earwig

Order Plecoptera   Stonefly

Order Orthoptera   Grasshopper

Order Orthoptera   Cricket

Order Phasmatodea   Walkingstick

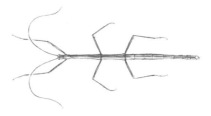

Order Embioptera   Webspinner

Order Psocoptera   Wingless Book Louse

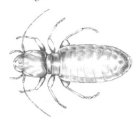

Order Psocoptera   Winged Book Louse

Order Mallophaga   Chewing Louse

Order Anoplura   Sucking Louse

Order Hemiptera   Squash Bug

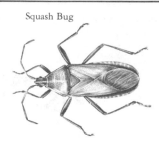

Order Hemiptera   Giant Water Bug

Order Homoptera   Leafhopper

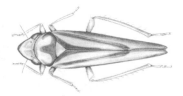

Order Homoptera   Cicada

Order Thysanoptera   Wingless Thrips

Order Thysanoptera   Winged Thrips

Order Megaloptera   Alderfly

Order Megaloptera   Dobsonfly

Order Raphidioptera   Snakefly

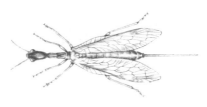

Order Neuroptera   Lacewing

Order Coleoptera   Beetle

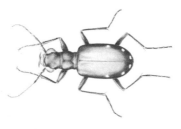

Order Strepsiptera   Male Twisted-winged Parasite

Order Strepsiptera   Female Twisted-winged Parasite

Order Mecoptera   Wingless Female Scorpionfly

Order Mecoptera   Winged Male Scorpionfly

Order Siphonaptera   Flea

Order Diptera   Mosquito

Order Diptera   Fly

Order Trichoptera   Caddisfly

Order Lepidoptera   Moth

Order Lepidoptera   Butterfly

Order Hymenoptera   Sawfly

Order Hymenoptera   Ichneumon

Order Hymenoptera   Wasp

Order Hymenoptera   Ant

Order Hymenoptera   Bee

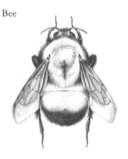

Major Orders of Arachnids

Order Araneae — Spider

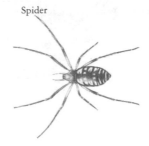

Order Scorpionida — Scorpion

Order Pseudoscorpionida — Pseudoscorpion

Order Opiliones — Daddy-long-legs

Order Acarina   Mite

Order Acarina   Soft Tick

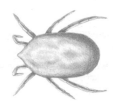

Order Acarina   Hard Tick

Order Uropygi   Whipscorpion

Order Amblypygi   Tailless Whipscorpion

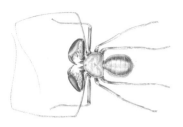

Order Solpugida   Windscorpion

INSECTS
(Class Insecta)

Primitive insects first appeared about 350 million years ago, and since then they have adapted to nearly every habitat on earth. Insects form the largest and most familiar class in the Phylum Arthropoda, with over a million species grouped in 34 orders. Of this large number, about 90,000–100,000 species are known in North America. Insects share with other arthropods a hard outer covering, or exoskeleton, and a body made up of segments, but they differ from all other arthropods in having the body segments grouped in 3 parts—a head, which bears a pair of sensory antennae; a thorax, with 3 pairs of jointed legs and often 2 pairs of membranous wings (sometimes reduced to a single pair); and an abdomen.

82 Proturans
(Order Protura)

Minute, unpigmented, wingless
insects, proturans were not discovered
until 1906. There are still no more
than 118 species known worldwide; 20
are found in North America. Proturans,
less than $\frac{1}{16}''$ (0.6–1.5 mm) long,
resemble other insects in possessing 3
pairs of jointed legs, a head, thorax,
and abdomen. The conical head has
piercing mouthparts and lacks eyes and
antennae. However, they hold their
single-clawed fore legs in front of them
like antennae. The cylindrical abdomen
has 12 segments at maturity, more than
in any other insect. Proturans have no
sensory appendages—hence the order
name, meaning "simple tail." Two
fingerlike organs called styli occur
below each of the first 3 abdominal
segments.
Proturans live in damp soil or vegetable
mold and feed on decayed organic
matter. Metamorphosis is simple:
adults and nymphs resemble each other
except that the number of abdominal
segments increases as the insect matures
and acquires sex organs. Proturans are
so tiny that they can be identified only
with the aid of a microscope.

81 Springtails
(Order Collembola)

These primitive, minute, wingless insects are widely distributed; about 2,000 species have been identified worldwide, including 315 species in North America. Springtails have elongate bodies, ¹⁄₁₆–¼″ (2–6 mm) long, that often are covered with hair or scales. They range in color from white, gray, or yellow to red, orange, purple, brown, or mottled hues. Uniquely, their abdomens have only 4–6 segments, less than any other insect—a curiosity that led some entomologists to suggest not treating collembolans as insects. Another unusual feature is a tube protruding from the abdomen. It is probably used for taking in water, although the Latin order name, meaning "glue peg," is derived from the belief that the tube was a means of holding the springtail on its perch. The common name derives from a special leaping organ, or furcula, that is folded forward under the insect and held by a catch. When this mechanism is released, the abdominal extension snaps back, tossing the springtail a few inches into the air.
Springtails occur almost everywhere, including regions beyond the snowlines of Antarctica. In North America some species swarm in the millions, peppering snowfields with their bodies. These "snow fleas" are sometimes cannibalistic. Some species also attack plants in greenhouses and spread plant diseases. Most springtails, however, are harmless, preferring to feed on molds, decaying vegetation, and algae. They are found under bark, among ground litter, and on the water. Metamorphosis is simple: nymphs resemble adults.

PODURID SPRINGTAILS
(Family Poduridae)

The elongate, wingless insects in this family have short antennae of no more than 4 segments, short legs, and a springing organ on the underside of the next-to-the-last abdominal segment. The well-developed prothorax can be seen from above. Most podurids have aggregated eyes. The different species are found in a wide variety of habitats, ranging from snowy ground to the seashore and water surfaces.

77, 80   **Snow Flea**
(*Achorutes nivicola*)

Description: $\frac{1}{16}''$ (2 mm). Cylindrical, somewhat broader past middle. *Dark blue.* Dull brown under legs and at forked abdominal tip. Antennae short. 16 minute eyes in cluster on each side of head.

Habitat: Leaf litter on forest floors and in holes in soil.

Range: Atlantic Coast to Pennsylvania, west to Iowa, north to Ontario and the Arctic.

Food: Pollen.

Life Cycle: Eggs are scattered in holes in soil in late spring. Nymphs feed during summer, maturing by early winter. Adults are active on snow surface on warm days and in early spring when they mate again.

On warm winter days these insects often swarm on the surface of snow, forming dark patches. Sometimes they are found floating in buckets of maple syrup.

79 Seashore Springtail
(*Anurida maritima*)

Description: ¹⁄₁₆–¹⁄₈″ (2–3 mm). Cylindrical. *Dark blue* to slate-gray. Antennae short. Springing organ short.

Habitat: Beach litter between tidemarks, under rocks, or in surface film of tide pools.

Range: Worldwide.

Food: Juices from decaying plant material.

Life Cycle: Eggs are scattered among beach litter, hatching in spring.

Huge clusters of these springtails are often found nestled in air pockets between the tides.

ENTOMOBRYID SPRINGTAILS
(Family Entomobryidae)

These primitive, wingless insects have elongate bodies with 4–6 segments, relatively long legs, and 4- to 6-segmented antennae. The pronotum is almost, if not entirely, concealed beneath the mesothorax. The last abdominal segment bears the well-developed springing organ called the furcula. These tiny insects are usually smooth skinned with hair or scales. Members of the largest family in the order, they inhabit leaf litter, soil, and fungi, where their brown and mottled bodies are difficult to detect.

64 Ainsley's Springtail
(*Orchesella ainsliei*)

Description: ¹⁄₈″ (3 mm). Cylindrical. *Brownish yellow,* sparsely covered *with erect, white-tipped long hair.* Bulging aggregated eyes on each side of head. *Antennae have 6 segments;* 4th abdominal segment twice as long as 3rd. Many have blackish-brown bands across back. No scales. Springing organ well developed, extending forward to bases of hind legs.

Habitat: Leaf litter on ground and among wet vegetation.

Range: New York to Missouri and Iowa.

Food: Juices from decaying plant materials.

Life Cycle: Male deposits sperm droplets on the ground. Female sweeps sperm up into slitlike genital opening. Fertilized eggs are deposited among ground litter. In spring nymphs appear in great numbers.

This springtail's mottled body so effectively blends in with ground litter that it is almost invisible. It runs and jumps with great agility.

72 GLOBULAR SPRINGTAILS (Family Sminthuridae)

These springtails have potato-shaped bodies, usually less than $\frac{1}{16}''$ (1 mm) long, and 6 short legs. A narrow necklike prothorax is connected to a globular head, which has elbowed antennae that are often longer than the head. The male has a hook at the bend of each antenna, which is used to grasp the antennae of the mate so male can be carried on mate's back until the female is ready to lay eggs. Females are larger than males and lack the antennal hooks. In both sexes the springing organ protrudes from beneath the main mass of the body. Nymphs and adults scavenge the surface of fresh water for pollen grains and other small particles of edible organic matter.

Diplurans
(Order Diplura)

These small, secretive, nocturnal insects
are known from about 400 species
worldwide, including 25 in North
America, in only 3 different families.
Most are less than ¼" (7 mm) long, but
a few are ⅜" (10 mm). They have
1-segmented tarsi, chewing mouthparts
that are concealed in a pouch, and 2
appendages called cerci at the tip of the
abdomen—Diplura means "two tails."
Usually these appendages are long and
filamentous, but the forcipate diplurans
have distinctively short, 1-segmented
cerci that resemble a pair of tiny
forceps. Campodeid diplurans are the
most common and widely distributed.
Their tail filaments have many
segments and are about the same length
as their long antennae. Anajapygid
diplurans are represented by only 1
species in North America. Their cerci
have fewer segments and are shorter
than the antennae. Metamorphosis is
simple: nymphs are diminutive versions
of adults. All diplurans live under
leaves, stones, or logs on the ground,
or under bark. They are easily
distinguished from immature members
of the Order Thysanura by their
concealed mouthparts, shallower
bodies, and the lack of compound eyes
and scales. Thysanurans are thicker and
have more exposed mouthparts and
3- or 4-segmented tarsi. Despite these
differences, diplurans formerly were
classified as part of Order Thysanura.

83 FORCIPATE DIPLURANS
(Family Japygidae)

Forcipate diplurans are whitish, slender, flattened, and wingless insects, ⅛–¼" (4–6 mm) long, with long legs and threadlike antennae almost as long as the body. Unlike other diplurans, members of this family have a distinctive pair of short 1-segmented cerci resembling tiny forceps at the tip of the abdomen. Like other diplurans, these live under leaves, stones, or logs on the ground, or under bark.

Jumping Bristletails
(Order Microcoryphia)

These brownish, medium-sized
jumping insects are represented by
approximately 350 species worldwide,
including 25 species in North America.
Like proturans, diplurans, and
thysanurans, jumping bristletails
evolved without wings. They differ
from other similar insects in that their
compound eyes are so large they meet
on top of the head. They also have
fingerlike movable projections called
styli on the bases of their legs. Like
other bristletails, they have 3 long
filament-like tails—the outer two are
cerci, while the central one is a
prolongation of the last abdominal
segment.
Jumping bristletails are about ⅝"
(15 mm) long not counting the
antennae or tails. They live under
loosely piled stones or rubble in
disturbed areas, among leaf litter or
bark in wooded and grassy regions, and
in cracks of rocky cliffs along the
seashore. These insects emerge at dusk
or on cloudy days to feed on algae
growing on stones. Metamorphosis is
simple: nymphs resemble adults. But
unlike other insects, jumping
bristletails often continue to molt
periodically after they mature.
Members of this order are distinguished
by microscopic differences in scales at
the base of the antennae and by the
number of pouches, called vesicles,
beneath each abdominal segment.

86 JUMPING BRISTLETAILS
(Family Machilidae)

Most members of this family are dark brown, highlighted by shimmering coppery reflections from the scales on the body. They have threadlike antennae, 1 long middle tail, and 2 shorter tails on the sides. Movable fingerlike appendages called styli are attached to the middle and hind coxae. All tarsi have 3 segments. They are about ⅝″ (15 mm) long.

Common Bristletails
(Order Thysanura)

These small, primitively wingless,
fast-running insects are represented by
about 250 species worldwide, including
40 in North America. Their spindle-
shaped, flattened bodies, ½″ (13 mm)
or less long, are covered with
overlapping scales. They have long,
threadlike antennae and 3 long, bristly,
tail-like appendages at the tip of the
abdomen. Thysanura means "tassel
tail." Bristletails either have small,
inconspicuous compound eyes and live
under stones, or they have larger
compound eyes (but never meeting
above the head, as in jumping
bristletails) and live under vegetation,
where there is more light. Most
thysanurans are found outdoors, but
indoor species, which feed on flour,
paste, bits of cloth, and paper, are better
known. Metamorphosis is simple: the
chief change is in size and the
development of functional sex organs.
A species of bristletail recently
discovered in California is one of the
few "living fossils." It is scaleless and
closely resembles insects preserved in
sedimentary rocks.

SILVERFISH
(Family Lepismatidae)

Silverfish have somewhat flattened
bodies, ⅜–½" (9–13 mm) long,
covered with minute, silvery scales.
Their compound eyes are small and
placed far apart. They have 3
pronounced tails and threadlike
antennae. These insects are usually
found in dark, warm places at night,
under sinks, stoves, and floors, and
around water pipes in homes. They can
become pests because they feed on
starchy substances, such as clothing or
book bindings, and dry foods.

87 Silverfish
(*Lepisma saccharina*)

Description: ⅜–½" (9–13 cm). Tapering, carrot-
shaped. *Silver-gray, coated with scales.*
Threadlike antennae and 3 tail filaments,
all shorter than body. Small black eyes.
Maxillary palps.

Habitat: Indoors in warm, dry or damp places,
including closets, bookcases, behind
baseboards, in partitions, or in
bathtubs.

Range: Worldwide in temperate climates.

Food: Dried cereals, flour, glue, and starch,
including stiffeners used in clothing
and bound books, and coated papers
used in magazines.

Life Cycle: Oval white eggs are dropped in a few
places week after week. Eggs hatch in
2–8 weeks. Nymphs, ⅛–¼" (4–
5 mm), have the same body form as
adults. In the South full size is attained
in about 2 years, longer in the North.

This insect has a scaly covering that
helps it to escape from the grip of ants
and spiders. Silverfish can survive
without food for months.

85 Firebrat
(*Thermobia domestica*)

Description: ⅜–½" (9–13 mm). *Mottled* gray and
tan or brown. *Threadlike antennae,*
longer than body, usually *swept back*
parallel to sides. 3 tail filaments.
Maxillary palps have 6 segments.

Habitat: Indoors near warm or hot places, such
as ovens, heating pipes, and furnaces.

Range: Worldwide.

Food: Crumbs and scraps of dry human food
near cooking and heating devices.

Life Cycle: At intervals female lays spherical white
eggs with soft shells in clusters of 50 or
more. Hatching and subsequent growth
take weeks or many months, depending
on temperature, humidity, and food
supply. Nymphs take up to 2 years and
about 40 molts to reach full size.
Maximum life-span in warm situations
is about 30 months.

Firebrats are alert and fast runners.
Probably once native to Eurasia, they
crawled into suitcases or cargo and have
been widely dispersed.

Mayflies
(Order Ephemeroptera)

May is not the only month when mayflies emerge in large numbers from lakes and streams. The order name, Ephemeroptera, means "living a day," and some adult mayflies do not survive that long—bursting from the water in the evening and dying before dawn. Many last a few days, but none is equipped to feed as an adult. They first appeared 350 million years ago. Today there are about 2,100 species of mayflies worldwide, including 550 species in North America.

These brownish or yellowish insects have a pair of large, triangular fore wings with many veins and crossveins. Usually they also have a pair of much smaller, rounded hind wings, although some have no hind wings at all. The soft bodies have small, bristlelike antennae, long slender fore legs, and 2 or 3 filamentlike tails that are frequently twice as long as the insect's abdomen. To mate, thousands of males perform a kind of dance, flying up and down in great swarms. They seize females that enter the swarm and mate in flight. Eggs are laid within an hour, attached by short filaments to aquatic plants or other supports.

Metamorphosis is simple. Naiads, unlike adults, have biting mouthparts and can feed on tiny plants and small aquatic animals. They resemble stonefly naiads but have 3, rather than 2, tail-like filaments, and gills on the abdomen rather than on the thorax and legs. Mayfly naiads sometimes live as long as 4 years. The last aquatic stage leaves the water, molts, and gains smoky wings. Called a subimago, or dun by fisherman, it soon sheds another skin to become a clear-winged adult, or spinner. Despite their common name, mayflies are seen in the spring and summer.

36, 37   SMALL MAYFLIES
(Family Baetidae)

These mayflies have diminutive hind
wings or none at all, 2 tail filaments,
and 3-segmented hind tarsi. Mostly
small, these insects measure ⅛–⅜"
(3–11 mm) long, excluding the tails.
Females have dark brown bodies, while
males are pale or transparent on
abdominal segments 2–6. The naiads,
to ⅜" (11 mm) long, have flat gills on
abdominal segments 1–7, 2–7, or 5–7.
They are most abundant in warm
regions, inhabiting shallow rapid water
and tolerating more water pollution
than naiads of other families.

396   Small Mayflies
(*Baetis* spp.)

Description: ⅛–⅜" (3–11 mm) excluding 2 tail
filaments. Dark brown to reddish
brown. *Male's abdominal segments 2–6
pale or transparent.* Fore wings clear;
hind wings much reduced, often
difficult to see.

Habitat: Near shallow flowing water.

Range: Throughout North America; most
species in central and eastern Canada.

Food: Adult eats nothing. Naiad feeds on
microscopic green plants and organic
particles in debris.

Life Cycle: Eggs hatch in 2–5 weeks. Naiads swim
upstream by wagging the abdomen side
to side. Subimagos transform to adults
in 7–12 hours. In warm streams adults
emerge any month, with 2 or more
generations a year. In colder water 1
generation matures in summer.

These mayflies are usually found near
the water edge among vegetation and
are relatively easy to catch with a net.
Adults of the first generation are larger
and darker than those of a second.

BUF
(Fam

Man\
Nort
famil
(25 r
filam
Adul
in m(
4-seg
naiad
fring(
2—7,
in th(
or ge
Amei
usual
yards

393 Coffi
(Ephe:

Description: ¾—⅞
filam(
Wing
dark
Habitat: Near
rivers
Range: Nova
north
Food: Adul
micro
matte
Life Cycle: Eggs
burro
grow
autun
comp
Subin
as late
adult:

Incre(
emerg
and d
formi

38 STREAM MAYFLIES
(Family Heptageniidae)

These small to medium-sized mayflies, ¼—⅝" (5—15 mm) long, have 2 tail filaments as adults and 5-segmented hind tarsi; they usually lack spots on their clear wings. Naiads, to ⅝" (15 mm) long, have 2 or 3 tail filaments. Their sprawling, flattened bodies bear single, usually leaf-shaped gills on abdominal segments 1—7. Naiads of most species live in streams, clinging to the undersides of stones, but a few are found in boggy ponds and sandy rivers.

391 **Brown Stream Mayflies**
(*Rhithrogena* spp.)

Description: ¼—⅝" (5—15 mm) excluding 2 tail filaments. *Body reddish brown to yellowish brown.* Fore legs of male somewhat longer than body, *tarsi 1½ times longer than tibiae.* Wings clear.

Habitat: Near swift streams and rivers with rocky bottoms, in mountains to elevation of 10,400' (3,120 m).

Range: Throughout North America; individual species more restricted.

Food: Adult does not eat. Naiad scrapes algae from rock surfaces.

Life Cycle: Female mates in late spring and drops eggs as it flies over water. Eggs hatch in late summer. Naiads grow slowly until water warms, then more quickly. Subimagos usually appear in evening, transforming to adults in 48 hours.

Naiads cling tightly to rocks or other underwater supports, remaining motionless or creeping quickly to the lower side if their support is lifted from the water and turned over.

SPINNERS
(Family Leptophlebiid₂

Spinners are small mayfli
(5–13 mm) long, exclud
filaments. The somewha
naiads have gills on abd₍
segments 1–6 or 1–7, a
segment of each long tai
whorl of short bristles. S
found near water with li

---

395 Early Brown Spinner
(*Leptophlebia cupida*)

Description: ⅜–½" (10–13 mm) ex₍
filaments. Dark brown ₁
paler below. Wings cle₂
brownish to brown, all
brown. *Middle tail filam*
shorter than 2 outer tails.
Habitat: Near quiet backwaters ₍
ponds, and lakes.
Range: Newfoundland to Geor₃
northern Florida, west ₁
north to southeastern M
Food: Adult does not eat. Na
and other microscopic ₂
and also skins shed by ₁
Life Cycle: Eggs may hatch in a fe
water bottom and rema
weeks. Naiads hide mₒ
crawl about in darkness
grown naiads emerge i₁
followed by subimago ₃
18–29 hours. Adults a
in the South and late M
southern Canada.

In spring naiads somet
streams formed by mel
crawling along the ban
water. They can travel
yards in 25 hours. Sub
from water as cold as 4

MIDBOREAL MAYF₁
(Family Ephemerellidaᵥ

Mostly brownish mayflies,
this family measure ¼–¾" ₍
long, excluding the 3 tail fila
They are distinguished from oᵥ
mayflies by details of wing venₐ
Naiads are either carrot-shaped ₍
flattened. They have broad, flatteᵢ
femora and often have tubercles on
head, thorax, and abdomen. Well-
developed gills appear on abdominal
segments 4–7 or 3–7, but those on
segment 1 are rudimentary. The one
North American genus is found near
rapid streams or small clear lakes.

Midboreal Mayfly
(*Ephemerella subvaria*)

⅜–½" (10–13 mm) excluding 3 tail
filaments. *Dark brown to reddish brown.*
Legs almost black. Wings clear.
Near small flowing streams.
Nova Scotia to Maryland, west to
Kansas, north to Manitoba.
Adult does not eat. Naiad feeds on
minute green plants and small insects.
Eggs hatch a few minutes after female
drops them into water. Naiads seek
shelter in water, moss, or crevices and
may overwinter in water, completing
full growth in spring. Subimagos
emerge April–May and transform to
adults 22–30 hours later.

Males swarm 6–50 feet above the
ground. Any female that flies into
swarm is seized by a male and the
drops downward to mate. Touch
chemical signals confirm that th
is of the same species.

The Brown Drake (*E. simulans*), same size, has translucent wings with dark mottling. It is the most widely distributed species of the genus, ranging throughout the United States and Canada.

392 Golden Mayflies
(*Hexagenia* spp.)

Description: ¾–1⅛" (18–30 mm) excluding 2 tail filaments. *Pale golden yellow* with brown midline on top of head, thorax, and tip of abdomen. *Pale brown band across abdominal segments 2–5.* Antennae, legs, and tails yellow. Wings clear to amber with yellow along front margins.

Habitat: Near large rivers and quiet lakes with sandy or silty bottoms.

Range: Widespread in eastern North America.

Food: Adult does not feed. Naiad scrapes diatoms and other algae from the bottom mud and submerged vegetation.

Life Cycle: Eggs are dropped on water surface and sink to the bottom, where they hatch. Naiads feed in bottom mud for 12 months, reaching full size by autumn or overwintering and completing growth in spring. Subimagos emerge May–August and in 12–24 hours transform to adults.

Naiads are important food for fish and dragonfly naiads. Adults are eaten by swallows and other insectivorous birds. For bait, fishermen use both wet flies that resemble the naiads and dry flies that look like adults.

Dragonflies and Damselflies
(Order Odonata)

Fossils resembling dragonflies and
damselflies date back 300 million years.
Today there are 5,000 species
worldwide, including 450 in North
America. These large, slender insects,
¾–5″ (18–127 mm) long, have evolved
as specialized hunters. Their freely
movable heads have large compound
eyes that in a dragonfly nearly cover the
head or in a damselfly bulge to the side.
Sharp biting mouthparts are used to cut
up insect prey. Their 4 powerful wings
move independently, enabling them to
fly forward and backward. Long legs,
unsuitable for walking, are used to hold
insects captured in flight. This order
cannot fold their wings flat against the
body—dragonflies extend them to the
sides horizontally and damselflies
hold them toward the rear vertically.
Both insects mate in flight. The male
curls the tip of its abdomen to deposit a
sperm packet in a chamber below its
second abdominal segment. Then, while
the male holds the mate by the neck
with special claspers called cerci, the
female picks up the packet using the
tip of her abdomen. Later eggs are
deposited in or close to water. The
naiads are aquatic predators that
capture insects, tadpoles, even small
fish, by extending a peculiar lower lip
at lightning speed. Bristles on the tip
grasp prey and pull it back to the
mouth. When not in use, this weapon
is folded under the naiad's head.
Although metamorphosis is simple, the
naiad and adult do not closely resemble
one another. Fully grown, the naiad
crawls out of water, splits the skin
along the midline of the thorax, and
releases the adult. Both naiads and
adults are highly beneficial predators,
destroying huge numbers of
mosquitoes.

43 DARNERS
(Family Aeschnidae)

Darners are among the largest and fastest-flying North American dragonflies, 2¼–4¾″ (57–120 mm) long. These brilliant blue, green, or brown insects have large, clear wings spanning up to 5⅞″ (150 mm). Their hemispherical compound eyes meet on top of the head. The female hovers above water and, using a well-developed ovipositor, thrusts eggs one at a time through the surface film of ponds and streams. Naiads, to 2½″ (65 mm) long, have a very flat lower lip, with no grasping bristles on the side, and relatively short legs.

343, 346 **Green Darner**
"Snake Doctor" "Darning Needle"
(*Anax junius*)

Description: 2¾–3⅛″ (70–80 mm); wingspan to 4⅜″ (110 mm). *Thorax green, abdomen blue to purplish gray.* Wings with pale yellowish area toward tips, darkening as insect ages. *Targetlike mark on face.* Compound eyes often color of milk chocolate. Naiad, to 1⅞″ (48 mm), is dark greenish brown.

Habitat: Near ponds and slow streams.

Range: Throughout North America, less common in the West.

Food: Adult preys on midges, mosquitoes, caddisflies, and other flying insects. Naiad feeds on tadpoles, small fish, and aquatic insects.

Life Cycle: Female inserts eggs singly into slit cut in stem of a submerged plant. Fully grown naiads crawl out of the water in early spring or late summer to transform into adults.

The Green Darner is one of the fastest and biggest of the common dragonflies. Other giants include the brown and

green Heroic Darner (*Epiaeschna heros*),
to 3⅝" (93 mm) with a wingspan to
5⅛" (130 mm), which is found from
Mexico to Quebec; and the green,
yellow, and brown Walsingham's
Darner (*Anax walsinghami*), 3½–4¾"
(90–120 mm) with wingspan to 5⅞"
(150 mm), found in the Southwest.

377 Brown Darner
(*Boyeria vinosa*)

Description: 2¾–3⅛" (70–80 mm), wingspan
3⅛–3⅞" (80–98 mm). Body brown
with *2 large pale yellow spots on each side
of thorax*. Wings have brown spot at
base and yellowish spot on front
margin. Naiad is black with pale
yellow middorsal line that widens
to spot.

Habitat: Shadowy edges of free-flowing streams.

Range: Nova Scotia to Florida, west to Texas,
north to Illinois.

Food: Adult eats small insects caught in
flight. Naiad preys on small aquatic
invertebrates.

Life Cycle: Mated pair rests on a lily pad while the
female lowers abdomen into water to
free eggs. Naiads crawl about 1'
out of water to transform to adults in
May in New England and in October in
Alabama.

Although they are good fliers, Brown
Darners are seldom seen very far from
home streams. Naiads live under stones
and trash near woodland streams.

GOMPHID DRAGONFLIES
(Family Gomphidae)

Mostly stream-haunting insects,
gomphid dragonflies, 1⅞–3" (48–
76 mm) long, rest on logs, stones, and
leaves, or dart from one resting place to

another, seizing prey along the way.
Only a few species frequent ponds. The
adult is easily recognized by the widely
separated compound eyes and sprawling
legs. The female lacks an ovipositor and
flies alone over the water, striking the
tip of her abdomen through the water
surface, each time discharging many
eggs. Naiads have wedge-shaped heads
and thick, 4-segmented antennae.
Using their stiff legs, they burrow in
bottom debris. There are about 10
genera in North America.

367   Small Western Gomphid
(*Octogomphus specularis*)

Description: 2–2⅛" (50–53 mm), wingspan 4"
(101 mm). Face yellowish. *Thorax black
with unbroken yellow markings.* Abdomen
black with yellow at base and tip and
thick yellow stripe narrowing to thin
line. Wings clear. Naiad, to 1"
(25 mm), is dark brown.

Habitat: Near small streams with numerous
riffles.

Range: Nevada to Baja California, north to
British Columbia.

Food: Adult eats flying insects. Naiad preys
on small aquatic insects and worms.

Life Cycle: Female drops her eggs in swift water of
riffles. Naiads stay among trash along
edges. Adults are active late April in
California to mid-August in British
Columbia.

Gomphids often fly far from the water,
sometimes along roads through shady
woods, and return to the stream to
mate and lay eggs.

344 Western Mountain Gomphid
(*Ophiogomphus bison*)

Description: 2" (50–51 mm), wingspan 2½–2¾"
(65–70 mm). *Thorax olive-green striped
with brown. Abdomen black and yellow,*
tail appendages yellow tipped with
brown. Head yellow with *2 large black-
tipped horns* projecting forward from top
of head of female. Wings clear.

Habitat: Small lowland trout streams and
vicinity.

Range: Nevada and California.

Food: Adult eats small insects. Naiad feeds on
small aquatic insects, worms, and mites.

Life Cycle: Female flies zigzag course over small
area of open water, dipping tip of the
abdomen to wash off eggs. Naiads stay
near sandy bottoms in places sheltered
from current, taking 4 years or more to
attain full growth. Adults fly May to
mid-August.

This dragonfly patrols the shorelines on
sunny days, settling only briefly on
logs, rocks, and low vegetation. It is
inactive on cloudy days.

BIDDIES AND FLYING ADDERS
(Family Cordulegastridae)

These large, hairy dragonflies, 2⅜–
3⅜" (60–85 mm) long, are brownish
black with yellow markings. They have
wide heads and eyes that meet at a
single point or are only slightly
separated. Wings range from clear to
smoky and have a black spot, or
stigma, along the front margin. The
stout and hairy naiads, which have
enormous spoon-shaped lower lips and
7-segmented antennae, cover
themselves with silt at the bottom of
small streams in woods or forests.
Adults fly slowly up and down these
waterways, hovering about a foot above
the water.

368 Western Flying Adder
(*Cordulegaster dorsalis*)

Description: 2¾–3⅜" (70–85 mm), wingspan to 5⅜" (135 mm). *Thorax chocolate-brown with yellow markings. Abdomen blackish with middorsal line of yellow patches;* segment 10 of male usually has a yellow spot on each side. Eyes green. Wings clear.

Habitat: Near woodland streams.

Range: Nevada and California to Alaska.

Food: Adult feeds on flying insects. Naiad eats small insects, tadpoles, and worms.

Life Cycle: Female uses long, strong ovipositor to press eggs singly into wet wood or other plant tissues near water's edge. Naiad burrows into silt or sand at stream bottom, then crawls out on some support and transforms to adult as early as late May in the North.

Between flights the adder often clings beneath a support, hanging at an oblique angle. Chiefly a forest dragonfly, it is not aggressive toward other dragonflies.

GREEN-EYED SKIMMERS
(Family Corduliidae)

These medium-sized or large dragonflies, 1⅜–3" (34–76 mm) long, have a distinctive lobe on the hind margin of each compound eye. Many species are black or metallic with bright green eyes; others are dark brown with yellow markings on the thorax and abdomen. Adults are common around wooded ponds and streams, where the flattened, long-legged naiads pursue small insect prey underwater.

372 Climber Dragonflies
(*Tetragoneuria* spp.)

Description: 1⅜–1⅞" (34–48 mm), wingspan 2–2⅞" (52–73 mm). Brown to greenish brown. Sides of thorax and top of abdomen have dull yellow markings. Face and top of thorax black; *tawny yellowish hair on thorax.* Wings usually clear, sometimes with dark brown areas at base of hind wings.

Habitat: Near broad slow streams and rivers.

Range: Throughout the United States and southern Canada; individual species more restricted.

Food: Adult eats small insects. Naiad preys on small aquatic insects.

Life Cycle: Eggs are washed off singly into water by flying female, alone or attended by male. Naiads sprawl on bottom or on aquatic plants. They emerge on stiff plants close to water.

Adults of both sexes are good fliers. Their common name refers to the way naiads climb up plant stems. Only a specialist can identify individual species, mostly on the basis of the adult male's genitalia or the spine pattern on naiads.

42 COMMON SKIMMERS
(Family Libellulidae)

Dragonflies in this large family have brightly colored bodies, ¾–2½" (18–64 mm) long, that are shorter than the wingspan, which is usually 1–4" (25–102 mm). The wings often have bands or spots. In some species males and females are different colors. Often the male does not attain full color until a few days after emergence. The short, squat naiads, to 1¼" (33 mm) long, inhabit shallower, warmer waters than naiads of other families and tend to be more active, maturing more rapidly.

361 Elisa Skimmer
(*Celithemis elisa*)

Description: 1⅛–1⅜″ (29–35 mm), wingspan
2¼–2⅜″ (56–60 mm). Patterned in red
and black. Prothorax fringed with long
brown hair; *thorax has middorsal black
stripe,* widest at front. *Abdomen black
with reddish triangular spots* on each
segment. Tail-like appendages orange.
Wings dark reddish at base, brownish or
black at tip; a small, vague, brown or
black spot in membrane about ⅔
distance to tip. Naiad, to ⅝″ (17 mm),
is greenish with many brown markings.

Habitat: Marshes, shallow bays, slow streams,
and ponds.

Range: Nova Scotia to Florida, west to Texas,
north to Minnesota.

Food: Adult eats small flying insects. Naiad
feeds on small aquatic insects and
worms.

Life Cycle: Female dips tip of abdomen through
water surface to wash off eggs. Naiads
probably overwinter. Adults begin to
emerge in early spring and fly April–
October.

This dragonfly perches on tall plants
rising above the marsh. Medium-sized
compared to other members of the
genus, this species is the most widely
distributed in northern states and
Canada, but seldom becomes abundant.

364 Brown-spotted Yellow-wing
(*Celithemis eponina*)

Description: 1⅜–1⅝″ (35–42 mm), wingspan
2¾–3⅛″ (70–78 mm). Head and body
amber, darkening with age. *Wings
yellow, banded and spotted with brown;*
veins initially yellow, later turning red.
Naiad is yellowish to greenish, mottled
with pale brown.

Habitat: Borders of ponds with weedy bottoms.

Range: Ontario and Massachusetts to Florida, west to Texas, north to Nebraska.

Food: Adult eats smaller flying insects. Naiad feeds on small aquatic insects.

Life Cycle: Accompanied by male, female washes off eggs below water surface. Naiads live among shallow weeds, climb well but swim poorly. They crawl up a stump before transforming to adults June–August.

Adults flutter like butterflies and are often carried by the wind some distance from ponds. This is the largest, most vivid member of the genus, which has about 10 species, all of which occur east of the Rocky Mountains.

345 Green Clearwing
(*Erythemis simplicicollis*)

Description: 1½–1¾" (38–44 mm), wingspan 2½–2¾" (65–70 mm). *Face and thorax bright green,* top of head darker. *Abdomen green, spotted with dark brown* on top segments 3–10, spots progressively larger toward tip; tip yellow in male. Legs black with long black spines on tibiae, also on middle and hind femora. Wings clear. Naiad is striped green and dark brown.

Habitat: Near borders of clear ponds.

Range: Quebec and Maine to Florida, west to Mexico, north to Washington.

Food: Adult eats smaller insects. Naiad feeds on small aquatic insects.

Life Cycle: Unattended female descends to touch water surface at widely separated intervals, washing eggs from tip of abdomen. Naiads transform to adults beginning in June.

This vigilant species often captures damselflies. It frequently rests on bare earth or floating trash and is active year-round in the South.

371 Black-faced Skimmer
(*Libellula cyanea*)

Description: 1⅝–1¾" (42–45 mm), wingspan 3⅛" (80 mm). *Mostly blue* with brownish-black streaks near base of both wings. Sides of thorax pale brown, crossed by a black stripe that widens below to leg bases. *Wings* black on outer ⅓; *yellowish-white spot* (stigma). With age color may be concealed by a whitish bloom (especially in males).

Habitat: Near ponds and slow streams.

Range: Maine to Georgia, west to Texas, north to Michigan.

Food: Adult eats small flying insects. Naiad preys on aquatic insects.

Life Cycle: Eggs are dropped into pond. Naiads sprawl on silty bottom, slowly pursuing prey. They crawl only a short distance out of water before transforming to adults. Adults fly June–October in Texas, April–September in the North.

Males pursue females and sometimes other males flying low over the water. They perch frequently on nearby twigs. This species' common name comes from the shiny black face of the old male.

365 Widow
(*Libellula luctuosa*)

Description: 1⅝–2" (42–50 mm), wingspan 1⅛–3½" (30–90 mm). Body dark brown, sometimes with yellow stripes on sides; thorax somewhat hairy. *Fore and hind wings blackish brown up to halfway toward tip;* wing tips clear or smoky-brown, especially in female.

Habitat: Near ponds, small lakes, and marshes.

Range: Ontario and Atlantic Coast to Georgia and Gulf Coast, west to Texas and northern Mexico, north to South Dakota.

Food: Adult eats smaller flying insects. Naiad feeds on small aquatic insects.

Life Cycle: Female, often unattended by male, drops eggs into pond water. Naiads crawl a few feet or more from water, cling to vegetation, and emerge as adults in April in the South, in late summer in the North.

These slow-flying dragonflies are easy to catch with a net. Their wings are disproportionately large, making these insects look bigger than they are. Adults hang below bare twigs of pondside shrubbery.

348 Twelve-spot Skimmer
(*Libellula pulchella*)

Description: 1¾–2¼" (45–58 mm), wingspan to 3⅛" (80 mm). Head and thorax chocolate to light brown. Abdomen gray-brown to whitish. *Each wing has 3 brown spots;* male develops milky cast on wing between spots, female clear in these areas. Sometimes *abdomen shows whitish bloom with age.*

Habitat: Near ponds or broad, slow regions of rivers.

Range: Most of the United States.

Food: Adult eats small insects caught in flight. Naiad feeds on aquatic invertebrates.

Life Cycle: Female drops eggs singly into water or settles on plants to attach eggs to stems close to water surface. Adults fly late May–September near the Canadian–United States border.

This skimmer often rests on lily pads or plants overhanging the water.

373 Four-spot Skimmer
(*Libellula quadrimaculata*)

Description: 1⅝–1¾" (42–45 mm), wingspan 2¾–3⅛" (70–80 mm). *Body olive-brown* with yellow stripe on each side. Face yellow, upper lip marked with black. *Wings have yellow streak along front margin,* a small dark area halfway along front margin, and a larger dark area next to body on hind wing but not extending to front margin. Naiad, to 1⅛" (28 mm), is orange-brown, somewhat hairy.

Habitat: Near ponds or broad, slow parts of rivers, sometimes straying to open woods.

Range: Labrador to New Jersey, southwest to Arizona, north to British Columbia, Alberta, and Alaska.

Food: Adult eats small flying insects. Naiad feeds on small aquatic insects.

Life Cycle: Male stands guard, driving away other males while female rests on a plant within reach of the water. Female dips the tip of abdomen into the water to drop eggs one at a time. Adults fly late spring—autumn.

The Four-spot Skimmer is the most common member of this genus. Mature males often develop a grayish bloom on the abdomen.

360, 376 **Red Skimmer**
(*Libellula saturata*)

Description: 1⅛–2⅜" (28–61 mm), wingspan 3⅜–3¾" (85–95 mm). *Red* with brown on top of head. *Thorax brownish red* without stripes. *Wings yellowish with reddish veins* and clear tips; brown streak near base of hind wing, far from front and hind margins.

Habitat: Stagnant small ponds and pools.

Range: Kansas to Mexico, northwest to California, north to Montana.

Food: Adult eats small insects caught in flight. Naiad feeds on small aquatic insects and worms.

Life Cycle: Male may defend territory while female places eggs in water. Hairy naiads creep and squat in dark ooze on bottom of stagnant pool, ambushing prey. Fully grown, they leave the water, crawl a short distance, then climb up any convenient support and transform into adults.

Primarily a southern species, this dragonfly is sometimes found near warm springs farther north. There are considerable differences in details of venation between sexes.

369 Bluebell
(*Nannothemis bella*)

Description: ¾" (18–20 mm), wingspan 1–1⅜" (25–35 mm). *Very small.* Body black and yellow, becoming powder blue with age. Wings clear or amber-tinged near base.

Habitat: Stagnant pools in marshy places.

Range: Maine to Florida, west to Louisiana, north to Ontario.

Food: Adult feeds on small flying insects, particularly flies. Naiad eats small aquatic insects and worms.

Life Cycle: Female washes off eggs in water, where naiads develop slowly. Adults may be found April–September.

This is the smallest North American dragonfly. Naiads inhabit narrow water-filled holes. Adults perch on grass stems in the sunshine, making brief flights to catch midges and other prey.

349 Swift Long-winged Skimmer
(*Pachydiplax longipennis*)

Description: 1⅛–1¾" (27–45 mm), wingspan to 3½" (88 mm). Stout. Mostly blue; *thorax with pale yellowish-green sides and 3 brown stripes.* Abdomen blue-violet. *Face white; top of head metallic blue a few days after emergence.* Wings clear to slightly smoky, often with a brownish cloud beyond the middle. Naiad is dark, handsomely patterned with greenish brown, including crossbanding on femora.

Habitat: Large ponds and broad streams.

Range: Throughout North America.

Food: Adult eats small insects. Naiad feeds on small aquatic insects.

Life Cycle: Female generally flies parallel to water surface, flicking the tip of abdomen downward to wash off eggs. Naiads clamber over submerged objects, generally transforming into adults close to water's edge. Adults fly late April–late September in the North.

This swift-flying skimmer is hard to capture. When two males meet, they face each other and dart upward together, often out of sight. Females generally rest on trees away from the shore unless they are laying eggs. The short abdomen makes wings appear extra long.

375 Rusty Skimmer
(*Paltothemis lineatipes*)

Description: 1⅞–2⅛" (47–54 mm), wingspan 3½–3¾" (90–95 mm). *Male rusty red, female gray;* both have shiny bare area on face, which is pale, becoming red with age. Thorax olive-brown along sides with 3 brown stripes. *Abdomen reddish above,* blackish below, *segment 10 narrower than rest,* partly pale. Wings clear, broad at base, tapering to tip.

Habitat: Temporary ponds in arid lands.

Range: Texas and Mexico to California.

Food: Adult eats small flying insects. Naiad preys on small aquatic insects, mites, and worms.

Life Cycle: Female flies over water surface and dips abdominal tip into the water to wash off small clusters of eggs. Naiads creep only a short distance from water before transforming to adults. Adults fly from May in California to October farther south.

This skimmer flits over ponds, occasionally perching to rest on tall grasses and plants. The only way to distinguish the species in this genus from other skimmers is by examining the fine details of wing venation.

374 Low-flying Amber-wing
(*Perithemis tenera*)

Description: ¾–1" (20–24 mm), wingspan to 1¾" (45 mm). *Mostly yellowish brown.* Face yellow; thorax brown; abdomen brownish yellow. Legs yellowish with black spines. *Wings amber;* male's with brown area ⅓ way to tip, female's with 2 brown crossbands or clouds.

Habitat: Near ponds and slow streams.

Range: Massachusetts to Gulf states, west to Texas and northern Mexico, north to Nebraska.

Food: Adult eats small insects. Naiad feeds on small aquatic insects.

Life Cycle: Female, often unattended by mate, flies over water touching surface with tip of abdomen, freeing 10–20 eggs with each dip. Naiads stand up high on slender legs searching for prey. They cling to stumps while transforming to adults. Adults are active from April in Texas to October in South Carolina.

This low-flying skimmer usually flies in sunshine, avoiding other dragonflies. It perches horizontally with fore and hind wings lifted unequally.

Specialists base identification of the different species in this genus on the form of male genitalia.

358 Half-banded Toper
(*Sympetrum semicinctum*)

Description: 1⅝–1¾" (41–45 mm), wingspan
1⅝–2" (41–50 mm). Male's head and
thorax reddish brown. Abdomen bright
red with black spot on each side of last
2 or 3 segments. *Wings dark yellowish in
basal half, clear beyond;* fore wings
sometimes entirely clear. Female is
similar, but with body olive-green.
Naiad, to ⅝" (15 mm), is dark brown.

Habitat: Close to shallow ponds, shallow edges
of slow streams, and spring-fed
marshes.

Range: New Brunswick to Virginia, west to
Illinois.

Food: Adult eats small flying insects. Naiad
feeds on small aquatic insects.

Life Cycle: Mated pair flies in tandem while female
flicks abdominal tip through the water
surface, washing off eggs one at a time.
Naiads are fully grown by August or
September. Adults fly to late October
in northern parts of range.

This skimmer has become scarce,
perhaps because suitable unpolluted,
marshy places close to deeper water
have become too few.

363 Jagged-edged Saddlebag
(*Tramea lacerata*)

Description: 1¾–2¼" (45–58 mm), wingspan
3¾–4" (95–102 mm). Black. *Abdomen
has 2 yellow spots near tip* on upper
surface. *Hind wings with broad black
saddle mark.* Naiad is green marked
with brown.

Habitat: Slow broad rivers with quiet backwaters

full of submerged vegetation.

Range: Massachusetts to Florida, west to Texas and California, north to Washington, south to Mexico.

Food: Adult preys on small flying insects. Naiad eats small aquatic insects and worms.

Life Cycle: Female released from mate's hold darts down to water, washes off eggs, then flies upward to be seized again by male. Naiads clamber actively among submerged plants, transforming to adults in late summer in the North. Adults seen May–June in the North almost certainly emigrated from the South. Probably 2 generations a year in the South, 1 in the North.

Saddlebags are named for the large saddle-shaped mark on the hind wings. In late spring or early summer often so many adults travel together they have difficulty finding separate suitable perches for the night and must compete for a roost.

362 Red Saddlebag
(*Tramea onusta*)

Description: 1¾–2¼" (45–58 mm), wingspan 3¼–3½" (83–90 mm). *Body red to reddish brown* marked with olive-brown on thorax and enlarged base of abdomen. Face has short black hair. Fore wings clear, *hind wings clear with dark saddle mark, outer boundary jagged; veins are red* even in dark band.

Habitat: Near broad, slow rivers with thick submerged vegetation.

Range: North Carolina to Florida, west to Texas, California, and Mexico, northeast to Ontario.

Food: Adult feeds on small flying insects. Naiad eats aquatic insects.

Life Cycle: Female settles on water plant while reaching below water to attach eggs to plant stems. Naiads clamber actively among underwater greenery, sometimes on floating algae, seeking prey.

This skimmer flies low over the water, but usually perches high up on bare branches, beyond the reach of collectors and photographers. In the Carolina Saddlebag (*T. carolina*), which has a wingspan of 3½–4″ (90–100 mm), the male's face is metallic violet at maturity and the female's is dark blue with some orange spots. It ranges from Ontario to Massachusetts, south to Florida, west to Texas.

BROAD-WINGED DAMSELFLIES
(Family Calopterygidae)

These large damselflies, 1–2″ (25–51 mm) long, differ from members of related families in that their wings broaden gradually from the base, rather than appearing stalked. When the insect is at rest, the wings are held together vertically above the slender body, which is usually metallic green or black. The slender, long-legged naiads stalk small prey among plants in shallow streams and occasionally in ponds. Damselfly naiads, unlike dragonfly naiads, have external gills on the tip of the abdomen.

366 Black-winged Damselfly
(*Calopteryx maculata*)

Description: 1⅝–1¾″ (40–46 mm). *Male's body metallic green;* female's dark brown, nonmetallic. *Male's wings black; female's brownish with glistening white spot* (stigma) near end of fore wing front margin. Naiad is pale brown with dark brown markings and yellow-brown or orange legs.

Habitat: Along edges of slow streams in forests.

Range: Throughout North America.
Food: Adult eats small insects, including aphids. Naiad feeds on small aquatic insects.
Life Cycle: Female forces eggs singly into soft plant stems. Male often remains nearby but usually is not in contact. Naiads live in slow water.

This damselfly flits about almost like a butterfly. It is often assigned to genus *Agrion*. The Western Dark-winged Damselfly (*C. aequabile*) has dark wings with a clear basal half. It is found from Nova Scotia and New York to Nevada and the Pacific Coast, northeast to Manitoba.

357 Ruby Spot
(*Hetaerina americana*)

Description: Male 1⅝–1¾" (40–46 mm), female 1½–1⅝" (38–40 mm). Head and thorax dark. Male's abdomen greenish bronze to dark brown; female's greenish. *Wings clear; male's ruby-red at base,* female's brown at base.
Habitat: Near streams in woods and forests, sometimes along marshy borders.
Range: Throughout the United States and southern Canada.
Food: Adult eats small soft-bodied insects. Naiad feeds on small aquatic insects.
Life Cycle: Female clings to plant just above water and forces eggs into soft stems. Male usually stays close by. Fully grown naiads climb up plant stem and out of the water to transform to adults.

When the male Ruby Spot flies through a shaft of sunlight, its red color shows spectacularly. If it suddenly flits into the shade, it seems to vanish.

SPREAD-WINGED DAMSELFLIES
(Family Lestidae)

These slender damselflies, 1¼–2"
(32–50 mm) long, have clear wings
that narrow to stalks at the base. They
are named for the way that their wings
diverge or spread when at rest. Naiads
have long gills with rounded ends
rather than leaflike gills, as in the
narrow-winged damselflies. Spread-
winged damselflies are found around
ponds and swamps, where they alight
on grasses and plant stems.

378 Dark Lestes
(Lestes congener)

Description: Male 1⅜–1⅝" (35–42 mm), female
1¼–1½" (32–38 mm); larger in the
West. *Male bronzy black above with
yellowish markings; last 2 abdominal
segments grayish.* Female paler, abdomen
reddish-brown. Wings clear; female's
longer than male's. Naiad, to 1"
(25 mm), is pale brown.

Habitat: Permanent and semipermanent ponds
and flooded stream margins.

Range: Southern Ontario to New Jersey, west
to California, north to British
Columbia.

Food: Adult eats smaller insects, such as
aphids. Naiad feeds on small aquatic
insects.

Life Cycle: Eggs are deposited on plants high above
the water or rather indiscriminately on
live or dead dry foliage in fall. When
vegetation collapses in winter, eggs are
submerged underwater, where naiads
hatch. Adults emerge late in July,
flying until mid-September.

This dark-colored, late-flying damselfly
produces eggs that can survive frost as
well as dry air.

Life Cycle: Female, unattended or held by male, lays eggs in wet wood. Sometimes both female and male enter water and remain submerged ½ hour or more. Naiads are usually found in gentle water.

Adults alight on bare sunny places or on boulders projecting from the water. They are alert, seldom resting long in one place. This genus is primarily native to South and Central America. About 20 species can be found in the United States—8 range into southeastern Canada and 2 into western Canada.

---

351 **Violet Tail**
"Violet Dancer"
(*Argia violacea*)

Description: Male 1⅛–1¼" (28–32 mm), female 1¼" (32–33 mm). *Male's* head black; *thorax and abdomen violet.* Female is dark brown to black. Wings clear.
Habitat: Along slow streams, coves of large fast rivers, or shallow weedy lakes.
Range: Throughout the United States and southern Canada.
Food: Adult eats small soft-bodied insects. Naiad preys on aquatic insects.
Life Cycle: Male clasps female by the neck with a pair of pincerlike appendages and beats wings to perch vertically upward from mate. Female dips abdomen into water to lay eggs on underwater leaf. When eggs are deposited, male takes flight like a helicopter, lifting female from the water.

These striking damselflies are often seen flying in tandem over streams and ponds. This is the most common species in the genus.

354 Circumpolar Bluet
(*Enallagma cyanigerum*)

Description: Male 1⅛–1½" (28–37 mm), female
1⅛–1⅜" (28–34 mm). *Blue,* marked
with black on top of head, pronotum,
and sides of abdomen. Wings clear.
Naiad, to 1⅛" (28 mm), is brown.

Habitat: Ponds, marshy lake shores, and other
quiet waters but not ponds with acidic
bottoms.

Range: Newfoundland to Connecticut, west to
Nevada and California, north to
southern Alaska. More abundant in the
West.

Food: Adult eats small insects, such as
aphids. Naiad feeds on small aquatic
insects.

Life Cycle: Female settles on plant alone or with
mate. Curling the tip of the abdomen
underwater, female attaches eggs to
plants. Adults fly May–November.

In the far North mated pairs are seen
flying much later in the season than
other damselflies, often by the
hundreds. These northern naiads live
under ice in winter.

353 Doubleday's Bluet
(*Enallagma doubledayii*)

Description: Male 1⅛–1¼" (28–31 mm), female
1¼" (31 mm). Male's head blue with
yellow and black markings; antennae
brown. *Thorax black, blue, or yellowish
green. Abdomen blue* with black
markings. Legs black with yellow
markings. Female has broad dark
stripes along each side of blue back.
Wings clear.

Habitat: Sand-bottomed ponds, especially in
pine barrens and glaciated regions;
found mostly near the coast.

Range: Cape Cod, Massachusetts, to Florida.

Food: Adult eats small soft-bodied insects.
Naiad feeds on aquatic insects.

Life Cycle: Male holds female by neck with pincers at the tip of the abdomen while female rests on plant above water surface. Backing into the water, female inserts eggs into submerged soft plant tissues. Later male helps mate out of the water. Naiads creep slowly over bottom, maturing June–August.

The body of this damselfly is noticeably blue, hence the common name Bluet. Details of male reproductive organs and of female spots and stripes aid a specialist in distinguishing this species from others with similar color and dimensions.

355 Red Bluet
(*Enallagma pictum*)

Description: Male 1⅛–1¼" (28–31 mm), female 1" (25–26 mm). Head, thorax, and abdomen *bright red or reddish yellow;* or brown and yellow, marked with black. Wings clear and narrow, reduced at base to a short stalk.

Habitat: Pond margins and wet meadows.

Range: Massachusetts to Maryland, particularly in New Jersey pine barrens, wherever ponds have water lilies.

Food: Adult eats small soft-bodied insects, such as aphids. Naiad feeds on small aquatic insects.

Life Cycle: Female rests on water lily and deposits eggs in soft underleaf tissues. When fully grown, naiads climb up plant stem out of water to release adults.

This unusual bluet does not have the characteristic bright blue and black color that has inspired this species' common name. One of the most common small damselflies, it alights fearlessly on a person's shoulder and sometimes harmlessly nibbles clothing.

Cockroaches
(Order Blattodea)

Cockroaches are among the oldest winged insects, appearing 350 million years ago; 3,500 species are known worldwide, 57 are found in North America. These dark brown to reddish-brown insects have flattened oval bodies and long swept-back antennae. The pronotum extends so far forward that it conceals most of the head. Although they have wings, cockroaches almost never fly, relying on well-developed legs to scurry about when they are disturbed. During the day their flat bodies slip easily into cracks and crevices. At night they forage for food. Metamorphosis is simple: nymphs resemble adults. Female cockroaches lay packets of 12–25 eggs in inaccessible places. Cockroaches infest buildings, contaminating food and producing unpleasant odors. Efforts to destroy them have made some species virtually immune to pesticides. Certain types are known to live for months on little more than dust. Tropical insects, they are most abundant in the southern regions of North America, but they flourish in any environment where there is sufficient food and warmth. Formerly these insects were included in Order Orthoptera because they share many features of grasshoppers and crickets. Contrary to popular belief, cockroaches do not transmit human diseases.

BLATTID COCKROACHES
(Family Blattidae)

The blattid cockroaches, ¾–2″ (19–50 mm) long, have flat oval bodies and long slender antennae. Their heads are concealed under the front of the pronotum. Their conspicuous cerci, composed of several segments, are important as sensory organs. Both the fore wings and pronotum are bare or only slightly covered with silky hair. Wings, when present, are held flat over the back, overlapping one another, but the tips of the hind pair do not fold.

---

126 Oriental Cockroach
"Asiatic Cockroach" "Shad Roach"
(*Blatta orientalis*)

Description: Male 1″ (25 mm), female 1⅜″ (35 mm). *Black or very dark reddish brown.* Adult male has short wings; flightless female has only wing stubs. *Rear side corners of abdomen project,* giving body scallop-shaped outline.

Habitat: Buildings, particularly basements and ground-floor storage areas. Rarely outdoors except during very hot, humid summer weather.

Range: Worldwide.

Food: Moist human and pet food, garbage, and other organic matter.

Life Cycle: Female carries capsules containing about 10 eggs, extruded partway, for a day or more, then cements capsule in dark cavity and abandons it. Nymphs mature in about 1 year.

This cockroach is known as the "Shad Roach" because its young appear in great numbers early in the year, just when shad are swimming into fresh waters to breed. It is also called "Black Roach" and "Black Beetle" because of its color and hard body.

125 American Cockroach
"Waterbug"
(*Periplaneta americana*)

Description: 1½–2″ (38–50 mm). *Reddish brown
with yellowish edges and markings on
pronotum.* Yellowish stripe along front
margin of each fore wing. Slender
antennae are longer than body.

Habitat: Crevices of tropical and subtropical
vegetation; indoors in hospitals,
warehouses, and homes.

Range: Florida to Mexico and, by introduction,
indoors in warm buildings from
Antarctica to Greenland.

Food: Moist human and pet food and other
organic matter.

Life Cycle: Female carries egg capsule containing
about 12 eggs and abandons it after
1–2 days, usually in some dark crevice.
Nymphs require about 12 months to
attain maturity. Females may survive a
year or more.

Adults fly well, occasionally attracted
to artificial lights at night. They often
crawl into grocery packages and are
transported to new locations. The
related Australian Cockroach
(*P. australasiae*), ¾–1″ (19–25 mm), is
dark red to brown with yellow
markings. It is well established as
far north as San Francisco and is a
greenhouse pest in Connecticut.

BLATTELLID COCKROACHES
(Family Blattellidae)

These small slender cockroaches, rarely
more than ⅝″ (16 mm) long, have
long, thin spiny legs and small spines
beneath the first segment of all tarsi.
The fore wings are densely covered by
silky hair. The tips of the hind wings
fold over when the insect is at rest.
Cerci do not taper markedly but are
clearly segmented. The adult female has

a broad rounded plate below its last abdominal segment. Unlike those in other families, blattellid females retain their egg capsules a month or more, until the embryos inside are ready to hatch. Some species are found outdoors in woods in the South.

124 German Cockroach
"Croton Bug"
(*Blatella germanica*)

Description: ½–⅝" (12–16 mm). *Pale to dark brown with 2 longitudinal brown marks* on pronotum. Wings of adult usually conceal continuation of dark stripes from pronotum onto thorax, but these marks are visible in nymphs.

Habitat: Homes, food factories, restaurants, breweries; rarely seen outdoors.

Range: Worldwide, wherever there are frost-free storage areas for food.

Food: Anything edible by people or household pets.

Life Cycle: Female carries a large, partly extruded egg capsule for 4 or more weeks, then abandons it. As many as 30 nymphs soon emerge, taking 4–6 months to develop. Several generations a year.

Adhesive pads on this insect's last tarsal segment enable it to crawl on vertical surfaces as smooth as glass. It leaves an offensive odor on packages and their contents if it gets inside. Its common name "Croton Bug" derives from the fact that the insect first became a household pest in 1890, when water from the Croton Reservoir began augmenting New York City's municipal supply.

Mantids
(Order Mantodea)

Mantids are easy to recognize. They have relatively long bodies, measuring ⅜–5⅞" (10–150 mm), and 4 thin legs. Their elongated pronotum bears fore legs that are held in a characteristic prayerful pose, inspiring the name praying mantis. The neck is so flexible that the mantid can "look over its shoulder"—a feature unique among insects. The strong prominent mouthparts of these predators can easily cut through the armored heads of insects, such as wasps. The fore legs, which have spiny teeth, can be moved very rapidly to snatch prey. The long, slender middle and hind legs are used to grasp twigs, which helps the mantid remain motionless while waiting to ambush victims. All are voracious eaters. The female often devours the male while mating. Both male and female feed on a broad range of insects, and large species even stalk small frogs, lizards, and hummingbirds. In autumn females lay hundreds of eggs in large oval masses resembling papier-mâché— each mass is coated with a foamy, hardening bird-repellent and attached to twigs. Only the eggs seem to tolerate frost. Metamorphosis is simple. In spring soft, cream-colored creatures squirm out, quickly expanding into tiny mantids. They immediately begin eating smaller insects and sometimes each other as well. Because mantids help control pests, they are considered highly beneficial and are often deliberately placed in gardens. Some entomologists assign mantids to Order Orthoptera along with grasshoppers, crickets, and cockroaches. Mantids are represented by 11 species in North America and 1,800 worldwide.

MANTIDS
(Family Mantidae)

These slender insects, ⅝–5⅞″ (15–150 mm) long, have a mobile triangular head, prominent compound eyes, and threadlike antennae. The elongated fore legs are adapted for grasping and holding prey. Long, slender middle and hind legs are used for walking, standing, and leaping.

301 Obscure Ground Mantid
(Litaneutria obscura)

Description: ⅝–1⅛″ (15–30 mm). Gray. *Male has sooty gray wings with a brown spot near base of fore pair.* Female has only small wing pads, leaving all but the 1st abdominal segment exposed. Middle and hind legs long, slender.

Habitat: Arid lands, among dry grasses.

Range: Colorado to Mexico, northwest to California, occasionally to Texas.

Food: Ants and other insects moving on ground and in low vegetation.

Life Cycle: Eggs are laid in masses on low shrubs in fall. They overwinter and hatch in spring. Mantids sometimes go through a dormant stage during extended droughts, common in southern parts of the range.

This mantid is most active during wet periods, when prey is plentiful and conspicuous. In the same range the Agile Ground Mantid (*L. minor*), 1–1⅛″ (25–30 mm), has a brown spot only in males, and females have a rough pronotum. The Pacific Ground Mantid (*L. pacifica*), ⅝–1⅛″ (15–30 mm), lacks the brown spot in males and is found only in California. The Boreal Ground Mantid (*L. borealis*), same size, has the brown spot in males, but females have a smooth pronotum. It occurs in California, rarely in Colorado.

299, 300 Praying Mantis
"European Mantid"
(*Mantis religiosa*)

Description: 2–2½" (50–65 mm), including wings, which extend beyond abdominal tip. Green to tan. Compound eyes tan to chocolate-brown, darker at night. *Fore coxae bear black-ringed spot beneath,* which is lacking in the Chinese Mantid and Carolina Mantid.

Habitat: Meadows, on foliage and flowers.

Range: Eastern United States into Ontario.

Food: Diurnal insects, including caterpillars, flies, butterflies, bees, and some moths.

Life Cycle: Eggs overwinter in flat mass attached to exposed twigs above snow. They hatch almost simultaneously in late spring. Nymphs are dispersed by wind or eat one another. Survivors are solitary. 1 generation matures in late summer or early autumn.

This mantid was accidentally introduced in 1899 on nursery stock from southern Europe. At a time when Gypsy Moth Caterpillars were burgeoning in the eastern states, it was recognized almost immediately as a beneficial predator. However, mantids are so cannibalistic that they are rarely numerous enough to have much effect in depleting caterpillar populations.

302 Carolina Mantid
(*Stagmomantis carolina*)

Description: 2⅜" (60 mm) including wings. Head and thorax almost as long as rest of the body. *Antennae about ½ as long as middle legs. Pale green to brownish gray,* often inconspicuous on vegetation. Wings do not extend to tip of abdomen, especially on female.

Habitat: Meadows and gardens, on herbs, low shrubs, and flower heads, including goldenrod.

Range: Virginia to Florida, west to Mexico and California, northeast to Indiana.

Food: Butterflies, moths, flies, small wasps and bees, true bugs, and caterpillars.

Life Cycle: Masses of 30–80 elongate eggs, packed in parallel rows, overwinter attached to plant stem. They are coated with a tan frothy material that dries like a hard meringue. Nymphs appear in spring and begin ambushing prey.

This voracious predator can make 2 separate strikes with its fore legs in a fraction of a second, often before a fly can spread its wings and escape. The adult female often captures and devours parts of her mate even as he continues to transfer sperm.

298 Chinese Mantid
(*Tenodera aridifolia*)

Description: 2½–3⅜″ (65–85 mm), including wings. Tan to pale green. *Fore wings tan with green along front margin.* Compound eyes chocolate-brown at sunset, pale tan soon after sunrise and during the day.

Habitat: Meadows and gardens, on tall herbs, flower clusters, and shrubs.

Range: Massachusetts to New Jersey, west to Ohio.

Food: Large caterpillars, butterflies, flies, bees, wasps, and day-flying moths.

Life Cycle: Overwinters in egg masses, about ¾″ (20 mm) wide and extending 1″ (25 mm) along tree stem exposed above snow. Nymphs hatch in late spring, disperse in wind, and thereafter are solitary. Mating pairs are seen mostly in September.

The largest mantid, this insect was introduced from China around 1896 as a beneficial insect. The Narrow-winged Mantid (*T. angustipennis*), introduced from southern Asia in 1933, is smaller and found in Delaware and Maryland.

Termites
(Order Isoptera)

Termites are small, soft-bodied, social insects with biting mouthparts. Because of their pale color and social habits, they are often called white ants. Like ants, they live in colonies, where various functions are performed by different castes, some of which are sterile and wingless. But termites differ from ants in several ways. Metamorphosis in termites is simple instead of complete as in ants, and while worker ants are all sterile females, in termites all the castes include both sexes. The reproductive caste consists of heavily pigmented, sexually active males and females, whose 4 finely veined wings are all the same size. They leave the colonies in swarms at certain seasons. After a short mating flight, a queen and its mate, or king, shed their wings and found a new colony that may last for many years. Their offspring include sterile workers that collect and distribute food and maintain the nest. The sterile soldier caste has enlarged jaws and defends the colony. A few species have a nasute caste: its members have a nozzle on the head which can be used to spray a repellent fluid at ants and other intruders. Some species have no worker caste, and the work of the colony is done by other castes. In others, there are no soldiers. Most species feed on wood, which they digest with the aid of microorganisms that live in their intestines. Their colonies, which may contain millions of workers, are found in logs or in standing dead timber, in buildings, in the soil, or in mound nests on the surface. Out of a total of 2,100 species, 41 are North American. Most are useful as scavengers but a few species attack buildings and are among the most destructive insects.

ROTTING-WOOD TERMITES
(Family Hodotermitidae)

These termites are ½–1″ (13–25 mm) long and live along the Pacific Coast. Winged forms are dark brown with orange heads; wingless forms and nymphs are paler. Unlike the subterranean termites, they have teeth on their jaws and no pale depression, or fontanelle, on top of the head between the eyes. The hind femora are swollen, and the tibiae bear long spines. Rotting-wood termites have no worker caste; the work of the colony is carried out by young of the soldier and reproductive castes. Colonies are found in dead trees, rotting wood, and especially in damp areas.

308 Pacific Coast Termite
(*Zootermopsis angusticollis*)

Description: Soldiers ¾″ (19 mm), reproductives 1″ (25 mm). Soldiers have large brownish heads, long dark mandibles, at least 23 segments in antennae. Secondary reproductives are grayish or whitish, lack wing pads but have compound eyes. All members of the colony have *prominent cerci* with 3 or more segments *and a groove on the midline above upper lip*.

Habitat: Wooded areas on rotting stumps and moist fallen trees, mostly above ground.

Range: West of the Rocky Mountains from northern California to British Columbia; rare in Southern California.

Food: Wet wood.

Life Cycle: Colony has no worker caste, only soldiers and 3 reproductive forms— fertile males, "first form" queens with small wing stubs, and secondary reproductives. Young serve as workers.

These termites occasionally enter and destroy the rotting heartwood of old trees or wet pine timbers in buildings.

SUBTERRANEAN TERMITES
(Family Rhinotermitidae)

These are small termites, ¼–⅜" (5–9 mm) long, that inhabit warmer parts of North America. Winged forms are black, while other castes are white or yellowish. Unlike rotting-wood termites, they have a pale depression, or fontanelle, on top of the head between the eyes. They have no teeth on their jaws; the hind femora are not swollen and there are no spines on the tibiae. Most members of this family live in colonies underground. A few establish colonies in damp wood.

324 Subterranean Termite
(*Reticulitermes hesperus*)

Description: Worker ¼" (5 mm), soldier ¼" (5 mm), male and female ⅜" (9 mm). *Worker and soldier have flat pronotum* without elevated ridge behind head. *Head of soldier twice as long as broad* with narrow long mandibles. Winged reproductives black to dark brown with compound eyes and wings, or, after wings are shed, with wing scales. Workers are grayish white and eyeless.

Habitat: Wooded areas or soil and moist wood.

Range: Baja California and Nevada north to British Columbia.

Food: Decaying roots, trees, even moist rugs. Winged reproductives may eat pollen.

Life Cycle: Female lays large eggs that hatch, then minute eggs that serve as food for nymphs. Once nymphs begin chewing wood, queen resumes laying large eggs. Queen doubles in size and is unable to leave the colony. Workers appear first; soldiers develop later. Secondary reproductives appear when royal pair age and egg production slackens. Founding queen may live more than 3 years.

These termites attack wood where humidity is high. Their tunnels follow the grain of wood until it collapses. The similar Eastern Subterranean Termite (*R. flavipes*) is paler, and its soldier's head is less elongated. It is the single most important termite, ranging from Massachusetts to Florida, west to the Mississippi River Basin, north to southern Ontario.

Zorapterans
(Order Zoraptera)

This very small order has only 22
species worldwide, including 2 in
North America. Zorapterans are minute
slender insects, measuring no more
than ⅛″ (3 mm) long. They have
chewing mouthparts, threadlike or
beadlike antennae composed of 9
segments, and short legs each having 2
tarsal segments. Their broad abdomens
are as wide as the thorax, sometimes
wider, and bear a pair of short, bristly
appendages, called cerci, at the tip.
Some have no wings at all. Others may
have 4 wings that are shed soon after
nuptial flights, leaving short stubs on
the thorax. Winged zorapterans and
those with wing stubs have compound
eyes as well as 3 simple eyes. Wingless
zorapterans lack eyes.
Like termites, zorapterans live in
colonies under bark and rotten wood
but they do not harm wooden
structures and have no caste system.
Using glands on the ends of their
abdomens, they spin silken tunnels and
feed chiefly on mites and smaller
insects. Metamorphosis is simple:
nymphs are identical to wingless
adults. Female zorapterans stand guard
over their eggs and sometimes attempt
to conceal them with chewed-up bits of
food. The scientific name of the order,
which means "purely wingless," was
given before the discovery of winged
adults.

307 **ZORAPTERANS**
(Family Zorotypidae)

This is the only family in Order
Zoraptera, and it contains just one
genus, *Zorotypus*. Its members are
minute insects, measuring less than
⅛" (3 mm) long. Found from Texas to
Illinois, east to Maryland, and south to
Florida, they are seldom noticed and
have no economic importance.

Ice Insects
(Order Grylloblattodea)

These small wingless insects were once called "living fossils." Because they slightly resemble crickets (gryllids) and cockroaches (blattids), their scientific name is a composite taken from the order names of both insects. Only 10 species are known worldwide; the 6 North American species are limited to icy mountains in California, Oregon, Washington, Montana, and northwest Canada. Ice insects are very primitive dark brown to black creatures with thin flattened bodies, ⅝–1⅛" (15–30 mm) long. The glossy convex head has biting mouthparts, small compound eyes or none at all, and long antennae. Widely spread legs have 5-segmented tarsi. Long antennae-like appendages called cerci trail from the tip of the abdomen. The female has a distinctive sword-shaped ovipositor used to thrust eggs into the soil. Ice insects feed on hardy plants and other small insects found under logs and stones in the ice. Because they live in remote places, these insects were not discovered until 1914 in the Canadian Rockies. For most of the year, ice insects remain frozen. If placed upon a human hand, some species will perish from heat in only a few minutes.

Earwigs
(Order Dermaptera)

This relatively small order contains less
than 1,100 species worldwide and only
about 20 in North America. Earwigs
have slender, flattened bodies, $\frac{1}{4}$–$1\frac{3}{8}''$
(6–35 mm) long, with beadlike
antennae and a pair of very large
pincers, or cerci, at the tip of the
abdomen. These pincers are used in
defense and also to fold the soft fan-
shaped hind wings beneath the short
fore wings. Dermaptera means "skin
wings," referring to the leatherlike
texture of the fore pair. But few earwigs
fly, and many species are wingless.
Female earwigs are especially maternal.
They protect eggs laid in burrows on
the ground, rush to collect those that
are dispersed, and feed nymphs until
they are strong enough to fend for
themselves. Metamorphosis is simple:
the only discernible changes are the
increased number of antennal segments
and the progressive development of the
wings until sexual maturity. Earwigs
hide during the day under debris or in
dark spaces, emerging at night to feed
on plants, organic wastes, and smaller
insects. When disturbed, they emit a
liquid that smells like creosote. The
common name comes from a totally
unfounded superstition that these
insects crawl into people's ears at night
to bite them. Earwigs are harmless,
only occasionally damaging flower
blossoms.

LONG-HORNED EARWIGS
(Family Labiduridae)

Long-horned earwigs, ⅜–1″ (10–26 mm) in length, have very long antennae that are about half as long as the body, which has 14–24 segments. Unlike other families, the 4th, 5th, and 6th antennal segments combined are usually shorter than the first segment. Some species have wings, others are wingless. Members of this family are found on both the Atlantic and Pacific coasts.

89 Ring-legged Earwig
(*Euborellia annulipes*)

Description: ⅜–1″ (10–26 mm). Brown to black. *Brown spots at tip of pronotum and brown rings on yellow femora and tibiae.* Antennae black with white ring near tip of each segment. Wingless as adults. Antennae have 15–16 segments. *Cylindrical 2nd tarsal segment.*

Habitat: Gardens, fields, also sometimes indoors.

Range: Throughout the United States, primarily in the South and Southwest.

Food: Aphids and other small insects, plants and ground litter, and perhaps foodstuffs found in homes and grain warehouses.

Life Cycle: White cylindrical eggs, to 1/16″ (2 mm), are deposited in autumn in cup-shaped cavities excavated under debris, where female remains until they hatch in spring. Nymphs grow to maturity in 1 summer in the North or fewer months in the South, where there are often 2 generations a year.

At night both adults and nymphs climb plant stems to reach aphids and other small insects.

90 Riparian Earwig
(*Labidura riparia*)

Description: ¾–1″ (18–26 mm). Pale brown. Head and *2 lengthwise stripes on pronotum dark.* Fore wings and projecting hind wings dark. *Abdominal forceps of male slightly curved, widely separated at base, symmetrical.* Antennae with 14–24 segments. *2nd tarsal segment cylindrical.*

Habitat: Under litter on sea beaches and river banks, and in irrigated fields.

Range: North Carolina to Florida, west to California.

Food: Small insects, mites, and similar prey.

Life Cycle: Egg clusters are laid in nest cavity prepared in surface soil, usually under a light-proof covering, such as a board or a rock. Female remains with eggs and nymphs until they are able to go off on their own. Growth requires 1 summer or less in warmer parts of range. Often 2 generations a year.

Adults sometimes fly to artificial lights at night. They produce an unpleasant odor when captured.

COMMON EARWIGS
(Family Forficulidae)

These brownish, shiny insects, ⅜–⅝″ (9–15 mm) long, have relatively long antennae with 12–15 segments. Adults usually have wings but seldom fly. The pincerlike cerci of the male are strongly curved, while the female's are straight and almost parallel.

88 European Earwig
(*Forficula auricularia*)

Description: ⅜–⅝″ (9–15 mm) including abdominal forceps. Body reddish brown to almost black. *Antennae, legs, and elytra yellow.* Underside yellowish brown. *Forceps reddish brown;* male's curved, female's almost straight and parallel. Short wings do not cover abdomen. Antennae have 15 or fewer segments; *2nd tarsal segment lobed beneath.*

Habitat: Dark damp crevices and ground litter; grasses, herbs, shrubs, trees, and even buildings.

Range: Eastern Canada and southern New England; also the Pacific Northwest.

Food: Vegetables, orchard fruits, garden flowers, garbage, as well as mites and insect larvae and pupae.

Life Cycle: Female digs cup-shaped nest in upper soil, deposits mass of up to 30 oval grayish-white eggs, and stays with them until a few days after they hatch. Nymphs mature in about 10 weeks. Eggs and adults overwinter in soil or under boards and stones. 1 or 2 generations a year.

Gardeners often lure these tiny insects by spreading poisoned bran sweetened with molasses. In California a parasitic tachinid fly has been introduced from Europe to control this minor pest.

Stoneflies
(Order Plecoptera)

Members of this small order, which has 1,600 species worldwide, including about 400 in North America, are found near streams or lakes. The common name for the order derives from the fact that stoneflies spend most of their brief adult lives (about 2–3 weeks) crawling among stones in or near fresh water. Stoneflies have 2 pairs of wings. The broader hind pair folds flat, fanlike under the fore wings to cover almost the entire abdomen. Plecoptera means "folded wings." Many species have long antennae-like cerci at the tip of the abdomen. Mouthparts are usually vestigial. Although a few adults eat algae and pollen, most cannot feed at all. Stoneflies are such poor fliers that they usually run when disturbed. They are sometimes seen fluttering weakly around lights at night.
Metamorphosis is simple. Naiads are predators in well-aerated water and need 2–3 years to develop into adults. They resemble mayfly naiads but can be distinguished by longer antennae, 2 rather than 3 tail-like appendages, and gills located on the thorax and base of the legs rather than on the abdomen. An absence of stoneflies in streams, where they would ordinarily thrive, indicates a lack of oxygen and usually pollution.

340 GIANT STONEFLIES
(Family Pteronarcidae)

These brown or gray stoneflies are the largest in the order, often measuring to 2½" (64 mm), with a wingspan of 1⅝–3¼" (40–84 mm). They have relatively short cerci. Most adults apparently never eat. Nocturnal, they often fly to artificial lights in late spring and early summer. Naiads bear gills on the first 3 abdominal segments and live mostly in medium to large rivers, where they generally feed on plant materials.

341 Californian Pteronarcys
(*Pteronarcys californica*)

Description: Male 1¼–1⅝" (31–40 mm), wingspan of 2¼–2⅝" (58–66 mm); female 1⅝–1¾" (40–46 mm), wingspan of 2⅞–3¼" (72–84 mm). *Pale brownish gray,* paler beneath. Head blackish with a sharp erect tooth on head plate above base of antennae. *Prothorax has reddish midline above.* Cerci brown in middle, paler at ends. Legs dark brown. Wings have rows of blackish cross veins.

Habitat: Near fast streams in mountains.

Range: Montana and Kansas to New Mexico, west to California, north to Washington.

Food: Adult eats nothing. Naiad feeds on algae.

Life Cycle: Eggs laid in masses on the water drop to the bottom to hatch. Naiads crawl on the underside of stones and molt several times over 3 to 4 years before becoming adults.

The largest North American stoneflies belong to this genus. Mostly nocturnal, adults fly to lights but are poor fliers and usually spend their time crawling over rocks near the water.

339 GREEN-WINGED STONEFLIES
(Family Perlodidae)

Stoneflies in this family are usually greenish or yellowish with greenish wings and measure ¼–⅝" (6–15 mm) long. They can be recognized by the rectangular pronotum which has sharp or barely rounded corners and the tail-like cerci, which are longer than the width of the pronotum. There is no trace of gills beneath the thorax of adults. These diurnal insects live on foliage near streams and feed on pollen. The naiads are aquatic predators.

40, 41 COMMON STONEFLIES
(Family Perlidae)

This is a carnivorous family of stoneflies with many species. Its members are yellowish or brownish and have 2 tail-like cerci distinctly longer than the width of the thorax and remnants of gills close to the leg bases. From tip to folded wing, they measure ⅜–1⅝" (10–40 mm) in length with a wingspan of ⅞–2⅜" (22–60 mm). Adults are seen in late spring and summer. Naiads, to 1" (25 mm), are aquatic predators.

342 Californian Acroneuria
(*Acroneuria californica*)

Description: Male ⅞" (23 mm), wingspan of ⅞–1⅛" (23–30 mm); female 1¼" (31 mm), wingspan of 1⅝–2⅛" (40–54 mm). Flat pronotum wider at front than rear. Prothorax narrower than head, broader than long. *Brown, mottled with yellow.* Abdomen yellowish at tip, darker near rear. Cerci yellowish brown. 1st antennal segment brown, pale toward tip. Legs brown with a

dark band near tip of femora, tarsi
blackish. Wings have dark veins.

Habitat: Near rapid streams.

Range: California, Oregon, and Washington.

Food: Adult eats virtually nothing. Naiad
preys on smaller insects.

Life Cycle: Masses of eggs are deposited in water.
Naiads take 1 year or more to become
adults, depending on the water
temperature. Full-grown, they climb
onto stones near water's edge, complete
final molt, and fly away.

This common stonefly is found near
mountain streams in late spring and
summer. Naiads can survive in near-
freezing water. Like most stoneflies, it
is a poor flier and rarely strays from the
water's edge.

Grasshoppers and Crickets
(Order Orthoptera)

These insects are easily recognized by the long, powerfully muscled hind legs that they use for jumping. Both grasshoppers and crickets have large flat-sided heads with big compound eyes, large chewing mouthparts, and antennae of various sizes depending on the species. They have a prominent saddle-shaped pronotum and 2 pairs of wings. The fore wings, called tegmina, are leathery, long, and narrow and are not used for flying. The order name means "straight wing," referring to this fore pair. The hind wings are broad and membranous, having many radiating veins that permit the entire wing to be folded flat, fanlike, hidden under the fore wing. Although in some species the appendages, or cerci, at the tip of the abdomen are long, most cerci are short and fingerlike. Females of many species have long ovipositors, which are used to cut slits inside plants, where eggs are deposited. Metamorphosis is simple: the nymphs resemble adults except they have no wings or sex organs.

Males are known for the musical sounds made when they rub together roughened portions of their wings or legs. So distinctive are these "songs" that grasshoppers and especially crickets can be recognized from their characteristic sounds. This music-making serves as a form of warning, a way to establish territory, or a move in courtship. Both grasshoppers and crickets can be troublesome agricultural pests. Crickets damage plants by cutting them with their ovipositors, and some grasshoppers eat foliage, destroying crops when they descend upon fields in huge swarms. The order includes 1,000 species in North America and over 23,000 species worldwide.

PYGMY GRASSHOPPERS
(Family Tetrigidae)

Pygmy grasshoppers, measuring less than ¾" (20 mm) long, can be recognized by the pronotum, which extends far back over the abdomen, and tapers to a point. Most are brown. Some have long functional wings, while others have reduced wings and cannot fly. Females are usually larger and heavier than males. Adults and nymphs feed on algae and other organic matter in wet soil. They are among the few grasshoppers that survive the winter as adults and are frequently seen in early spring and summer. These insects are also called grouse locusts.

253   Aztec Pygmy Grasshopper
(*Paratettix aztecus*)

Description: ¼–½" (5–12 mm). Dull gray, brown, or reddish, *sometimes edged with yellow along sides of pronotum, which extends backward beyond tip of abdomen.* Femora of middle legs extremely long. Side ridges of pronotum weak toward front. Front projection of head no wider than 1 compound eye.

Habitat: Desert soil, especially gravelly edges of intermittent streams.

Range: Mexico into California, Nevada, and Arizona.

Food: Algae and decaying plant material.

Life Cycle: Eggs are concealed in soil, usually in spring. Female overwinters. Life-span 2 or more years.

Pygmy grasshoppers are found throughout North America. They are usually seen in spring and summer. All *Paratettix* species have a narrow ridge and 12–15 segments in antennae. Some species include both long-winged and short-winged members. A few species are mostly females and produce asexually by parthenogenesis. Species previously included in genus *Telmatettix*.

SHORT-HORNED GRASSHOPPERS
(Family Acrididae)

Short-horned grasshoppers get their name from their short, horn-shaped antennae, which are usually less than half the length of the body. They are ½–3⅛″ (12–80 mm) long. These common grasshoppers have 3-segmented tarsi and flat, round hearing organs called tympana on each side of the first abdominal segment. The pronotum does not extend over the abdomen. These grasshoppers produce a low buzzing sound by rubbing the roughened surfaces of their hind wings against the fore wings. Females lay eggs in large masses below the surface of the ground. Depending on the species, a mass contains 8–25 eggs, which usually hatch in summer. Many members of this large family attack crop plants. Some are called locusts from the Latin word for grasshopper.

273 Creosote Bush Grasshopper
(*Bootettix argentatus*)

Description: ¾–1″ (20–26 mm). Green, *somewhat marked with silvery white;* some brown and black. Female's antennae rather short, no longer than fore femora. Male's fore wings have many transparent cells near margin.

Habitat: Arid lands with creosote bush.

Range: New Mexico, Arizona, and southern California.

Food: Foliage of creosote bush.

Sound: Scratchy, gritty noises at night.

Life Cycle: Eggs are concealed a few inches below surface of soil, hatching after spring rain, when nymphs find soft expanding foliage to eat. Rate of growth varies greatly. Adults are found throughout the year.

This well-camouflaged species stays in shade when possible, flying only if approached closely.

262 Lubber Grasshopper
(*Brachystola magna*)

Description: 1½–3⅛" (38–79 mm). Brown to gray-green with *brown, green, or blue and black markings.* Hind tibiae have both inner and outer spines at tip. Pronotum has 3 smooth longitudinal ridges. Fore wings pink with black spots, shorter than pronotum. Small hind wings, useless for flight, concealed by fore wings.

Habitat: Rocky and gravelly soil and sparse vegetation.

Range: Minnesota to Arizona and Mexico, north to Montana.

Food: Vegetation and dead insects on roads.

Life Cycle: Eggs are concealed in soil, hatching after a soaking rain. Adults are active August–October. Probably 1 generation a year.

The Lubber Grasshopper eats and sometimes destroys the scanty vegetation of its habitat. This insect retains so much internal moisture that after death it may rot.

259 Painted Grasshopper
(*Dactylotum bicolor*)

Description: ¾–1⅜" (20–35 mm). *Boldly patterned* in red-orange, yellow, blue-black, and white. Hind tibiae spiny.

Habitat: Desert grasslands; sometimes in alfalfa fields.

Range: Northern Mexico north to Arizona, Texas, and the foothills of the Colorado Rocky Mountains.

Food: Foliage of desert vegetation, especially grasses; also alfalfa.

Life Cycle: Female works abdominal tip into soft soil, depositing up to 12 egg masses of about 100 eggs each. Eggs overwinter, hatch in late spring, and nymphs become adults by midsummer. 1 generation a year.

This grasshopper is named for its bright coloring. The Rainbow Grasshopper (*D. pictum*), same size, has narrower wings and black and yellow bands, shading to orange or pink. It ranges from northern Texas and northeastern New Mexico to Montana.

268 Carolina Locust
(*Dissosteira carolina*)

Description: 1⅜–2" (35–50 mm). Cinnamon-brown to grayish tan, matching color of dry soil. *Hind wings black with broad pale yellow border.* Pronotum has a high, narrow middle ridge.

Habitat: Roadsides and dry fields.

Range: Throughout North America.

Food: Grasses, other herbaceous plants, and sometimes beans.

Sound: Very fast purring or beating, then a fluttering sound; made only in flight.

Life Cycle: Egg masses, each containing 20–70 eggs, are deposited in soft soil. Nymphs' speckled bodies blend inconspicuously with ground. Usually 1 generation a year.

This locust is less destructive than most other species in the genus. The related Long-winged Locust (*D. longipennis*), 1⅜–1¾" (35–44 mm), is brownish yellow with brown-spotted fore wings and black hind wings with a narrow yellow border. It is quite destructive, eating plains grasses at moderate elevations from southwestern Kansas to Texas, west to New Mexico, north to Idaho.

256 Dragon Lubber Grasshopper
"Gray Dragon"
(*Dracotettix monstrosus*)

Description: ¾–1¾" (18–44 mm). Ash gray. Head bulging. *Pronotal crest has 3 deep notches.* Wings short, flightless.
Habitat: Arid soil and sparse vegetation.
Range: Southern California into Mexico.
Food: Desert plants and dead insects.
Life Cycle: Eggs are laid in soil but do not hatch until rain, which allows nymphs to find newly sprouted vegetation to eat. Nymphs develop rapidly and gradually disperse.

These grasshoppers hop on roads, feeding on dead insects, including their own kind. They are often killed by passing vehicles, especially the females, which are relatively heavy and cannot leap far.

260 Three-banded Grasshopper
"Banded Range Grasshopper"
(*Hadrotettix trifasciatus*)

Description: 1–1¾" (25–44 mm). Brown. Antennae thick, black. Hind tibiae red, *inner femora blue-black on basal ⅔.* Fore wings have blackish crossbands. Hind wings yellow, narrowly bordered and tipped with black (not visible at rest).
Habitat: Sandy and rocky areas, open grassy plains.
Range: Manitoba to Texas and Mexico, northwest to Alberta.
Food: Grasses and herbaceous foliage.
Life Cycle: Up to 12 egg masses, each containing about 100 eggs, are deposited in soft

soil. Eggs overwinter and hatch in late spring. Nymphs become adults by June in the South or October in the North. 1 generation a year.

This handsome grasshopper sometimes has pink on its body and legs.

264 Differential Grasshopper
(*Melanoplus differentialis*)

Description: 1⅛–1¾" (28–44 mm). Shiny brownish yellow. Antennae brownish yellow or red. Hind femora yellow with a black herringbone pattern. *Hind tibiae yellow with black toothlike spines,* all tarsi yellowish. Glossy pronotum and fore wings. Male's cerci have a prominent projection at side.

Habitat: Grasslands and open woods.

Range: Throughout the United States.

Food: Grasses if available; crop plants and fruits.

Life Cycle: Female presses 8 egg masses, each with up to 11 eggs, into soft soil. 1 generation in early summer.

These grasshoppers do not migrate and are often destructive.

271 Red-legged Locust
(*Melanoplus femur-rubrum*)

Description: ¾–1" (18–25 mm). Dark brown to greenish, yellow, or red-brown. Hind femora have herringbone pattern or black spots. *Hind tibiae bright red to yellowish with black spines.* Pronotum has no appreciable crest. Fore and hind wings of male at rest project beyond abdominal tip. Colors paler in the South.

Habitat: Fields, vacant lots in cities and suburbs, open woods, or along irrigation ditches in more arid areas.

Range: Atlantic Coast to Florida, west to
Arizona, north to Alberta.

Food: Native grasses, introduced weeds,
alfalfa, grains, and crops in the
Southwest, soybeans in the
Midwest.

Life Cycle: Female thrusts several egg masses, each
containing about 20 eggs, into soft
soil, where they overwinter. Nymphs
appear in spring and become adults by
June in the South, where they feed
until December; they mature later in
the North.

Both sexes of these locusts transmit
poultry tapeworms and also parasites
that mature in quail, turkeys, and
guinea fowl. The similar Rocky
Mountain Grasshopper (*M. spretus*),
1⅛–1½" (29–38 mm), is brownish
above with dark markings. It reached
plague proportions in the West before
1900, but now is probably extinct. The
"Grasshopper Glacier" near Cooke,
Montana, contains millions of
embedded Rocky Mountain
Grasshoppers, presumably from
swarms that settled on the glacier and
froze.

265 Spur-throated Grasshopper
(*Melanoplus ponderosus*)

Description: 1–1⅜" (25–34 mm). Mostly yellowish
to pale brown. *Hind tibiae* often orange
or reddish *with yellowish and brown rings
near base. Blunt projection below prothorax*
between bases of fore legs.

Habitat: Open fields and grasslands.

Range: Tennessee and Mississippi to New
Mexico, north to Iowa and Illinois.

Food: Native grasses.

Life Cycle: Egg masses, each containing about 20
eggs, overwinter in soft soil. Nymphs
appear in spring and become adults by
June in the South, where they feed
until December.

The Migratory Grasshopper (*M. sanguinipes,* formerly *M. bilituratus*), 1–1⅜" (25–34 mm), also has a spur-throat. It is pale brown and yellow with red hind tibiae. This species ranges from the Atlantic Coast to Georgia, southwest to Mexico, north to Alaska. Spectacular mass flights were noted in 1938 and 1940 but have not been repeated since then.

278 Two-striped Grasshopper
(*Mermiria bivittata*)

Description: 1⅛–2⅛" (29–55 mm). Pale green to brown with *2 dark brown stripes* extending backward from compound eyes over head and thorax, becoming vague on fore wings. Hind tibiae have 15 or more spines in outer row; tibiae reddish.

Habitat: Grasslands.

Range: South Carolina to Florida, west to Texas, north to Nebraska.

Food: Grasses.

Life Cycle: Female works abdominal tip into soft moist soil, deposits eggs in compact mass as deep as possible. Nymphs work way to surface, where they begin feeding. Adults are active in summer. Probably 1 generation a year.

This grasshopper flies as well as leaps to avoid disturbance. The Three-striped Grasshopper (*M. maculipennis*) has an additional dark stripe along its midline and ranges from Mexico to Alberta and Saskatchewan.

276 Panther-spotted Grasshopper
(*Poecilotettix pantherina*)

Description: 1–1⅜" (25–34 mm). Green, greenish brown, or greenish yellow. *Head and pronotum have small black and white dots.*

Antennae black, ringed with white on alternate segments. Hind tibiae pale blue-green, legs otherwise brownish or greenish yellow with rows of small black spots.

Habitat: Prairies, on herbs such as sunflower, daisy, and other composites, especially groundsel tree.

Range: Mexico to Arizona.

Food: Foliage of composite plants.

Life Cycle: Mass of cylindrical eggs overwinter in soft soil. Nymphs work their way to the surface and attain full size in a few weeks. Adults are active April–November. 1 generation a year.

This grasshopper's bright colors fade rapidly after death. The related Long-winged Grasshopper (*P. longipennis*), same size, is greener and has red stripes on its head and pronotum. It feeds on baccharis and other shrubby composites in Arizona, Nevada, and California.

---

267 Southeastern Lubber Grasshopper
(*Romalea microptera*)

Description: 2–2¾″ (50–70 mm). Short winged. *Hind wings pink to red with black borders.* Two different color forms: either dull straw-yellow with black markings and, in the South, a reddish stripe on fore wing; or mostly black with a yellow stripe around edges of pronotum, along back ridge, and on rear edge of abdominal segments.

Habitat: Roadsides, field edges, and gardens.

Range: North Carolina to Florida, west to Louisiana, northeast to Tennessee.

Food: Herbaceous plants of many kinds.

Life Cycle: Eggs are pressed into soft soil. Nymphs usually emerge in early spring. Adults often appear in June and are active through November.

This slow-moving grasshopper cannot fly. When handled, it gives off a foul-smelling liquid, perhaps to repel predators. Its spiny hind tibiae can cut human skin.

266, 269 **Alutacea Bird Grasshopper**
"Leather-colored Bird Grasshopper"
(*Schistocerca alutacea*)

Description: 1⅛–1¾" (30–45 mm). *Greenish yellow to dark greenish brown with yellow midline stripe.* Fore wings blackish yellow without obvious spots. Hind tibiae red to green, or reddish yellow, bright yellow, or black.

Habitat: Tall grass and open sandy woods.

Range: New England to Virginia, southwest to Texas and New Mexico, north to Canadian Northwest Territories.

Food: Grasses.

Life Cycle: Female thrusts mass of about 25 eggs 1" or more into the soil. Nymphs emerge in a week or less, unless soil is very dry. They hatch when it rains.

Because it can fly rapidly over great distances, this insect has earned the name bird grasshopper.

---

263 **American Bird Grasshopper**
(*Schistocerca americana*)

Description: 1⅝–2⅛" (40–55 mm). *Brown and beige along back and sides with brownish-yellow midline stripe.* Yellow on sides and underneath prothorax. *Hind wings clear,* hind femora without crossbands, tibiae brownish red.

Habitat: Grasslands and forest edges.

Range: Florida to Mexico year-round; north to Ontario and southern parts of Canadian prairie provinces in summer only.

Food: Grasses, other herbaceous plants, tree leaves.

Sound: Whistling buzz of wings in flight.

Life Cycle: Eggs are laid in a mass in soil. Nymphs

work their way to the surface. Cool weather reduces activity and growth in temperate regions.

A particularly alert insect, the American Bird Grasshopper will fly into a tall tree if disturbed. It cannot tolerate heat, migrating north in summer.

270   Green Valley Grasshopper
(*Schistocerca shoshone*)

Description: 1½–2¾" (38–70 mm). Green with yellow midline stripe on head and pronotum. *Hind tibiae red-pink.*

Habitat: Tall grasses and open, sandy woods.

Range: Colorado to Texas and Mexico, northwest to California.

Food: Grasses.

Life Cycle: Egg masses are thrust into soft soil, hatching in a week or less except when soil is very dry. Often nymphs do not hatch until rains soften soil.

One of the largest grasshoppers, the Green Valley Grasshopper sometimes appears in devastating hordes and severely damages range grasses. The related Vagrant Grasshopper (*S. vaga*), approximately same size, is brown with a pale brown midline stripe. It occurs in the same range and feeds on grapes as well as grasses.

281   Toothpick Grasshopper
(*Stenacris vitreipennis*)

Description: 1–1⅛" (24–29 mm). Green, usually with a faint yellowish stripe on each side. Hind tibiae green. Fore wings pointed at tips, extending to or beyond end of abdomen. *Spine between fore legs.*

Habitat: Vegetation close to water.

Range: North Carolina to Florida.

Food: Foliage of swale and marsh plants.
Life Cycle: Eggs overwinter in soil and hatch in late spring. Nymphs become adults by midsummer. 1 generation a year.

A wary grasshopper, it flies away or dodges behind upright stems when approached.

275 Horse Lubber Grasshopper
(*Taeniopoda eques*)

Description: 1½–2½" (38–64 mm). Stout. Shiny black with orange or yellow. Veins on fore wings yellow on black background. *Hind wings red with black borders,* large enough in male to permit flight.
Habitat: Grasslands and woods with live oak.
Range: Texas and Arizona to Mexico.
Food: Desert annuals and foliage of perennial shrubs, including mesquite.
Life Cycle: Eggs overwinter in soil and survive mild drought, hatching after spring rains. In good years adults appear August–November. In very dry years eggs do not hatch.

These grasshoppers flutter about and leap into bushes when disturbed. Males snap their fore wings together noisily. They are occasionally destructive.

261 Pallid-winged Grasshopper
(*Trimerotropis pallidipennis*)

Description: 1¼–1⅝" (31–41 mm). Grayish to brownish or reddish. *Fore wings with 1 or 2 black crossbands.* Hind wings whitish to yellow except for dark edge. Hind tibiae yellow or yellowish brown.
Habitat: Many different habitats, sea level to 9,150' (2,789 m), including areas of alkaline soil.
Range: Manitoba to Oklahoma, Mexico to British Columbia.

Food: Herbaceous plants on range and arid
         land.
Life Cycle: Up to 12 egg masses, with 100 eggs
         each, are deposited in soft soil. Eggs
         overwinter, hatch in late spring.
         Nymphs become adults by June in the
         South or October in the North. 1
         generation a year.

The related Seaside Grasshopper
(*T. maritima*), ¾–1⅜" (19–34 mm),
has a pale brown to pale gray body,
finely speckled with black; buff to pink
hind tibiae; and pale yellow hind wings
edged with dark gray. It often snuggles
into beach sand along the Atlantic
Coast and the Great Lakes.

---

274   Great Crested Grasshopper
"Dinosaur Grasshopper"
(*Tropidolophus formosus*)

Description: 1½–1¾" (37–45 mm). *Toothed crest
         along full length of pronotum.* Green with
         brown markings. Hind wing orange,
         partially edged with black. More than
         12 spines in outer row on hind tibiae.
Habitat: Meadows.
Range: Kansas to Texas and Mexico, west to
         Arizona, north to Wyoming.
Food: Low-growing mallows and horse nettle,
         some grasses.
Sound: High-pitched rattling produced in
         flight.
Life Cycle: Up to 12 egg masses, with 100 eggs
         each, are deposited in soft soil. Eggs
         overwinter and hatch in late spring.
         Nymphs become adults by June in the
         South or October in the North. 1
         generation a year.

The distinctive crest on this
grasshopper reminds some collectors of
the fin-backed dinosaur (*Dimetrodon*).
The crest is so pronounced on the
nymphs that it has inspired the name
"Dinosaur Grasshopper."

LONG-HORNED GRASSHOPPERS AND KATYDIDS
(Family Tettigoniidae)

These large insects, ½–3″ (14–75 mm) long, have extremely long antennae and 4-segmented tarsi. Males have flat, round hearing organs called tympana located at the base of the front tibiae. Females have flat, swordlike ovipositors. Long-horned grasshoppers range from dark brown to greenish hues; most katydids are green. Many members of this large family are known for their songs, especially the katydid with its shrill *katy-DID-katy-DIDN'T*. Most live in forest trees and shrubs, where they feed on leaves. A few species prey on insects. Eggs are inserted into plant tissues, where they usually overwinter.

---

287 Oblong-winged Katydid
(*Amblycorypha oblongifolia*)

Description: 1⅝–1¾″ (40–45 mm). Leaf-green or rarely pastel pink or tan. Hind femora shorter than fore wings. *Fore wings* are rounded at tips and *at least 3 times as long as broad.* Ovipositor flat, upcurved, shorter than abdomen. Head rounded. Antennae longer than body.

Habitat: Deciduous woods and gardens, on trees and bushes.

Range: Quebec to Pennsylvania, southwest to Texas, north to Nebraska.

Food: Foliage of deciduous woody plants.

Sound: A series of lisping chirps.

Life Cycle: Eggs are thrust into plant tissues, overwinter, and hatch in late spring. 1 generation a year.

The Round-winged Katydid (*A. rotundifolia*), 1¼–1⅜″ (31–34 mm), appears humpbacked; it is also leaf-green or pink, and has hind femora about the same length as the oval fore wings. It ranges from Nova Scotia to southern New Jersey, southwest to Texas, north to Minnesota.

257   Mormon Cricket
(*Anabrus simplex*)

Description: 1–2⅜" (25–60 mm). Dark brown to bluish black, sometimes with pale markings. Large pronotum extends backward, concealing female's vestigial wings and almost covering male's reduced wings. Ovipositor upcurved, swordlike, as long as body. Antennae as long as body.

Habitat: Open fields.

Range: Missouri River to northern Arizona, west to southeastern California, north to Alberta.

Food: Lupine, sagebrush, and many plants, including grains and vegetables.

Sound: A hoarse chirp, repeated at intervals.

Life Cycle: Dark brown eggs are deposited below soil surface in midsummer. They turn gray, overwinter, and hatch in spring. Up to 100 nymphs emerge, maturing in about 60 days.

This common cricket got its common name after thousands suddenly attacked the Mormon pioneers' first crops in Utah in 1848. Fortunately, many California gulls arrived in time to devour the crickets and save the crops.

255   Short-legged Shield-back Katydid
(*Atlanticus testaceus*)

Description: ¾–2" (18–50 mm). Pale brown. Front margin of fore wings black. Fore wings more than ½ length of prothorax. *Flat shieldlike prothorax* extends back over 1st abdominal segment. Ovipositor as long as body. Long antennae reach tip of ovipositor.

Habitat: Ground and low vegetation in
dry upland woods.

Range: New England to Kentucky, north to
Minnesota and Ontario.

Food: Foliage.

Sound: Song alternates between about 3
seconds of trill and an equal interval
without any sound.

Life Cycle: Eggs overwinter in plant tissue or soil.
Adults seen June–September. 1
generation a year.

Several related species occur east of the
Rocky Mountains, differing in the form
of the pronotum and in preferred foods.

---

285 California Katydid
(*Microcentrum californicum*)

Description: 1¾–2⅜" (45–60 mm). Green. Hind
femora short, no more than ⅔ length of
fore wings. Pronotum incurved in front
at midline without any blunt toothlike
projection. *Rear border of fore wings form
an obtuse angle,* giving silhouette a
humped appearance.

Habitat: Oak woods.

Range: Arizona to California.

Food: Oak foliage.

Sound: Clicking.

Life Cycle: Eggs are cemented overlapping one
another on a tree branch. Nymphs
usually mature by summer.

The green nymphs are seldom seen
because they feed high up in treetops.

---

286 Angular-winged Katydid
(*Microcentrum retinerve*)

Description: 2–2½" (50–65 mm). Pale green, head
often yellowish green. Front edge of
pronotum straight. Hind femora short,
usually less than ⅔ length of fore
wings. Antennae separated at base.

Broad, obtusely angled fore wings.
Ovipositor curves or bends abruptly
upward.

Habitat: Woods and meadows, on foliage of
trees and shrubs.

Range: Massachusetts to northern Georgia,
west to Arkansas, north to Ontario.

Food: Foliage of trees and shrubs, especially
oak, willow, cottonwood, and citrus
trees.

Life Cycle: Female cements flat, oval gray eggs to
leaf or plant stems in late summer.
Eggs overwinter, hatch, and mature by
midsummer. 1 generation a year.

The related Broad-winged Katydid
(*M. rhombifolium*) is darker green. Its
pronotum has a blunt tooth at the
middle of the front edge, and its fore
wings are at least 3 times as long as
wide. It ranges from New York to
northern Florida, west to California,
north to Oregon.

254   Keeled Shield-back Katydid
(*Neduba carinata*)

Description: ¾–1½″ (19–39 mm). Brown to gray-
brown, usually with yellowish-brown
mottling or stripes on pronotum and
sides of thorax. *Pronotum narrows at
front.*

Habitat: Woods and meadows.

Range: Coastal areas from California to British
Columbia.

Food: Foliage of trees and shrubs.

Life Cycle: Flat, oval gray eggs are cemented to
plant stems in late summer. They
overwinter and hatch the following
spring. Growth is slower in the North
than in the South. 1 generation a year.

The larger Splendid Shield-back
Katydid (*N. ovata*), 1–1⅝″ (26–
40 mm), has an oval pronotum and
green markings. It ranges from Nevada
to the southern California deserts.

282 Nebraska Cone-head
(*Neoconocephalus nebrascensis*)

Description: Male 1⅛–1¼″ (28–31 mm); female
1⅛–1⅜″ (28–36 mm), ovipositor 1⅛–
1½″ (29–39 mm). Body and wings
leaf-green or tan, with narrow yellowish
edging. Black beneath *conical projection
of head.* Pronotum hind edge rounded.
Ovipositor curved, swordlike.

Habitat: Marshes, thickets, and cornfields.

Range: Maine to Virginia and Tennessee, west
to Nebraska, north to Minnesota.

Food: Flowers and foliage of trees.

Sound: *Tsip-tsip,* almost continuous by day,
sometimes at night. At close range a
loud monotonous buzz.

Life Cycle: Eggs are thrust into soft plant tissues,
where they overwinter. 1 generation a
year in the North, 2 in the South.

The Robust Cone-head (*N. robustus*), to
2⅜″ (60 mm), has a narrow yellow line
on each side of the flat-topped
prothorax. It ranges from the
Appalachians to the Great Lakes and
Atlantic Coast, and is also in
California. The eastern Sword-bearer
Cone-head (*N. ensiger*), to 2″ (50 mm),
is either green with yellow edging on
its prothorax, or dark tan with black
speckling on its fore wings.

272 Gladiator Katydid
(*Orchelimum gladiator*)

Description: 1⅜″ (34–36 mm). Green with tan or
brown markings. Undersurface of
prothorax has several long slender
spines. Ovipositor curved, almost as
long as abdomen. Wings slender, hind
pair at rest projecting beyond tips of
fore pair.

Habitat: Grasslands and roadsides with tall
grass.

Range: Quebec and New England to
Tennessee, west to northeastern

California, north to Washington.

Food: Grass.

Sound: Trilling alternating with staccato clicks.

Life Cycle: Female cements eggs among stiffer grass stalks, where they overwinter in the North. Green nymphs mature by late summer. Probably 1 generation a year.

Each species in this genus has its own sound and range. Identification of the different species is based on a comparative study of male genitalia, of the projecting conical midpart on the head in both sexes, and of the ovipositor.

258 Mescalero Shield-back Katydid
(*Plagiostira mescaleroensis*)

Description: ¾–2" (20–50 mm). Brownish green marked extensively with ivory-white. Black-speckled fore wings less than ½ as long as extended *shieldlike top of prothorax*, which reaches 1st abdominal segment. *2 white lines extend forward from rear edge of pronotum*, running parallel, then diverging. Ovipositor longer than body. Antennae reach halfway to ovipositor.

Habitat: Dunes.

Range: Mescalero dunes of New Mexico.

Food: Dune vegetation.

Life Cycle: Eggs are thrust deep into dune plants, hatching after heavy rains. They matur rapidly. Adults cannot survive drought

The Gillette's Shield-back Katydid (*P. gillettei*), 1–2⅝" (25–68 mm), is yellowish brown with black wings and heavy ridges along each side of the pronotum. It ranges from California to Colorado and Nevada, feeding on various desert plants.

288 True Katydid
"Northern Katydid"
(*Pterophylla camellifolia*)

Description: 1¾–2⅛" (45–55 mm). Leaf-green.
Fore wings convex, oval, *crossed by many conspicuous veins* between front margin and longitudinal vein. Head pointed at front in midline above. Side plates of prothorax about as long as high.

Habitat: Woodlands and forests.

Range: Massachusetts to Florida, west to Texas and Kansas, northeast to Ontario.

Food: Foliage of deciduous trees.

Sound: Both sexes make sounds. The most common call is the loud two-part *katy-DID;* less often is the three-part *katy-DIDN'T*.

Life Cycle: Eggs are laid on bark and young stems in fall, overwinter, and hatch in spring. 1 generation a year.

Nymphs are seldom seen near the ground, but adults perch on low shrubs, sometimes along roadways and forest margins. This katydid is known as the True Katydid because it was the first species to have its call transcribed. It can be heard easily when chirping from treetops to ground level.

283, 284 Fork-tailed Bush Katydid
(*Scudderia furcata*)

Description: 1⅝–2" (40–50 mm). Leaf-green, fore wings long and narrow. *Male has forked sex organs* (claspers) *at tip of abdomen,* hence common name.

Habitat: Woodlands and forests.

Range: Throughout North America from Mexico to limit of deciduous trees in Canada.

Food: Foliage of bushes and trees.

Sound: 1 or 2 strong rasping pulses close together, separated by pauses of varying length. Female may chirp in response.

Life Cycle: Oval flat eggs are attached like overlapping shingles to leaves and twigs.
They hatch in spring. 1 generation a
year.

This katydid sometimes destroys young
oranges in California by nibbling holes
through the rind. The northeastern
Broad-winged Bush Katydid
(*S. pistillata*), slightly larger, has
tapering fore wings resembling willow
leaves. The eastern Curve-tailed Bush
Katydid (*S. curvicauda*), same size, has
narrower, tapering fore wings.

CAMEL CRICKETS AND KIN
(Family Gryllacrididae)

Camel crickets and their kin have
humped, tan or gray bodies, ⅜–2″
(10–50 mm) long, and huge hind
femora. Most are wingless. Their
threadlike antennae are as long as their
bodies or longer, and all tarsi have 4
segments. Most males do not produce
sounds, and both sexes usually lack
hearing organs, or tympana. All
members of this family are nocturnal,
living in dark caves and basements or
under loose bark and rocks.

248 Spotted Camel Cricket
"Cave Cricket"
(*Ceuthophilus maculatus*)

Description: ⅜–¾″ (10–19 mm). *Humpbacked.* Very
long antennae. Glossy dark brown
spotted or streaked with yellowish to
reddish brown on thorax and hind
femora. Ovipositor ½ length of
abdomen with 5–7 spines below near
tip.
Habitat: Dark places under logs, stones, or bark
or in soil, damp basements, and humic
caves.

Range: Maryland to Arkansas, north to South
Dakota and Manitoba.
Food: Fungi, roots, foliage, fruits, or dead
insects, even its own species.
Life Cycle: Eggs are inserted singly into soil or into
bat droppings inside of caves.

This cricket's long antennae and coarse
leg bristles help alert it to nearby
predators, such as spiders or centipedes.
Only a specialist can reliably identify
species.

249 Secret Cave Cricket
(*Ceuthophilus secretus*)

Description: ⅜–1″ (10–25 mm). *Humpbacked*. Very
long antennae. Yellowish gray to dark
brown. Ovipositor ½ length of body.
Habitat: Dark places, primarily caves; also in
deciduous forests.
Range: Texas.
Food: Fungi, roots, foliage, fruits, or dead
insects, even its own kind.
Life Cycle: Eggs are inserted singly in loose soil in
caves in summer; overwinter there.
Adults emerge late spring–early
summer.

These crickets come out on warm
humid nights, feed, then return to
their caves. Their waste products are
recycled by other cave-dwellers and are
an important source of energy for the
entire cave ecosystem.

247 Jerusalem Cricket
(*Stenopelmatus fuscus*)

Description: 1⅛–2″ (30–50 mm). *Humpbacked*.
Very long antennae. Shiny amber-
brown with darker brown crossbands on
abdomen. Wingless. Head large with
wide space between small compound
eyes and jaws. Legs short; hind tibiae

have 2 rows of spines with inside row shorter and flattened on inner surfaces.

Habitat: Hillsides and valley slopes, under rocks.

Range: Nebraska to New Mexico and Mexico, north along the Pacific Coast to Washington, east to Montana.

Food: Other insects, plant roots, decaying vegetation, and potato tubers.

Sound: A scratchy noise.

Life Cycle: Female prepares depression in soil for masses of oval white eggs. Nymphs and adults are extremely slow-moving. Adults are particularly slow during spring mating season. Female often devours mate. 1 generation a year.

Active both day and night, these crickets often leave distinctive smooth tracks on dusty roads by dragging their bulky abdomens. Other species of the genus are distinguished by the number and form of tibial spines.

TRUE CRICKETS
(Family Gryllidae)

True crickets have broad bodies measuring ⅜–1″ (9–25 mm) long. They resemble long-horned grasshoppers but have only 3 tarsal segments. Their long, tapering antennae are ⅓–½ as long as the body, and long cerci are conspicuous at the end of the abdomen. Females have elongate cylindrical or needlelike ovipositors, which are used to lay eggs singly in the ground or in rows in plant tissues. Males produce rapid chirping or trilling sounds with a stridulating organ, similar to that of katydids, located at the base of the fore wings. Flat, round hearing organs, or tympana, are located at base of fore tibiae. Crickets' songs are much higher pitched than those of grasshoppers, and their fore wings are shorter.

251 House Cricket
"Cricket on the Hearth"
(*Acheta domestica*)

Description: Male ⅝–¾" (15–19 mm), female
⅝–⅞" (15–21 mm). *Stout,* almost
cylindrical. Yellowish brown. *Hind
wings at rest project beyond abdomen and
cerci.*

Habitat: Mostly indoors, in bakeries, breweries,
and kitchens.

Range: Almost worldwide.

Food: Crumbs and food scraps.

Sound: Common song is a triple chirp.
Courtship song is a continuous trill.

Life Cycle: Eggs are deposited singly in cracks and
crevices during almost any season.
Nymphs and adults are present year-
round. Females emit a chemical
repellent that causes other females to
disperse and start families elsewhere.

The House Cricket is often an indoor
nuisance because of its continuous
chirping. It emerges most often at
night to seek vegetable refuse.

252 Field Cricket
(*Gryllus pennsylvanicus*)

Description: ⅝–1" (15–25 mm). Black to dark
reddish brown. Black antennae longer
than body; cerci hairy, longer than head
and prothorax combined. *Wings do not
project beyond cerci.*

Habitat: Undergrowth where there is moderate
humidity and protection from night
winds and cold.

Range: Throughout North America to Alaska.

Food: Plant materials outdoors, including
seeds and seedlings of wild and crop
plants, small fruits, and when
available, dying and dead insects.

Sound: Common song is a series of triple
chirps. Courtship song is a continuous
trill at a pitch near the upper limits of
audibility for the human ear.

Life Cycle: Female inserts eggs singly deep into the soil. Eggs overwinter in the North, where all unmatured nymphs and adults die of the frost. In the South nymphs and adults may overwinter and produce 3 generations a year.

This cricket enters houses in autumn, attracted by the warmth. In courtship the male dances about and "sings" to excite the female.

---

277 California Tree Cricket
(Oecanthus californicus)

Description: ½–⅝″ (13–15 mm). Pale green, female sometimes brownish green. Reddish dots on 1st and 2nd antennal segments and pronotum. *Male's fore wings paddle-shaped.* Nymph is whitish.

Habitat: Deciduous woods and forests.

Range: Nevada to California; sometimes Texas and Oklahoma to Utah, north to Idaho.

Food: Adult feeds on aphids and caterpillars. Nymph eats foliage, flowers, and young fruits.

Sound: Sustained trilling.

Life Cycle: In darkness male courts female with song until female nudges male and nibbles secretion from gland in its back. After mating, female cuts pinlike holes into thin bark of woody branches, dabs a little excrement in each hole, and then embeds eggs. Eggs overwinter and hatch in spring. Nymphs grow slowly and mature by midsummer.

This cricket is often mistaken for the Snowy Tree Cricket in California, where both species occur.

280 Snowy Tree Cricket
(*Oecanthus fultoni*)

Description: ½–⅝" (13–15 mm). Pale green.
Wings transparent. Antennae with
black spot on 1st and 2nd segments.
Male's fore wings paddle-shaped, held flat
over back, concealing all but tips of
folded hind wings and short cerci.
Female's fore wings narrow, curved
closely around body. Nymph is almost
white.

Habitat: Deciduous woods and forests.

Range: Most of North America, except
southeastern states.

Food: Adult feeds on aphids and caterpillars.
Nymph eats flowers, foliage, and young
fruit.

Sound: A continuous trill. Pitch and harmonics
are higher in warmer temperature.

Life Cycle: Female selects mate and nudges male
until it stops singing. After mating,
female cuts pinlike holes in thin bark of
woody branches, dabs excrement inside,
embeds eggs, and coats them with
secretion. They overwinter and hatch in
spring. Nymphs attain maturity by
midsummer.

Often several males in the vicinity
make the same sound, giving a
stereophonic effect. Formerly known as
O. niveus.

279 Black-horned Tree Cricket
(*Oecanthus nigricornis*)

Description: ½" (13 mm). Pale greenish yellow to
very dark green, almost black.
Antennae with a few black or brownish
marks on 1st segment, rest of antennae
black. Legs mostly black. *Males fore
wings paddle-shaped.*

Habitat: Deciduous woods and forests.

Range: Nova Scotia to North Carolina and
Tennessee, west to California, north to
Oregon.

Food: Adult feeds on aphids and caterpillars. Nymph eats leaves, flowers, and young fruit.

Sound: A continuous trill. Pitch higher in warmer weather.

Life Cycle: Female selects and nudges male until it stops singing. Female then nibbles on secretion from gland on male's back while they mate. Eggs inserted in bark overwinter and hatch in spring. Nymphs mature by midsummer.

These crickets sometimes damage slender stems of grapevines, rose bushes, and fruit trees by incising rows of egg pits in their bark. The various species of tree crickets are distinguished mostly by the form and spotting of the first 2 antennal segments.

MOLE CRICKETS
(Family Gryllotalpidae)

This is a small family of elongate, brown crickets that spend most of their time burrowing in the ground. They are usually ¾" (20 mm) or longer and have short antennae and spadelike front legs, which they use for digging. A dense coat of fine hair prevents soil from clinging to the body. The broad fore wings are only about ½ as long as the abdomen, but the hind wings are longer, extending beyond the tip of the abdomen when folded. Mole crickets lay their eggs in clusters in tunnels among the roots of plants, on which both nymphs and adults feed. They are most common in moist soil.

250 Northern Mole Cricket
(Gryllotalpa hexadactyla)

Description: ¾–1⅜" (20–35 mm). Earth-brown with a coating of short fine hair. *Fore legs broad and spadelike.* No evident ovipositor in female.

Habitat: Moist soil.

Range: East of the Rocky Mountains, from southern Canada to Mexico.

Food: Plant roots.

Sound: A guttural growling note.

Life Cycle: Eggs are laid in chamber at end of underground tunnel, where roots can be reached. Female guards eggs and nymphs until they are partly grown. 1 generation a year.

This cricket spends most of its time underground, but it can fly well. It often damages crops, garden plants, and orchard trees. The larger European Mole Cricket (*G. gryllotalpa*), 2" (50 mm), has shorter hind wings. It is found in the East.

Walkingsticks and Timemas
(Order Phasmatodea)

Walkingsticks and timemas blend in so effectively with the twigs and leaves of their environment that they become invisible. Most of the approximately 2,000 species are tropical. Only 10 live in North America—8 usually wingless walkingsticks and 2 wingless timemas. Most walkingsticks are brownish and have extremely long, thin bodies, up to 5⅞" (150 mm) long. These insects resemble the leafless twigs and branches that they cling to with their long legs. However, unlike mantids, walkingsticks do not use their camouflage to ambush other insects but to protect themselves from predators. They sway with the wind or remain motionless all day, waiting for darkness before moving to feed on foliage. Timemas, up to 1" (25 mm) long, have stouter bodies shaped like leaves and leaflike markings on their limbs. They rely on this camouflage for protection. Reproduction in many phasmatodeans takes place by parthenogenesis— females lay eggs without mating. Dozens of seedlike eggs are dropped one at a time to the ground and sometimes take up to 2 years to hatch. Metamorphosis is simple: nymphs resemble adults except that they change in color, sometimes develop wings, and become sexually mature. Formerly, walkingsticks and timemas were included in Order Orthoptera because their mouthparts, wings, and some internal features resemble those of grasshoppers.

WALKINGSTICKS
(Family Phasmidae)

The walkingstick has a greatly elongated, almost cylindrical body with a long thorax and a long abdomen. It has long, threadlike antennae on a tiny head; long, slender legs on the front, middle, and rear portions of the thorax; and short, 1-segmented cerci at the tip of the abdomen. The tarsi are 5-segmented. Nymphs are green, becoming a twiglike brown at maturity. They show an amazing ability to regenerate lost legs. These insects are found mostly in the South, where some species grow to a length of 5⅞" (150 mm).

| 295, 297 | **Northern Walkingstick** |
| | (*Diapheromera femorata*) |

Description: Male 3" (75 mm), female 3¾" (95 mm). *Very elongated,* wingless. Male brown, female greenish brown. Antennae ⅔ length of body. Cerci, with 1 segment, often resemble palps at tip of abdomen.

Habitat: Deciduous woods and forests.

Range: Atlantic Coast to northern Florida, west to New Mexico, north to Alberta.

Food: Foliage of deciduous trees and shrubs, especially oaks and hazelnuts.

Life Cycle: Female drops eggs singly. Eggs overwinter among ground litter and hatch in spring, when nymphs push open domelike ends of the eggs. Nymphs crawl up woody vegetation at night to reach edible foliage.

The Northern Walkingstick's resemblance to slender twigs camouflages it from predatory birds during the day. When many females are dropping eggs, the sound is like the pitter-patter of light rain.

296 Giant Walkingstick
(*Megaphasma dentricus*)

Description: 3–5⅞" (75–150 mm). *Large, very elongated.* Greenish to reddish brown, legs often paler. Antennae ½ as long as body, threadlike. Middle femora swollen; both middle and hind femora have a spiny ridge along lower surface.

Habitat: Woods, forests, and grasslands.

Range: Indiana to Louisiana, west to Texas, north to Iowa.

Food: Foliage of grasses and woody plants, especially grapevines and oaks.

Life Cycle: In the fall, females drop eggs individually. They overwinter among leaf litter and hatch in late spring. Nymphs clamber up plants to feed in darkness and are seldom active by day. 1 generation a year.

The largest North American walkingstick, this insect so closely resembles leafless twigs and leaf stalks that it is generally overlooked.

TIMEMAS
(Family Timemidae)

These stout, wingless insects are greenish to pinkish and resemble leaves. Their legs arise from below the thorax, and coxae are invisible from above. Unlike walkingsticks, timemas have 3-segmented tarsi. Females, measuring up to 1" (25 mm) long, have tapering abdomens that terminate in a pair of short, straight cerci held close together. Males, to ⅝" (15 mm) long, have abdomens that end more abruptly with a pair of forcepslike cerci. Adults and nymphs feed on the foliage of West Coast trees and shrubs, including oaks, ceanothus, and firs.

84 California Timema
(*Timema californica*)

Description: Male ½″ (12–14 mm), female ¾–⅞″ (20–22 mm). *Stout, leaflike;* body tapering to pair of forcepslike cerci, male's curved, female's pointed. Wingless. Long legs with 3-segmented tarsi. Male often green with pink legs; female leaf-green or bright pink. Antennae threadlike, ½ as long as body.

Habitat: Oak forests.

Range: California to British Columbia.

Food: Oak foliage.

Life Cycle: Eggs are dropped singly on the ground, overwinter, and hatch in late spring when each nymph bursts out through trapdoor at end of egg. Nymphs mature in May of the following year.

These insects are well camouflaged; those found on green leaves are green, those on reddish buds are pink. They somewhat resemble earwigs but run quickly when disturbed, emitting a repellent odor. The male commonly rides on the female, as if ready to mate.

Webspinners
(Order Embioptera)

Uncommon insects, webspinners are
represented by only 10 species in North
America, restricted to the Gulf states
and the West Coast. There are about
150 species known worldwide. These
brownish or yellowish insects, ⅛–⅜"
(3–10 mm) long, live in colonies inside
the silken galleries they spin among
mosses, or in cavities in the soil. They
have slender, cylindrical bodies with
threadlike antennae, chewing
mouthparts, short stout legs, and 1 or
2 asymmetrical appendages called cerci
at the tip of the abdomen. Only males
have 4 downy, brownish wings. The
fore wings are longer than the hind
wings and well separated from them.
Webspinners spin silken webs from
glands on the tarsi of their fore legs.
Females eat mostly decayed plant
matter, but males are carnivorous.
When disturbed, webspinners run
rapidly backward to their nests, or
sometimes "play dead." Females lay
clusters of elongated, curved eggs in
the silk-lined tunnels, which they
carefully guard. Metamorphosis is
simple: both males and females become
sexually mature, and males gain wings.

Booklice and Barklice
(Order Psocoptera)

Some 150 species of booklice and barklice have been identified in North America, out of a total of about 1,100 worldwide. These small insects, rarely more than ¼" (5 mm) long, have heads that are quite large in proportion to their bulbous bodies. Huge compound eyes protrude from the sides of the head, and long, threadlike antennae with 13–50 segments sweep back toward the abdomen. Many species are wingless; others have 4 wings that fold above the back like a steeply pointed roof when the insect is resting. Although a few psocopterans superficially resemble lice, accounting for their common name, they are neither parasitic nor true lice. Barklice, the most numerous and common, are found on the bark of trees, where they may spin irregular webs. Booklice are wingless and sometimes infest houses, where they feed on the sizing, paste, and glue of book bindings. Metamorphosis is simple. Eggs are deposited singly or in clusters, sometimes covered by silk. Nymphs resemble adults. Some are gregarious; most feed on dry plant materials and bark.

Chewing Lice
(Order Mallophaga)

There are about 320 species of these minute, flat-bodied, wingless insects in North America and nearly 2,700 species worldwide. Their bristly bodies, $\frac{1}{16}$–$\frac{1}{4}$" (2–5 mm) long, are mostly abdomen. The large, toothed mandibles are quite different from the piercing mouthparts of sucking lice. Unlike sucking lice, chewing lice have heads as large as or larger than the thorax, small compound eyes, and short or hidden antennae. Each leg has 1 or 2 claws, rather than the 1 claw of sucking lice. All chewing lice spend their entire lives on their host. Different species have adapted to specific areas. Although the order name means "wool-eating," few chewing lice eat wool. Some feed on bits of animal hair and feathers. Female chewing lice lay 50–100 eggs, which they cement individually to hairs or feathers. Metamorphosis in chewing lice is simple: nymphs are similar to adults. Chewing lice are divided into 6 families, distinguished according to adaptive features such as claws and mouthparts. The minute size of these insects makes identification possible only with a microscope.

78 BIRD LICE
(Family Philopteridae)

This family has more species than any other in the order. Its members have 2 claws on each tarsus. The threadlike antennae have 5 segments and are not concealed in grooves. The mandibles are vertical rather than horizontal, and there are no maxillary palps. Unlike other families of chewing lice, the mesothorax and metathorax are fused together with no apparent division. They are small, usually measuring less than $\frac{1}{16}''$ (1.5 mm).

Sucking Lice
(Order Anoplura)

These flattened, wingless insects, usually measuring less than ¼" (6 mm) long, are external parasites on mammals. About 250 species are known worldwide, including 62 in North America. Sucking lice are distinguished from other lice by their broad, flattened bodies, a single powerful claw on each leg, and sucking mouthparts, which pierce skin to draw blood. They have small heads with short antennae and minute compound eyes or none at all. Each species lives on a particular mammalian host, including humans. Almost all sucking lice lay their eggs directly on the host and cement them to hair, or in the case of human lice, to clothing threads. When the eggs hatch, the nymphs resemble the adults, indicating a simple metamorphosis. Like adults, the nymphs are parasitic. Human head and body lice and pubic lice are the only species that regularly attack people. Their presence causes great irritation and in some cases transmits typhus. Since sucking lice feed repeatedly, making a fresh puncture each time, they cause irritation and itching disproportionately greater than the actual amount of blood withdrawn.

MAMMAL SUCKING LICE
(Family Haematopinidae)

These lice form a large group of external parasites that attack mammals other than humans, including cloven hoofed animals, rodents, horses, and even domestic dogs. They have vestigial, nonfunctional eyes or no eyes at all. Neither nymphs nor adults can survive many hours if separated from the host.

---

71 Hog Louse
(*Haematopinus suis*)

Description: ¼" (5–6 mm). *Head long and narrow.* Abdomen flat and broad, resembling pubic lice. Body mostly gray. Thorax reddish brown. Abdomen dark along sides. Distinctive point projects from each eye.

Habitat: Mostly in ears and folds of hog skin, particularly in winter.

Range: Worldwide, wherever pigs are kept.

Food: Hog blood.

Life Cycle: Female glues eggs to hog hair, where they soon hatch. Nymphs and adults feed frequently, making new punctures each time.

This is the only louse species that attacks hogs and is the largest member of the family. The smaller Horse Sucking Louse (*H. asini*), ⅛–¼" (3–5 mm), is yellow except for its brown head, thorax, and the sides of the abdomen. The Short-nosed Ox Louse (*H. eurysternus*), 1/16–⅛" (1.5–3 mm), has a head nearly as wide as long and a yellowish-gray body with 2 black spots on the last segment. All of these lice are capable of transmitting serious diseases.

HUMAN LICE
(Family Pediculidae)

Called "cooties" and "crabs," these
wingless parasites attack people. Their
flat, oval, or somewhat elongated
bodies, $\frac{1}{16}-\frac{1}{8}''$ (1.5–3 mm) in length,
may have legs with crablike claws and,
depending upon species, either
compound eyes or eye tubercles.
Human lice are classified by shape and
by areas of the body they invade.
"Crabs" inhabit pubic hair, "cooties"
head hair and other areas. Although
extremely annoying pests, these lice can
be thwarted simply by bathing and by
changing clothes and bedding
frequently. They can transmit typhus.

70 Human Body Louse
"Cootie" "Gray Back"
(*Pediculus humanus*)

Description: $\frac{1}{8}''$ (3 mm). *Grayish*. Head only
slightly narrower than thorax and often
$\frac{1}{2}$ as wide as abdomen. No side lobes
on abdomen. Legs have sharp tarsi, no
claws.

Habitat: Human bodies, and temporarily on
clothing and bedding.

Range: Worldwide, especially among people
who bathe infrequently or wear unclean
clothing.

Food: Human blood.

Life Cycle: Eggs are attached singly to clothing
and hatch in about 8 days. Nymphs
require less than 2 weeks to mature.

Human Body Lice hide in folds and
seams of clothing, crawling on skin
only to feed. Infected lice can spread
epidemics, such as typhus fever and
relapsing fever. The related Human
Head Louse (*P. h. capitis*), $\frac{1}{16}-\frac{1}{8}''$
(1.5–3 mm), glues its small white
eggs, called nits, to hair, especially
between the ears or on the neck.

True Bugs
(Order Hemiptera)

Members of Hemiptera, which number about 4,500 species in North America and over 40,000 worldwide, are called true bugs to distinguish them from other insects commonly called bugs, such as beetles. Almost all true bugs have a first pair of wings that fold flat over the back. These fore wings, called hemelytra, are usually leathery at the base and membranous at the tip, where they overlap. Hemiptera means "half wings," referring to the half-membranous fore wings. The hind pair, which are the flying wings, are uniformly membranous and slightly shorter than the fore pair. True bugs have sucking mouthparts in the form of a beak, which always arises from far forward on the head instead of at the back of the head, as in cicadas and their kin. Between the wings, the enlarged and modified back of the metathorax forms a distinctive triangular shield, called the scutellum. Antennae in terrestrial species may be long, but in aquatic species the antennae are commonly short and concealed. Females lay from a few to many eggs, which are usually inserted into plant tissues or glued onto plant surfaces. Metamorphosis is simple through 5 developmental stages. Nymphs resemble adults except that they may change colors several times and are wingless. Most true bugs are terrestrial, but a few are aquatic, using oarlike legs to skim through or over the water. Most have glands that exude strong odors to repel enemies. Some species feed on both insects and small vertebrates, while a few others are parasitic and can be disease carriers. Many species suck plant juices. This order was formerly called Heteroptera.

WATER BOATMEN
(Family Corixidae)

These common aquatic insects have elongate oval bodies, ⅛–½″ (3–13 mm) long, and are mottled brownish gray with yellowish crossbands. The feeble beak appears to arise at the rear of the head, rather than the front, and is unsuited for piercing or sucking. Instead, short fore legs have 1-segmented, scooplike tarsi, which are used to collect algae and other minute, submerged food particles. Middle and hind legs serve as oars in swimming. Adults can fly but usually are seen swimming in shallow freshwater ponds, pools, sewage tanks, and birdbaths, or browsing inconspicuously on aquatic vegetation. Many can produce squeaky sounds, important in attracting mates, by rubbing pegs at the base of the front femora against ridges on the head. These insects are the preferred food of many fishes. In Mexico, people eat one water boatman species in all of its life stages.

100 Water Boatmen
(*Corixa* spp.)

Description: ¼–½″ (5–13 mm). Elongate oval. Gray to brown, upper surface usually with fine crossbands. Head concave, concealing forward portion of prothorax and making *fore wings appear to arise immediately behind head.* Scutellum usually concealed. Fore legs short, tarsi scooplike. *Middle and hind legs are flattened, paddlelike.* Hind pair fringed with hair.

Habitat: Ponds, puddles, and birdbaths.

Range: Worldwide.

Food: Minute algae.

Life Cycle: Oval eggs are cemented to underwater supports, sometimes forming a dense crust. They hatch in 7–15 days.

Unless a birdbath is scrubbed almost daily, water boatmen will come to feed on algae. Some species are attracted to artificial light at night. Only a specialist can identify the many species and accurately distinguish between the genera *Arctocorixa, Callocorixa, Corixa,* and *Palmacorixa.*

BACKSWIMMERS
(Family Notonectidae)

Backswimmers propel themselves upside down. These elongate, torpedo-shaped water bugs, ¼–⅝″ (5–15 mm) long, swim in a jerky, erratic manner, using extended hair-fringed hind legs as oars. They can pause at any depth by controlling their buoyancy. Air trapped in fringed abdominal pockets enables them to remain submerged for 6 hours of inactivity, if necessary. Their coloration provides protective camouflage while they swim on their backs—wings and backs are pale, the undersides are dark. At all ages, backswimmers prey upon insects that get caught in the surface film, aquatic insects, and tadpoles. Their sharp mouthparts are used to inject digestive agents into the prey and to suck out liquefied food.

99 Kirby's Backswimmer
(*Notonecta kirbyi*)

Description: ⅜–½″ (10–13 mm). Black beneath, *white to green or black on back.* Fore wings yellowish white with dark tips. Compound eyes large, black. Legs brown and long, particularly *hind legs,* which are *flattened and bear a fringe of hair on tibiae and tarsi.*

Habitat: Ponds and slow-flowing shallows of streams.

Range: Middle Atlantic states.
Food: Insects and other small animals that fall into or live in water, as well as tadpoles, small fishes, and aquatic insects.
Life Cycle: Elongated white eggs are cemented to underwater plant stems, hatching in a few weeks. Nymphs often feed on their own species. Adults of 1st generation appear in July. 2 generations a year.

Master predators, backswimmers catch their victims by quietly swimming underneath them. If disturbed, backswimmers dive to the bottom and at intervals rise to the surface to get air. Occasionally they fly to other suitable habitats.

98 Common Backswimmer
(*Notonecta undulata*)

Description: ⅜–½″ (10–13 mm). Black underneath; *white to dark green on back*. Fore wings ivory-white with red markings and dark overlapping tips. Compound eyes large, black. Legs brown; fore and middle legs used for grasping, much longer *hind legs flattened, fringed with hair,* used for rowing. Color often paler in the West.
Habitat: Ponds and slow-flowing shallow streams.
Range: Throughout North America.
Food: Insects and other small aquatic animals.
Life Cycle: Elongated white eggs are attached to plant stems underwater, where they hatch in a few weeks. Nymphs are active predators. Adults of 1st generation appear in July and overwinter. 1 or 2 generations a year.

Occasionally, a backswimmer will attack a person's bare hand or leg in the water, earning its reputation as a water bee or water wasp.

GIANT WATER BUGS
(Family Belostomatidae)

The largest true bugs, these common brown insects are 1–2⅜" (25–60 mm). Their flattened hind legs are used for swimming. The fore legs grasp prey while the insect thrusts its powerful, sucking beak into a victim. To obtain air, giant water bugs raise the tip of the abdomen to the water surface and extend 2 tail-like breathing tubes. The larger species lay their eggs on water plants, whereas in the smaller species large batches of eggs are cemented to the back of the males, who carry them about and aerate them until they hatch.

102 Ferocious Water Bugs
(*Abedus* spp.)

Description: 1¼–1⅜" (28–35 mm). *Flat, oval, at least ½ as wide as long.* Dull brown. Compound eyes prominent, strong beak. Fore legs adapted for grasping and holding prey. Hind tibiae and tarsi flattened.

Habitat: Slow streams and ponds, even temporary ones; usually under rocks.

Range: Mexico to central California and Arizona.

Food: Small aquatic animals, including fish.

Life Cycle: Female cements elongated pale brown eggs to the exposed surface of male's fore wings. Keeping fore wings above water, male swims into shallows, then rocks from side to side, washing and aerating eggs. After nymphs hatch, mass comes free, letting male approach another female. Nymphs develop quickly. Several generations a year.

Among the most ferocious water predators, members of this genus live in streams. Other giant water bugs inhabit standing water.

97 **Eastern Toe-biter**
"Electric Light Bug"
(*Benacus griseus*)

Description: 1¾–2⅜" (45–60 mm); wingspan 4⅜"
(110 mm). *Broad, flat.* Brown. Beak
short and stout. Fore femora heavy,
without groove to accommodate tibiae
when folded tightly. Hind legs
flattened, fringed, used in swimming.
Fore wings leathery.

Habitat: Shallow freshwater ponds and pools,
among bottom vegetation.

Range: East of the Mississippi River.

Food: Other insects, tadpoles, small fishes,
and salamanders.

Life Cycle: Eggs are dark brown to gray, striped
with a white ring at the end where
nymphs emerge. They are attached in
rows of 100 or more on plants or other
supports above the water. Nymphs
hatch in about 2 weeks. Adult female
lays 150 or more eggs in a lifetime. But
few nymphs survive cannibalism and
other predators.

Because of their attraction to artificial
lights, these insects are called "Electric
Light Bugs." Underwater, they often
stab or seize a person's bare foot,
earning the name "Toe Biter."

101 **Giant Water Bug**
"Toe-biter" "Electric Light Bug"
(*Lethocerus americanus*)

Description: 1¾–2⅜" (45–60 mm) long, ¾–1"
(20–25 mm) broad; wingspan 4⅜"
(110 mm). *Large, flat.* Brown. Beak
short and stout. Front femora heavy,
with groove to accommodate tibiae
when folded. Hind legs flattened,
fringed. Fore wings leathery.

Habitat: Shallow freshwater ponds and pools,
among bottom vegetation.

Range: Throughout the United States and
Canada.

Food: Other insects, tadpoles, small fishes, and salamanders.

Life Cycle: In late spring or early summer, eggs are attached in rows of 100 or more on plants or other supports above water. Nymphs emerge in about 2 weeks. Adult female alternates between feeding and mating and lays 150 or more eggs in lifetime, but few nymphs survive because of cannibalism and other predators. Adults usually overwinter.

Both the large nymph and the adult feign death if picked up, but they can stab suddenly with their beaks, injecting the anesthetic saliva used to subdue prey. The groove in each fore femur distinguishes this species from the Eastern Toe-biter.

WATERSCORPIONS
(Family Nepidae)

Waterscorpions are easily recognized by their scissorlike fore legs and the long tail-like breathing tubes at the tip of the abdomen, which are thrust through the surface film for air. They have large, compound eyes and short, stout beaks capable of piercing a person's skin if carelessly handled. Three pairs of oval disks below the abdomen, called false spiracles, are used to gauge depth and compensate for changes in water pressure. Some species have long, slender bodies with thin stick legs, while others are flat and oval-shaped. They range from ¾" to 1¾" (20–43 mm) long. Despite well-developed wings, they fly rarely and only at night. All prey upon other insects, tadpoles, and small fish.

293 Western Waterscorpion
(*Ranatra brevicollis*)

Description: 1" (25 mm) with 1" (25 mm) *paired tail-like breathing tubes.* Very elongate, *sticklike.* All brown. Pronotum narrow and necklike, no wider than head at compound eyes. Slender middle and hind legs, each about as long as body.

Habitat: Bottoms of shallow ponds, among debris and vegetation.

Range: California.

Food: Smaller insects, mites, and aquatic worms.

Life Cycle: Eggs are thrust into soft tissues of underwater plants or bottom debris. Nymphs are full grown in a few months and may overwinter before or after final molt to adulthood.

Swaying gently with the wind, this insect resembles a walkingstick. Both sexes produce scratchy sounds underwater by rasping together rough areas on the enlarged front coxae and the grooves into which the coxae fit.

294 Brown Waterscorpion
(*Ranatra fusca*)

Description: 1⅜–1¾" (35–43 mm) with ⅞" (21 mm) *paired tail-like breathing tubes.* Very elongate, *sticklike.* Pale to dark yellow brown. Pronotum narrow and necklike, no wider than head at compound eyes. Slender middle and hind legs each about as long as body.

Habitat: Bottoms of shallow fresh waters, among vegetation and debris.

Range: Throughout southern Canada and the United States.

Food: Body juices from other insects, tadpoles, salamanders, and fishes.

Life Cycle: Eggs are thrust into soft, living or dead tissues of aquatic plants, hatching in 2–4 weeks. Nymphs mature in about 5 weeks. Adults overwinter.

This waterscorpion can puncture a person's skin with its beak. It injects a salivary secretion rich in anesthetizing substances that tranquilizes prey and initiates digestion.

TOAD BUGS
(Family Gelastocoridae)

These small, heavy-bodied, hopping insects, ¼–⅜″ (7–10 mm) long, resemble tiny toads. The prothorax is relatively large, broad, and convex, partially concealing the head and compound eyes. The short antennae are concealed under the eyes. The femora of the fore legs are greatly enlarged having muscles used for grasping prey. The hind legs are adapted for jumping. Like toads, they hop along the edge of fresh water, preying on small insects.

129 Toad Bugs
(*Gelastocoris* spp.)

Description: ⅜″ (8–10 mm). *Toadlike.* Oval, ⅔ as wide as long; sides of prothorax projecting. Bulging compound eyes at sides of head. Mottled brown to yellowish brown. Legs yellowish with brownish bands.

Habitat: Shores of quiet lakes, ponds, and streams.

Range: United States and southern Canada; individual species more limited.

Food: Smaller insects.

Life Cycle: Eggs are laid near water's edge in mud or sand, or under stones. Nymphs grow slowly if food is scarce. Adults overwinter, but few eggs hatch before early summer.

Seen close to the water's edge, a toad bug can be mistaken for a small grayish-brown pebble until it hops or

suddenly runs. The Western Toad Bug
(*G. variegatus*), ¼″ (7 mm), is mottled
pale and dark gray and ranges from
Texas to California and Nevada, south
into Mexico. More widespread is the
slightly larger Big-eyed Toad Bug
(*G. oculatus*), ranging from New
England to Virginia, west to Manitoba
and Oregon, south to Mexico.

WATER STRIDERS
(Family Gerridae)

Water striders are slender, elongate
insects with dark brown bodies, ⅜−1″
(10−25 mm) long. Some species have
spherical bodies that are much shorter
than their legs. All have short fore legs
and longer, slender middle and hind
legs that arise close together. They dart
about on the surface film of fresh and
salt water, feeding on small animals
that fall into the water or float up from
below. Some adults are winged, others
wingless.

292 Common Water Strider
"Skater"
(*Gerris remigis*)

Description: ½−⅝″ (12−16 mm). Flattened,
elongate. Dark brown to black. Short
fore legs. *Long slender middle and hind
legs.* Mostly wingless.
Habitat: Surfaces of ponds, slow streams, and
other quiet waters.
Range: Throughout North America.
Food: Aquatic insects, including mosquito
larvae that rise to the surface, and
terrestrial insects that drop into water.
Life Cycle: Courtship and mating involve
communication by ripples in the surface
film. Female lays parallel rows of
cylindrical eggs on an object at water's
edge. Nymphs mature in about 5

weeks. Adults live many months, and in northern parts of their range they overwinter under fallen leaves on land near water.

These insects are called "Skaters" in Canada and "Jesus Bugs" in Texas because they "walk" on the water.

BED BUGS
(Family Cimicidae)

Bed bugs are flat, oval, almost wingless insects, usually less than ¼" (7 mm) long. These reddish-brown bugs have moderately long, slender antennae, thin legs, and vestigial wings in the form of stubs. Notorious pests, they can run at a surprising speed. At night they hunt for sleeping mammals and birds. Adults have been known to survive without food for a year or more. Bed bugs were brought to North America by early colonists.

67 Bed Bug
(*Cimex lectularius*)

Description: ⅛–¼" (4–7 mm). *Very flat,* usually ⅔ as wide as long. Rusty red to brown or purplish. Antennae slender, 4 segments. Sides of pronotum flangelike. Legs relatively short with 3 tarsal segments. Wings vestigial, represented by stubs.

Habitat: Human habitations, particularly bedrooms.

Range: Worldwide.

Food: Human blood.

Life Cycle: During lifetime, female lays up to 300 eggs, which hatch within a week. Depending on food supply, nymphs take 6 weeks–12 months to become adults. Many generations a year.

When the Bed Bug pierces skin, it takes a small amount of blood, then withdraws its mouthparts. If not disturbed, it moves a few steps and repeats this process 2 or 3 times more, leaving wounds marked by tiny blood clots. Lacking human blood, this insect will feed on the blood of rats, mice, rabbits, or chickens. It can survive without food for up to 15 months. The Bed Bug is not known to transmit any human disease. Its remote ancestors probably preyed upon bats in caves that people subsequently inhabited.

PLANT BUGS
(Family Miridae)

This family has the most species in the order and includes the largest number of native North American true bugs. Most are small, soft-bodied insects. Their elongate or oval bodies, less than ⅜" (9 mm) long, are often brightly colored. The head is shorter than the prothorax and much shorter than the antennae. The beaklike mouthparts project downward or forward when used to suck plant juices. Compound eyes are present, but not simple eyes. Tarsi have 3 segments. Females insert eggs into plant tissue using a bladelike ovipositor. Most members of this family live on plants, and some are crop pests.

122 Adelphocoris Plant Bugs
(*Adelphocoris* spp.)

Description: ¼–⅜" (6–9 mm). Green, orange-red, yellowish brown, or blackish, with *bold markings or stripes*. Antennae and legs often dark. Nymph is paler than adult.

Habitat: Crop fields, pastures, gardens, and open lands.

Range: Throughout North America; individual
species more limited in distribution.

Food: Juices of alfalfa and other herbaceous
plants; each species has favorite host.

Life Cycle: Eggs are embedded individually in soft
tissues or crevices of food plant.
Nymphs grow rapidly. Winged adults
disperse to new areas in mid- to late
summer, aided by the wind. 1 or more
generations a year.

Many species injure alfalfa plants,
including the western Superb Plant
Bug (*A. superbus*), ⅜" (8 mm) long,
formerly known as *Lygus pratensis*.

121 Scarlet Plant Bugs
(*Lopidea* spp.)

Description: ¼–⅜" (6–9 mm). Slender, flattened.
Brownish red with *bright red on
prothorax, head, and outer portions of fore
wings*. Membranous portion of fore
wings blackish. Antennae, legs, and
undersurface are black. Nymph is
brownish, red, and black.

Habitat: Meadows and forest edges.

Range: Throughout North America.

Food: Plant juices; 1 host for each species.

Life Cycle: Minute, pale green eggs are thrust into
soft leaf stalks. Nymphs grow quickly,
maturing in a few weeks. Sometimes
several generations in a summer.

When disturbed, adults quickly fly
away or, like nymphs, drop to the
ground. Only a specialist can positively
identify the different species, but the
host plant usually provides a valuable
guide. Among the most common are
the Poplar Plant Bug (*L. cuneata*),
the Caragana Plant Bug (*L. dakota*),
the Phlox Plant Bug (*L. davisi*), the
Goldenrod Plant Bug (*L. media*),
the Black Locust Plant Bug (*L.
robiniae*), and the Willow Plant
Bug (*L. salicis*).

235 Tarnished Plant Bug
(*Lygus lineolaris*)

Description: ¼" (6–7 mm). Dark brown to pale green with *reddish-brown, black, and yellow markings.* Head yellow with 3 narrow dark stripes. Pronotum has yellowish margins and lengthwise stripes. Legs pale. Nymph is green or pale yellow.

Habitat: Meadows, gardens, and crop fields.

Range: Eastern North America.

Food: Plant juices from foliage, soft stems, and fruits, including strawberries, wild mustard, goldenrod, aster, and on vegetable, fruit, and alfalfa crops.

Life Cycle: Tiny, pale green, seedlike eggs are inserted into leaf stalks and soft stems. Nymphs grow rapidly. In the North only adults overwinter, mostly in litter on the ground. Often 5 generations a year in the South, where nymphs as well as adults overwinter.

This genus probably includes the most common leaf bugs in the United States and Canada, with about 40 species.

156 Four-lined Plant Bug
(*Poecilocapsus lineatus*)

Description: ¼" (6–7 mm). Greenish yellow except jet-black on front of head, *4 jet-black lines on pronotum and fore wings.* Nymph is bright red with a black spot on thorax and yellow stripes on wing pads.

Habitat: Meadows, gardens, and crop fields.

Range: East of the Rocky Mountains.

Food: Plant juices, nectar, and occasionally liquids from a freshly killed insect.

Life Cycle: Eggs are thrust into soft tissues of plant stems, where they overwinter. As nymphs grow, their wings lengthen; wing stripes, separated by a black line, also lengthen. When nymphs molt to adults, 2 additional black lines appear on fore wings. 1 generation a year.

White or dark spots appear on plants where these insects have sucked juices, often causing the whole leaf to wither and drop. These plant bugs will attack garden flowers and small fruits.

ASSASSIN BUGS
(Family Reduviidae)

Assassin bugs get their name from the way they attack and inflict sharp stabs on their victims. Most are elongate, often nearly parallel-sided bugs, ½– 1⅜" (12–36 mm) long. Some species have long, slender bodies like walkingsticks. The fore femora have powerful muscles for grasping and holding prey while its body fluids are sucked out. In most species the elongate head has a cross groove between the eyes. The short, strong beak is usually curved; at rest, it tucks into a groove, which is cross-ridged, between the bases of the fore legs. Sounds are produced when the rough tip of the beak is moved back and forth along these ridges. Most assassin bugs prey on other insects, although some suck blood from vertebrates, and a few transmit diseases to people.

118, 119 Bee Assassins
(*Apiomerus* spp.)

Description: ½–⅝" (12–15 mm). Variably colored: red with blackish-brown markings or brown with yellowish markings. Dense short hair on head, thorax, and legs. Distance between simple eyes greater than the distance between compound eyes. 2nd antennal segment rather comblike, not subdivided into small ringlike units. Nymph is dark and reddish.

Habitat: Meadows, fields, and gardens.

Range: North America; most common in the West.

Food: Other insects, especially bees.

Life Cycle: Eggs are attached to foliage. Nymphs, like adults, are voracious predators. 1 generation or more a year in the North.

This insect pounces on Honey Bees and other pollinating insects. It holds the captive in its powerful legs, thrusts its cutting beak into the victim's back, injects an immobilizing digestive agent, then sucks out the body juices.

105 Wheel Bug
(*Arilus cristatus*)

Description: 1⅛–1⅜" (28–36 mm). Gray-brown to black. *Prothorax raised into a curved series of coglike teeth along midline,* inspiring the common name. Nymph is blood-red with black markings.

Habitat: Meadows and crop fields.

Range: East of the Rocky Mountains, from southern Canada south to the Gulf states.

Food: Caterpillars, even large ones including hornworms, Japanese Beetle larvae, and other insects.

Life Cycle: Eggs are laid in clusters on foliage. 1 generation a year.

This handsome predator can give a painful stab when it defends itself from a careless handler.

120 Eastern Blood-sucking Conenose
"Big Bed Bug" "Mexican Bed Bug"
(*Triatoma sanguisuga*)

Description: ¾" (19–20 mm). *Black or dark brown with 6 orange spots* above and below each side of abdomen. *Head prolonged* in front of eyes, *tapering to a blunt point,* concealing nontapering beak.

Habitat: Nests of small mammals.
Range: New Hampshire to Florida, west to Texas, north to Ontario.
Food: Blood of mammals.
Life Cycle: After a meal, female scatters many oval whitish eggs. Nymphs grow slowly, taking up to 2 years to become adults.

If hungry, the adult may fly, but after a full meal it is unable to get off the ground. Bites cause severe allergic reactions in some people. Like other members of the genus, this bug can transmit parasites that cause a debilitating but rarely fatal disease that is common from Mexico into South America. The lingering illness suffered by Charles Darwin may well have been Chagas' disease, acquired from an attack of the related Blood-sucking Conenose (*T. infestans*) in 1835 in western Argentina.

AMBUSH BUGS
(Family Phymatidae)

These small, greenish-yellow or brown-and-yellow insects are usually seen on flowers. Well camouflaged by shape and color, they wait there to ambush insect prey, often much heavier than they are. Most are $\frac{3}{8}-\frac{1}{2}''$ (8–12 mm) long. Their fore femora, swollen with muscles, are used to snatch and hold prey, which may be a bumble bee, wasp, or butterfly. Ambush bugs then immobilize their victims by injecting saliva through their short beaks. The 4-segmented antennae are slightly thickened at the end. The wings at rest cover only the midzone of the broad abdomen, leaving the sides exposed.

127 Jagged Ambush Bug
(*Phymata erosa*)

Description: ⅜–½″ (8–12 mm). *Greenish to bright yellow with a wide dark band across abdomen.* Antennae slightly clubbed. Fore legs adapted for seizing and holding prey. Fore tibiae when folded fit into grooves in underside of femora. *Flaring sides of abdomen* extend beyond closed fore wings.

Habitat: Gardens and meadows on flowers.

Range: Throughout the United States and southern Canada.

Food: Flies, bees, butterflies, day-flying moths, and other true bugs.

Life Cycle: Black oval eggs are coated with adhesive secretion and glued to plant. Nymphs emerge through uncoated cap at tip of egg. They attack small insects, but the adults often seize prey much larger than themselves.

This bug is almost invisibly camouflaged on greenish-yellow flower clusters, such as goldenrod. Because Honey Bees are killed so frequently, beekeepers regard ambush bugs as pests rather than as beneficial insects. The common name refers to the jagged spines at the rear side corners of the prothorax.

LACE BUGS
(Family Tingidae)

These small, flat, grayish bugs, ⅛–¼″ (3–6 mm) long, have lacy patterns on the head, thorax, and fore wings. Most of the head is concealed beneath the expanded pronotum, which extends backward in a triangle between the wings. Simple eyes are absent. Females cut slits in the underside of leaves for their eggs. The wounded plant oozes juices that often cover the eggs, drying to form miniature stalactite-like projections from the leaf. Nymphs are covered with spines and are so much darker than adults that they seem unrelated even when feeding side by side with them. Lace bugs suck the juices from foliage of trees and shrubs, sometimes causing death of tissues and considerable damage.

56 Oak Lace Bug
(*Corythuca arcuata*)

Description: ⅛″ (3–4 mm). *Prominent hood on front of pronotum,* usually covering head. Body pale, crossed by *2 dark bands;* black below. Antennae and legs yellow. Wings clear.

Habitat: Oak forests and isolated trees.

Range: Wherever oaks grow in North America.

Food: Sap from oaks and from some other trees, sucked from foliage and young stems.

Life Cycle: Female forces eggs into leaf tissues close to midrib and coats them with hard, resinlike secretion. Nymphs pierce leaf cells from below, then deposit fecal pellets, which adhere to leaf, giving it peppered appearance. Adults overwinter beneath ground litter, fly and mate on warm spring days. Usually 1 generation a year, more in the South.

Nymphs cling to leaves and remain motionless if approached, but adults drop off or fly away. The similar Alder Lace Bug (*C. pergandei*), ⅛″ (3 mm), feeds on hazel and birch as well as alder in most of the United States and eastern Canada. The Goldenrod Lace Bug (*C. marmorata*), ⅛″ (3–4 mm), is yellowish white above, speckled with brown, and has 4 narrow brown lines across the fore wings. It is common on plants of the daisy family, including asters, horticultural chrysanthemums, and goldenrod, and sometimes invades greenhouses.

SEED BUGS
(Family Lygaeidae)

This is the second largest family of true bugs in North America. Most of its members suck juices from developing or dry seeds or the sap of grasses, but some are predators. Many are brown and have camouflaging colors; others have reddish markings. Their elongate or oval bodies are ⅜–¾″ (8–18 mm) long. Unlike similar insects in other families, most have simple eyes as well as compound eyes. Fully-developed fore wings have 5 veins. All have 4-segmented antennae low on the head.

---

116 Small Eastern Milkweed Bug
(*Lygaeus kalmii*)

Description: ⅜–½″ (11–12 mm). Black with red spot on top of head and red band across prothorax. *Fore wings have red X-shaped mark* on thickened part *and gray or white spots* on black membranous part.

Habitat: Meadows and fields.

Range: Throughout the United States.

Food: Maturing and mature seeds of milkweeds.

Life Cycle: Eggs are laid on milkweeds, when spring foliage and flower buds show. Nymphs, resembling adults, develop rapidly. Adults overwinter. 1 or more generations a year.

This insect is immune to toxic chemicals in milkweed but is itself toxic to other insect predators.

---

123 Long-necked Seed Bug
(*Myodocha serripes*)

Description: ⅜″ (8–10 mm). *Head very elongate, constricted to necklike region behind eyes.* Collarlike front lobe of prothorax narrower than rear lobe. Head and undersurface glossy black. Antennae long. Fore wings are reddish brown with yellowish mottling on thickened part and along front margin.

Habitat: Old fields among leaf litter.

Range: East of the Rocky Mountains; scarce in the North.

Food: Seeds, occasionally plant sap.

Life Cycle: Female inserts eggs into plant tissues. Nymphs are solitary. They overwinter if not fully grown, feeding again in early spring. Adults also overwinter and emerge on warm winter days.

The Banded Long-necked Seed Bug (*M. annulicornis*), ⅜" (9–10 mm), occurs in Florida. Its fourth antennal segment is black with a broad yellow ring, and the fore wings are black with pale yellow on the lower margin.

115 Large Milkweed Bug
(*Oncopeltus fasciatus*)

Description: ⅜–⅝" (10–15 mm). Elongate oval. Black with *red* (older) *or orange* (younger) *markings* on all edges of thorax and scutellum. Wings banded in red or orange and black. Nymph is red with black antennae and legs.

Habitat: Meadows and fields.

Range: East of the Rocky Mountains, scarcer in the North.

Food: Maturing and mature seeds of milkweeds.

Life Cycle: Small elongate eggs are bright red with 3 downcurved projections near tip. They hatch in spring. Adults overwinter, appearing in great numbers on warm winter days. 1 or more generations a year.

The Large Milkweed Bug sometimes sips nectar from flowers of other plants in gardens and crop fields but seems to do no significant damage.

STILT BUGS
(Family Berytidae)

These slender insects, ¼–⅜″ (5–9 mm) long, have distinctive, long threadlike antennae and slender, stiltlike legs, each ⅓–⅔ as long as the body. The first antennal segment is about half as long as the body and articulates with the second segment like an elbow. The last, or fourth, antennal segment forms a short spindle-shaped club. Stilt bugs have simple eyes and a 4-segmented beak. Stilt bugs resemble small crane flies except for the distinctive antennae, mouthparts, wing structure, and scutellum. They live on vegetation, feeding on plant juices and occasionally on smaller insects. Some species live on spiny plants and themselves have a covering of spines on the pronotum. This family was previously known as Neididae.

383 Stilt Bugs
(*Jalysus* spp.)

Description: ⅜″ (8–9 mm). *Slender body and appendages.* Pale brown. Last segment of antennae black. Nymph has dark green body with pale marks.

Habitat: Weedy meadows, crop fields, and gardens.

Range: Throughout the United States.

Food: Juices of nettle, potato family plants, alfalfa, and various grasses.

Life Cycle: Eggs are laid on host plant in spring. Nymphs feed and grow rapidly. Adults overwinter.

Nymphs and adults puncture stems and ovaries of flowers, and young fruits, often causing them to drop prematurely. The Spined Stilt Bug (*J. spinosus*), ⅜″ (8 mm), is the most common species in the East, ranging from New England south to Virginia, west to Colorado, north to Manitoba. Western species are distinguished from each other and from eastern species only by inconspicuous features.

LEAF-FOOTED BUGS
(Family Coreidae)

These medium- to large-sized insects, generally more than ⅜" (9 mm) long, have leaflike expansions of the hind tibiae. Their hind femora are often also swollen and bear heavy spines. These bugs have 4-segmented antennae, large compound eyes, a pair of simple eyes, and many parallel veins on their wings. All suck plant juices, and give off a foul-smelling secretion when disturbed.

103 Florida Leaf-footed Bug
(*Acanthocephala femorata*)

Description: ¾–⅞" (18–23 mm). *Black with red on tips of antennae, fore legs, and tarsi.* Antennae arise from short projecting tubercles. *Hind femora very wide,* much broader in male than in female; in both sexes bearing sharp spines beneath. Hind tibiae are only slightly expanded.

Habitat: Meadows, unmowed ground, and adjacent areas.

Range: Florida.

Food: Plant juices.

Life Cycle: Elongate orange eggs are cemented to foliage. 1 or 2 generations a year, more in southern Florida.

When handled, the Florida Leaf-footed Bug gives off a foul-smelling secretion. Adults fly away if rapidly approached. Many similar species are widely distributed in the East.

104 **Squash Bug**
     (*Anasa tristis*)

Description: ⅝" (15—17 mm). Dark to pale grayish
              brown above, pale below. *Sides of
              abdomen are orangish or striped with orange
              or dark brown.* Wing tips reddish-
              brown. *Hind tibiae cylindrical,* not
              leaflike. Nymphs are green with pink
              antennae and legs; often coated with
              white powder.
     Habitat: Crop fields and gardens.
       Range: Throughout North America.
        Food: Juices of cultivated cucumber, squash,
              melon, pumpkin, and other gourds.
  Life Cycle: Shiny, somewhat 3-sided eggs are laid
              singly or in groups of 15—50 below
              leaves or on stems. Feeding nymphs
              cause leaves to wilt, blacken, and dry to
              a crisp. Adults overwinter among
              withered leaves.

              Since adults require the shelter of debris
              to survive frost, infestations can often
              be reduced by gathering up and
              burning all refuse after the harvest.

SCENTLESS PLANT BUGS
(Family Rhopalidae)

These insects resemble leaf-footed plant
bugs but are smaller and do not have
scent glands. Mostly pale colored, they
are less than ⅝" (15 mm) long. They
have both compound eyes and 2 simple
eyes and 4-segmented antennae as long
as or longer than the head. The
membranous part of the fore wing has
many veins. These bugs are usually
found on weeds and grasses, where
adults and nymphs feed in late summer
and early fall.

117 Eastern Boxelder Bug
(*Leptocoris trivittatus*)

Description: ⅜–⅝″ (10–15 mm). Elongated oval.
Dull black to gray-brown. *Marked
narrowly with brick-red* along each side,
midline of pronotum, and on fore
wings. Wing veins red. Nymph is red,
adding black as it grows.

Habitat: Deciduous woods, forests, and gardens.

Range: East of the Rocky Mountains.

Food: Juice in foliage of boxelder, other
maples, and deciduous fruit trees.

Life Cycle: Eggs are hidden in bark crevices or left
on leaves and sometimes on seed pods
in spring. Nymphs appear in late
spring or early summer and develop
rapidly. Adult females overwinter. 1 or
2 generations a year.

Although this insect does little damage
to trees, it causes deformities and
blemishes in fruit. In autumn, huge
swarms of females can be seen near
buildings, looking for a place to
overwinter. The Western Boxelder Bug
(*L. rubrolineatus*), same size, is found
west of the Rocky Mountains.

STINK BUGS
(Family Pentatomidae)

Stink bugs get their common name
from the copious amounts of foul-
smelling fluid they discharge when
disturbed. Both adults and nymphs
possess large stink glands on the
undersurface, opening through
conspicuous slits. All of these medium-
to large-sized bugs, usually ¼–¾″
(6–20 mm) long, have 5-segmented
antennae—the first segment thick, the
second slender. The membranous part
of the fore wing has many branching
veins. A minority of stink bugs prey
upon caterpillars and herbivorous larvae
of beetles, sucking their juices instead

of plant sap. Eggs of stink bugs are often shiny, barrel-shaped, and armed with spines. They are attached to leaf surfaces in small masses resembling the cells of honeycomb.

111   Green Stink Bug
"Green Soldier Bug"
(*Acrosternum hilare*)

Description: ½–¾" (13–18 mm). Shield-shaped. *Bright green with narrow yellow, orange, or reddish edges.* Black spots on sides below. Outermost 3 antennal segments marked with black. Nymph has oval green body with red, yellow, and black markings.

Habitat: Crop fields, orchards, and gardens.

Range: Throughout North America.

Food: Juices of foliage, flowers, and fruit from a wide variety of plants.

Life Cycle: Keg-shaped eggs are attached in double rows of 12 or more on underside of foliage. They hatch in spring. 1 generation a year in the North, 2 in the South.

This pest damages apple, cherry, orange, and peach trees, and eggplant, tomato, bean, pea, cotton, corn, and soybean crops.

128   Brochymenas
(*Brochymena* spp.)

Description: ½–⅝" (14–16 mm). Brown to mottled gray. Surface dull, roughly pebbled. *Head elongate.* Antennae arise far in front of compound eyes. Beak long, directed toward base of hind legs. *Side of pronotum extends into toothlike projections.*

Habitat: Deciduous fruit orchards, woods, and isolated trees.

Range: Throughout the United States.

Food: Juices of caterpillars and other soft
insects.
Life Cycle: Clusters of pearl-white eggs are
attached to twigs and leaves in spring
and soon hatch. Nymphs grow slowly.
Adults overwinter in crevices and under
trash on ground. 1 generation a year.

These well-camouflaged insects
match the color of tree bark. The
Four-blistered Brochymena (*B.
quadripustulata*) is widespread, while
Van Duzee's Brochymena (*B. affinis*)
ranges from Idaho to northern
California, north to British Columbia.

175 **Two-spotted Stink Bug**
"Conspicuous Stink Bug"
(*Cosmopepla conspicillaris*)

Description: ¼–⅜" (6–8 mm). Shiny blue-black
with *orange band across prothorax that
often surrounds 2 black spots,* and narrow
white line on scutellum and sides of
abdomen.
Habitat: Fields and meadows.
Range: East of the Rocky Mountains in the
United States and Canada.
Food: Plant juices of weeds.
Life Cycle: Eggs are laid on undersurface of foliage,
where they hatch in 1–2 weeks.
Nymphs disperse progressively in about
10 weeks, feed in groups when small
but singly later, and may overwinter,
maturing in late spring.

The related Small Red-black Stink Bug
(*C. bimaculata*), same size, has red or
yellow stripes on the pronotum, sides of
the abdomen, and tips of the wings as
well as 2 red spots on the scutellum. It
feeds on many kinds of weeds
throughout most of the United States.

152 Harlequin Cabbage Bug
"Calico Bug" "Fire Bug"
(*Murgantia histrionica*)

Description: ⅜" (8–10 mm). Smoothly convex,
proportioned like a heraldic shield with
rounded corners. Jet-black with *bright
yellow or orange marks above.* Antennae
and overlapping portion of fore wings
black. Nymph is shiny red and black.

Habitat: Crop fields, orchards, gardens, and
meadows.

Range: New England and adjacent areas of
Canada to southern states, west to
California, south into Mexico.

Food: Juices of cruciferous plants, including
commercial cabbage, kale, and broccoli
crops; also turnip, horseradish, potato,
beet, bean, grape, squash, sunflower,
ragweed, and citrus foliage.

Life Cycle: Female lays barrel-shaped eggs in
double rows of 12 or more, cementing
them to undersurface of foliage.
Nymphs are active all summer. 3 or
more generations a year in the South,
only 1 in the North, where adults
overwinter and become active on warm
days in early spring.

This insect causes white and yellow
blotches on the foliage of infested
plants, ruining their commercial value.

236 Spined Soldier Bugs
(*Podisus* spp.)

Description: ⅜–½" (10–12 mm). Shield-shaped.
Pale brown to yellowish tan, peppered
above with minute black dots; pale
below with dark area at rear. *Thorax
projects in a sharp spine on each side.* Fore
wings project beyond abdominal tip.
Fore tibiae have a conspicuous spine
about midway below.

Habitat: Crop fields, gardens, and open woods
on wild and cultivated plants,
herbaceous or woody.

Range: Throughout North America.

Food: Caterpillars, sawfly larvae, grubs of leaf beetles, and other insects, including nymphs of its own species.

Life Cycle: During 5–6 week season, female lays up to 40 clusters of metallic bronze eggs, each cluster containing well-spaced arrays of 20–30 eggs. Nymphs at first stay together, perhaps feeding on leaf juices. After molting, they are predators.

Although spined soldier bugs destroy many harmful insects such as caterpillars, they are not totally beneficial because they also prey on useful insects, such as ladybug beetles. The widespread Underspotted Spined Soldier Bug (*P. maculiventris*), same size, has a conspicuous black spot underneath on its pale rear.

Cicadas and Kin
(Order Homoptera)

Approximately 45,000 species of homopterans are known worldwide, including 6,000 in North America. Although some entomologists group them with true bugs, the differences are enough to place homopterans in a separate order. Cicadas and their kin have sucking, beaklike mouthparts that arise far back beneath the head, whereas the mouthparts of true bugs usually arise from the front of the head. Homopteran wings are uniformly membranous (homoptera means "similar wing"); they are not divided into membranous and leathery areas, as is usual in the fore wings of true bugs. In some families these wings are nearly equal in length. All homopterans are plant feeders, but their diets vary enormously, as do their means of reproduction. Some species reproduce sexually, while others do so without mating, or parthenogenetically. Female cicadas and leafhoppers mate sexually and most have relatively large ovipositors, which they use to cut slits into plant tissues for their eggs. But aphids can reproduce parthenogenetically and are among the most destructive agricultural pests—theoretically, one female is capable of producing many billions of offspring. In all homopterans, metamorphosis is simple. Many homopterans exude a sweet secretion called honeydew, while others produce waxy materials that help protect them from attack and rain. Because each species has distinct features, most of the larger homopterans are easily identified. Aphids, scale insects, and some other small homopterans are exceptions to this rule, and a microscope is needed to classify them.

CICADAS
(Family Cicadidae)

These large insects, 1–2⅜" (25–60 mm) long, have membranous wings; the fore pair are twice the length of the hind pair. Males have sound-producing organs below the base of the abdomen. Although sometimes called locusts, cicadas are nonjumpers, unlike grasshoppers and crickets. They are usually seen emerging or flying from empty nymphal skins that may hang by the thousands from trees during July and August after the large adults have emerged. Each species has its own distinctive song, which may be a loud buzz or a pulsating clacking sound. Some species, primarily eastern, have life cycles of 13 or 17 years. Most of this time is spent as subterranean nymphal stages feeding on the roots of trees that they later climb as adults. More commonly the nymphal period is 1–3 years.

291   Periodical Cicadas
(*Magicicada* spp.)

Description: 1⅛" (27–30 mm); wingspan to 3" (77 mm). Stout. Black to brownish. Eyes bulging, dark red. *Wings membranous with orange tinge and orange along basal ½ of fore wing front margin.* Legs reddish. Undersurface of abdomen primarily reddish brown to yellow.

Habitat: Deciduous and mixed forests, adjacent grasslands and pastures.

Range: Mostly east of the Mississippi River, from the Great Lakes to the Gulf of Mexico.

Food: Sap of tree roots.

Sound: An intense whining, rising and falling in pitch.

Life Cycle: Mated female uses ovipositor to slit tree branch lengthwise, then wedges a series of eggs into the fresh crevice. Nymphs burrow into soil to reach tree roots, where they feed and grow very slowly, requiring 13–17 years to complete development. Each nymph crawls to nearest upright support, splits its skin, and transforms to an adult. The simultaneous appearance of thousands of cicadas during a few weeks overwhelms predators, permitting the great majority to mate undisturbed.

Unlike cicadas of other genera, Periodical Cicadas emerge in a single locality only once every 13 or 17 years. Each synchronized population is called a brood. Some are large and occupy major areas of the United States, while others are small and cover less than 100 square miles. Only 14 broods of 17-year cicadas and 5 of 13-year cicadas are known to exist today. Broods that are separated by 4 years tend to overlap in geographic distribution, whereas those separated by only 1 year border each other geographically, without any overlap.

290 Dogday Harvestfly
(*Tibicen canicularis*)

Description: 1⅛–1¼" (27-33 mm), wingspan to 3¼" (82 mm). *Black with green markings.* Wings clear green along rear ½ of fore wing margin.

Habitat: Coniferous and mixed woods.

Range: Northeastern United States and adjacent Canada.

Food: Adult is not known to eat. Nymph feeds on root juices, especially pine.

Sound: A powerful call that sounds like a circular saw cutting through a board.

Life Cycle: Nymphs take 3 years before maturing to adult. A new generation hatches each summer in the same area.

Since this cicada disappears from mixed forests soon after all the pines are eradicated, it probably feeds on pine roots. It is seen during the hot, "dog days" of summer, hence its common name.

289 Grand Western Cicada
(*Tibicen dorsata*)

Description: 1⅜–1⅝" (35–40 mm), wingspan to 4⅜" (110 mm). *Brown-black or greenish.* Sometimes yellow on thorax and mouthparts. Often dusting of white powder above. Wings brownish green.

Habitat: Deciduous and mixed woods.

Range: Colorado and Arizona into New Mexico and adjacent states to the east.

Food: Adult does not eat. Nymph feeds on juices from roots.

Sound: A sudden loud whine, maintained steadily, then dropping in pitch and intensity as it dies away.

Life Cycle: A new generation appears in the same geographic range each summer. Probably nymphs mature in 3 years.

This insect is the biggest and most beautiful western cicada.

TREEHOPPERS
(Family Membracidae)

Small, strange-looking jumping insects, most treehoppers are ¼–⅝" (5–15 mm) long. The pronotum projects far back over the thorax and often in other directions as well, making the entire insect resemble a thorn of the plant upon which it rests. Treehoppers take sap from vegetation, particularly trees and shrubs. They press their eggs into slits in twigs, often killing the tip of the twig. Many treehoppers are brightly colored, some in intriguing patterns.

107 Thorn-mimic Treehopper
(*Campylenchia latipes*)

Description: ¼" (5–6 mm). Dark reddish brown over head and thorax, also beneath. Wings and most of high crest along midline cinnamon-brown. *Short sharp horn extends on each side of pronotum.*

Habitat: Deciduous and mixed forests, and adjacent grasslands.

Range: California.

Food: Juices of many plants.

Life Cycle: Eggs are inserted in slits of stems, where they overwinter. Nymphs go through 5 molts in about 6 weeks. 1 or 2 generations a year.

These treehoppers resemble sharp-edged thorns on plants and are easily overlooked when motionless.

109 Oak Treehopper
(*Platycotis vittata*)

Description: ⅜" (10 mm). Sea-green to bronze or olive-green; or pale blue marked with red dots or 4 red lengthwise stripes. Wings clear, yellow along front margin; veins with fine black lines. *Prothorax extends to sharp point on each side,* varies in how far median crest rises and projects forward as horn. Fine pattern of pits, resembling food plant.

Habitat: Woods and forests with deciduous or evergreen oaks.

Range: Arizona to California, north to British Columbia.

Food: Juice of oaks, especially from leaves.

Life Cycle: Eggs are inserted into slits in young twigs. Nymphs are mostly unmarked, resembling adults. 1 generation a year.

These attractive insects congregate in compact groups along twigs when newly hatched. Older nymphs tend to be more solitary. Related species occur on oaks in the eastern United States.

110 Buffalo Treehoppers
(*Stictocephala* spp.)

Description: ⅜" (9 mm). Bright green to yellowish above, yellowish below. Wings clear, tapering toward end of abdomen. *Pronotum projects forward at each side to short stout point,* suggesting horns of buffalo or bison, hence the common name.

Habitat: Woods, orchards, crop fields, and meadows.

Range: Throughout the United States and southern Canada.

Food: Alfalfa, if available, and low succulent plants; also willow, elm, cherry, locust, and orchard trees, sometimes potato, tomato, clover, goldenrod, and aster.

Life Cycle: Pairs of smooth white eggs are pressed into crescent-shaped slits cut in bark of young stems. Eggs overwinter, hatch in spring. Nymphs mature in about 6 weeks.

This is the best-known and most widely distributed treehopper. The genus is represented by almost a dozen species, each having minute distinguishing features and some differences in food and habitat. Formerly assigned to genus *Ceresa*.

106 Locust Treehopper
(*Thelia bimaculata*)

Description: ⅜–½" (9–13 mm). Male gray mottled with yellow, lemon-yellow stripe along each side of prothorax; female larger, mottled gray-black and brownish. *Pronotum projects in large thornlike crest,* sloping from front of eyes to wing tips.

Habitat: Woods, meadows, and gardens.

Range: New Hampshire to Florida, northwest to Ohio.

Food: Plant juices of locust and other leguminous trees and shrubs.

Life Cycle: Eggs are laid at roots of host tree, just below leaf litter. Nymphs remain close to adults. 1 or 2 generations a year.

Nymphs produce a sweet secretion that attracts ants, which probably protects them from predators.

SPITTLEBUGS OR FROGHOPPERS (Family Cercopidae)

Spittlebugs are small to medium-sized jumping insects, ⅛–½" (4–13 mm) long, that hop about like tiny frogs on plants and shrubs, where they feed. Adult spittlebugs are sometimes mistaken for leafhoppers but have only 1 or 2 spines on the hind tibiae instead of 2 rows of spines. Some species are coated with waxy white fluid. Spittlebug nymphs cover themselves with masses of bubbly, wet spittle, which accounts for their common name. Each frothy mass contains 1 or more tiny nymphs, effectively hiding them from predators as they suck plant juices. Nymphs usually feed upside down. Spittle secreted from the anus flows down over the body and mixes with an excretion from glands on the seventh and eighth abdominal segments. Air mixed into the spittle makes the bubbles long-lasting. The adults do not produce spittle and jump around freely.

65 Meadow Spittlebug (*Philaenus spumarius*)

Description: ⅜" (9–10 mm). *Near bubbly froth.* Elongated, pear-shaped. Rounded head, *very short antennae.* Gray to green, yellow, or chocolate-brown with pale spots. Wings short, at rest conform to body shape. Nymph is pale, scarcely pigmented.

Habitat: Meadows and croplands where forage plants are raised.

Range: New England south to Florida; West Coast to north central states and adjacent Canada.

Food: Juice of many different plants.

Life Cycle: Frost-resistant eggs are deposited in angle between leaf and stem, hatching in spring. Nymphs soon cover themselves with froth consisting of self-made bubbles, which provides protection from dry air, predators, and potential parasites. Nymphs move frequently, but maintain their cover until they mature. Winged adults crawl or fly into open. 1 generation a year.

Adults are often inconspicuously colored, and nymphs are overlooked because of their bubbly covering. This pest destroys alfalfa and clover crops.

LEAFHOPPERS
(Family Cicadellidae)

Leafhoppers are very common jumping insects, $\frac{1}{16}-\frac{5}{8}"$ (2–15 mm) long. There are over 2,500 species in North America. These tiny insects look like adult froghoppers but have more elongate bodies and 1 or more rows of slender spines on the hind tibiae. Many are brightly colored. They live and feed on all kinds of plants—from grasses, flowers, and low shrubs to trees—but each species requires a specific plant food and habitat. As they suck, many leafhoppers secrete a honeydew from the anus, composed of filtered plant sap. This sweet liquid attracts ants, flies, and wasps. Some leafhoppers are serious pests of crop plants. Their feeding causes wilted, discolored leaves and stunted growth. Some transmit diseases in plants.

113 Grape Leafhopper
(*Erythroneura comes*)

Description: ¹⁄₁₆–¹⁄₈″ (2–3 mm). Pale yellow or white with yellow, red, and blue marks. Overwintering adults often nearly all red. *Fore wings bear 3 small black dots,* 2 of them on front margin.

Habitat: Open fields, woods, and gardens.

Range: Throughout North America.

Food: Juice from foliage of vines and briars.

Life Cycle: Eggs are inserted singly into leaf tissues. Nymphs feed mostly on underleaf surfaces. Adults of last generation overwinter among plant trash on the ground. 2 or more generations a year.

Although this insect's principal food is grape, it also infests beech, blackberry, Boston ivy, burdock, catnip, currant, gooseberry, grass, maple, plum, and Virginia creeper.

112 Scarlet-and-green Leafhopper
(*Graphocephala coccinea*)

Description: ³⁄₈″ (8–9 mm). Slender, pointed head. Pale yellow to brownish; or bright green or blue. *Fore wings are mostly orange to red, edged in green or blue,* with a diagonal stripe of same color forming a deep V when wings are folded together over back.

Habitat: Meadows and gardens.

Range: Eastern United States and adjacent areas of Canada.

Food: Juices of weeds and cultivated plants.

Life Cycle: Eggs are thrust into soft plant tissues in early spring. Nymphs are mostly green. Adults overwinter in leaf trash on the ground. 1 or more generations a year.

Saliva injected into a plant by this leafhopper blocks plant tubes essential for the transport of sap, causing the plant to wither and drop leaves.

114 Sharpshooter
(*Oncometopia nigricans*)

Description: ⅜″ (8–9 mm). Tubular. Yellow-orange with fine black markings. *Front rim of pronotum yellow with black dots,* rest of pronotum and fore wings pale ash-gray with fine black markings along wing veins. Wings at rest are almost parallel to body.

Habitat: Meadows and undeveloped lands.

Range: Florida.

Food: Juices of various weeds.

Life Cycle: Eggs are inserted into tissues of soft stems. Nymphs disperse quickly to feed. 1 or more generations a year.

One of the largest leafhoppers, this treehopper is called Sharpshooter because it leaps rapidly from danger with the speed of a sharpshooter's bullet.

FULGORID PLANTHOPPERS
(Family Fulgoridae)

These wedge-shaped planthoppers, up to ⅜″ (8–10 mm) long, have a blunt, hornlike projection at the top of the head. Short antennae arise below the eyes on a narrow, flat "cheek." There are usually 2 simple eyes located immediately in front of the compound eyes on either side of the head. The slender, long legs have 3 tarsal segments and a few spines on the hind tibiae. Fore wings are either long or quite short. Some fulgorids are wingless. Almost all suck plant juices. They are sometimes called lanternflies, referring to the early belief, now disproved, that tropical species were luminous.

108 Partridge Scolops
(Scolops perdix)

Description: ¼–⅜" (6–8 mm). *Unusually long head, prolonged into a horn,* almost as long as distance from eyes to tip of abdomen. Body speckled tan and dark brown, paler below.

Habitat: Meadows and open woods.

Range: Pennsylvania to Missouri, north to Manitoba and Ontario.

Food: Juices of many native plants.

Life Cycle: Little known.

Remarkable for their body shape and especially the long head horn, these insects and related species have little impact on human activities.

PSYLLIDS
(Family Psyllidae)

Psyllids are small insects, ¹⁄₁₆–¼" (2–6 mm) long. They hold the membranous, round-ended wings rooflike over the body like tiny cicadas, which some species resemble. The fore wings are longer, thicker, and more heavily veined than the hind pair. All psyllids seem able to jump, although the hind femora are not noticeably enlarged. The tarsi have 2 segments; the long, threadlike antennae usually have 10 segments. Nymphs are oval and flat, scarcely suggesting the adult form. Many psyllid nymphs produce waxy filaments which may help conceal their bodies or make them unpalatable to predators. All species feed on plant juices. A few become pests on orchard and garden plants.

53 American Alder Psyllid
(*Psylla alni*)

Description: ¹⁄₁₆–¹⁄₈″ (1–4 mm). *Body brown or green;* wings often milky white. *Nymphs are* paler brown with dark crossband on base of abdomen; usually *concealed beneath long white cottony fibers of wax.*

Habitat: Stream edges and lakeshores, where alder grows.

Range: Nevada and California.

Food: Juices from alder stems and young leaves.

Life Cycle: Pale yellow eggs are laid on bark, hatching about 30 days later, when new leaves are opening. The last nymphs of the year are darker. They overwinter and complete development in early spring. 1 or more generations a year, depending on rainfall.

Nymphs of the similar Cottony Alder Psyllid (*P. floccosa*) are yellow and green, while adults are yellowish green to brown. The Cottony Alder Psyllid is common in many eastern states and also occurs in Colorado and California.

WHITEFLIES
(Family Aleyrodidae)

Whiteflies look like tiny, white specks on plants. These tropical insects are often found outdoors on citrus trees in the South and indoors on home or greenhouse plants everywhere. Only ¹⁄₁₆–¹⁄₈″ (1–3 mm) long, whiteflies have a powdery white covering on their body and wings and sometimes on their legs. Their disproportionately large wings are paddle-shaped; the fore wings, which are slightly larger than the hind pair, fold horizontally over the body when at rest. The threadlike antennae have 7 segments. Females lay eggs either after mating or asexually by parthenogenesis. The flattened, oval nymphs usually

have a fringe of white, waxy fibers
which attach them to the host plant.
This downy covering makes them
resemble the nymphs of soft scale
insects. When nymphs molt for the
next-to-the-last time before becoming
adults, they stop eating, a characteristic
similar to the pupal stage in insects
with complete metamorphosis.

58 Greenhouse Whitefly
(*Trialeurodes vaporariorum*)

Description: 1/16" (1–2 mm). *Body pale, resembling
white speck.* Wings white, unspotted.
Nymph is oval, flat, almost transparent
whitish green *with white waxy filaments
underneath.*

Habitat: Gardens in the South and tropics,
greenhouses, and homes in the North.

Range: Worldwide.

Food: Juice from host plants of many kinds.

Life Cycle: Elongated eggs are supported separately
on short, slender stalks. Nymphs molt
to a nonfeeding pupal stage. They
become coated with distinctive long
and short transparent waxy rods.
Winged adults emerge after a week or
more. Only cool weather limits
continuous new generations. They
cannot survive frost.

This common insect is a troublesome
pest in greenhouses, southern gardens,
and in some homes. It will quickly
move from one plant to the next,
covering the foliage with white specks.

APHIDS
(Family Aphididae)

Aphids feed on the juices of plant
stems, leaves, and flowers and are found
almost everywhere. Many are major
pests; they reproduce so rapidly and in

such great numbers that they can destroy entire crops. Most are green, red, or brown, but they may range from gray, yellow, and black to pink and lavender. Aphids have soft, pear-shaped bodies, $\frac{1}{16}$–$\frac{3}{8}$" (2–8 mm) long, with small heads and long antennae. Some adults have wings, while others are wingless. Usually a pair of slender, wax-secreting tubes, called cornicles, project from the back of the abdomen. Aphids also secrete from the anus a saplike liquid called honeydew that attracts ants and other insects; some ant species "milk" aphids, like cows, for this secretion. Most aphids have a complex life cycle. Overwintering eggs hatch in spring into wingless females, which give birth parthenogenetically to young that mature as wingless females. This continues until a generation of winged females is born in summer and flies to another species of plant. There they produce more wingless females until a last generation of winged females returns to the original plant. In fall these produce wingless males and egg-laying females, which mate, and the cycle begins again. In this way huge aphid colonies build up rapidly. They are among the most destructive plant pests of crop and orchard trees. Luckily they are kept in check by many predators, including ladybug beetles, fire beetles, and parasitic wasps.

59 Green Apple Aphid
(*Aphis pomi*)

Description: $\frac{1}{16}$–$\frac{1}{8}$" (2–3 mm). Pear-shaped. *Spring and summer forms bright green,* head sometimes yellowish, legs dark. Wingless males grayish yellow, females dark green. Antennae less than $\frac{1}{2}$ as long as body. Tubular projections from abdomen short, never with shelflike flange at tip.

Habitat: Orchards.

Range: Throughout North America.

Food: Sap of apple or pear trees, sometimes that of hawthorn, firethorn, cotoneaster, and pittosporum.

Life Cycle: An entire cycle can be completed on a single tree or shrub. Pale green eggs are laid on new branches in fall, overwinter, turn shiny black, and hatch when leaf buds open in spring. Nymphs enter unfolding buds, feed on leaves, mature in 2–3 weeks as wingless females. These reproduce asexually for several generations, including both wingless and winged female adults. Wingless adults and nymphs remain close together; winged adults disperse to found new colonies.

This insect was introduced from Europe in colonial times. It is second only to the Rosy Apple Aphid as a major pest in orchards.

55, 60 Rosy Apple Aphid
(*Dysaphis plantaginea*)

Description: $\frac{1}{16}$–$\frac{1}{8}$″ (2–3 mm). Pear-shaped. *Wingless form rosy brown to purplish.* Winged form has brownish to brownish-green body with dark head and thorax. Both often dusted with whitish powder. Antennae more than $\frac{1}{2}$ as long as body. Tubular projections from abdomen have shelflike flange at tip.

Habitat: Wherever apple and related trees and plantain grow.

Range: Throughout the United States and Canada.

Food: Juice of apple or plantain, according to the season.

Life Cycle: Eggs are laid on apple trees, hatching young that reproduce asexually for several generations. Winged females appear in July and fly to plantain, where several more generations of

wingless females develop. In fall
winged females reappear, fly to apple
trees, and produce wingless males and
females, which mate. Pale green eggs,
which turn shiny black, overwinter
in bark crevices.

Rosy Apple Aphids cause leaves to curl
and fruit to grow gnarled and ripen
early.

63 Rose, Pea, and Potato Aphids
(*Macrosiphum* spp.)

Description: ⅛″ (3–4 mm). Pear-shaped. Light to
medium green or pink, with darker
midline stripe, or mottled green and
pink. Antennae longer than body; legs
long. Abdomen tapers to narrow blunt
projection between pair of tubular
cornicles.

Habitat: Crop fields, gardens, and meadows.

Range: Throughout North America.

Food: Juices of various host plants; preference
varies according to species.

Life Cycle: In autumn, eggs are laid on woody
stems of a shrub or tree, where they
overwinter and produce wingless
females in spring. These feed, then
reproduce asexually for a month or so,
creating several generations of wingless
females. A winged generation appears
and moves to another host plant,
primarily annual herbs and grasses,
where asexual reproduction continues.
In autumn another winged generation
returns to the woody host shrub or tree,
then a wingless sexual generation
develops, mates, and repeats the cycle.

The best-known species in this genus is
the Potato Aphid (*M. euphorbiae*),
known also as the Pink-and-green
Tomato Aphid. It places overwintering
eggs primarily on roses. Summer hosts
include rose, apple, potato, tomato,
eggplant, corn, gladiolus, iris, and

various weeds. The related Rose Aphid (*M. rosae*) may overwinter on roses and stay on these plants all year.

61 Root Aphids
(*Pemphigus* spp.)

Description: ⅛″ (3–4 mm). Pear-shaped. *Wingless form checkered yellowish and dark*, sometimes with a tuft of cottony wax at rear. Winged form usually black. Both often covered with whitish dust.

Habitat: Trees in winter and spring; on roots of plants in meadows, including asters, goldenrod, and knotweed, and on crop plants such as alfalfa, beets, and wheat.

Range: Throughout North America.

Food: Plant juices.

Life Cycle: Eggs overwinter in crevices of poplar and other trees, hatching in early spring. Young females reproduce asexually for several generations. In summer, winged forms develop and fly to wild grasses and cereal crop plants, where they produce new generations of wingless females. Later, winged adults appear, fly back to trees, and produce wingless males and females, which mate.

Many members of this genus, such as the Sugar-beet Root Aphid (*P. betae*), induce galls on trees, in this case the leaf stalks of poplars. The first generations that feed on roots are often watched over by ants, which lap the sugary secretion, called honeydew.

57 Cloudy-winged Cottonwood Aphid
(*Periphyllus populicola*)

Description: ⅛″ (3–4 mm). Pear-shaped. Yellow-green to greenish black or blackish brown. *Wings have grayish or whitish cloudlike pattern* that does not obscure

veins. Nymph is translucent amber.
Habitat: Cottonwood and poplar groves.
Range: Montana to New Mexico, west to California.
Food: Sap of poplar or cottonwood.
Life Cycle: Eggs overwinter in bark crevices of cottonwood, hatching into wingless females that reproduce several generations of wingless females. In summer winged females migrate to poplar trees, where they give birth to more wingless generations until winged females return to cottonwood and produce males and females, which mate.

Aphid nymphs are often inconspicuous on dark twigs, but adults' cloudy wings are usually easy to spot. A mature female can often be seen surrounded by a cluster of young.

62 Giant Willow Aphid
(*Pterochlorus viminalis*)

Description: ⅛–¼" (4–5 mm). Pear-shaped. Brown, spotted with black, often appearing gray because of *covering of waxy white dust*. Abdomen has large black projection from midline and 2 very short black tubes surrounded by small black areas. Nymph is translucent amber.
Habitat: Willow groves.
Range: Throughout North America.
Food: Juice of willows.
Life Cycle: Several generations, including wingless and winged adults, are produced asexually all summer. In fall wingless males and wingless females appear and mate. Females lay frost-resistant eggs in bark crevices.

These aphids are called "giants" because of their unusually large size. They are generally found in huge compact colonies, often near ground level.

When disturbed, they kick energetically, raising the hind legs high above the abdomen.

WOOLLY AND GALL-MAKING APHIDS
(Family Eriosomatidae)

Most of these small, somewhat pear-shaped insects secrete waxy or woolly material that nearly covers their bodies. They resemble aphids but have very short tubular extensions near the tip of the abdomen or none at all. Like aphids, eriosomatids usually live on 2 distinct host plants and have bisexual and asexual generations. Members of the wingless, bisexual generation lack mouthparts. A mated female lays only a single egg. Females that reproduce without fertilization produce several generations that have thick woolly or waxy coverings during nymphal stages. Many species induce galls to form on buds, leaves, stems, or roots of the first host plants. All members of this family are serious pests.

51 Woolly Apple Aphid
(*Eriosoma lanigerum*)

Description: 1/16″ (2 mm). Short, broad, soft. *Purple or dark red with multiple strands of cottony wax* from abdomen. Antennae barely more than ½ as long as body.

Habitat: Wherever apple and elm trees grow.

Range: Worldwide.

Food: Juice from inner bark of young stems and roots of apple and elm trees.

Life Cycle: Overwintering eggs are deposited deep in elm bark crevices and hatch in spring as wingless females. They reproduce asexually for 2 generations, until winged females fly to apple or related trees, where they crawl down to roots and produce more asexual generations. Late in summer more winged females appear and fly to elms, where they produce a sexual generation.

Apple twigs become swollen and knobby, and roots develop nodules where these woolly aphids suck juices. They also attack hawthorn, mountain ash, pear, and quince. Elm leaves curl in groups where insects are clustered. The Woolly Elm Aphid (*E. americanum*) is very similar but causes individual leaves of elm to curl. Shadbush is the alternate host.

GIANT SCALE INSECTS
(Family Margarodidae)

Like all scale insects, members of this family are hidden under waxy scales. Females have large, rounded bodies, ⅛–1″ (4–25 mm) long, and no wings. They are especially noticeable when they produce cottony or waxy egg capsules with lengthwise fluting. Winged males are slender and have long feathery antennae and a pair of hair tufts at the back end. As a nymph grows, the number of antennal segments increases, and the insect becomes hidden under its waxy secreted scale. Southern species that feed on grass sap form wax cysts on roots. These thin, hard, bronzy or golden balls are known as "ground pearls" and are sometimes used for jewelry.

50 Cottony Cushion Scale
(*Icerya purchasi*)

Description: Female ⅛–¼″ (4–5 mm), male less than ¹⁄₁₆″ (1 mm). Female yellow, orange-red, or brown; partly or wholly covered by white or yellowish wax.

Female has large, white, fluted, cottony egg sac extruding from rear. No wings. Male red with 2 wings, 2 knoblike halteres similar to those of true flies, and 2 tufts of hair on rear. Nymph is bright red with brown legs and 6-segmented antennae.

Habitat: Woodlands and citrus orchards in the West, grasses in the South, greenhouse plants in the North.

Range: Outdoors from South Carolina to Florida, west to California; occasionally to southern Ontario and New England.

Food: Plant juices.

Life Cycle: Egg mass containing 600–1,000 red eggs bursts in 2 months during mild winter or in a few days in summer, releasing nymphs. They feed at first on foliage, then twigs, and become covered by secreted cottony wax. Male nymphs spin flimsy cocoons, then transform to adults. Females free themselves of waxy covering and molt to adults. Many generations a year.

These insects were introduced from Australia in 1866 on acacia plants imported to California. Twenty years later they had become so pervasive in citrus orchards that the predatory Australian ladybug beetles were introduced in 1888 and 1889 in the first successful use of biological pest control.

ARMORED SCALE INSECTS
(Family Diaspididae)

This largest family of scale insects includes many important pests. Their crusty, scalelike covering can be seen in huge assemblages on infested twigs and branches. Both males and females have small, flattened oval bodies that are hidden under a detached, waxy scale covering made of their own secreted wax and the cast-off skins of nymphal

stages. Scales differ in shape and color
in each species; males usually have
smaller, more elongate scales than
females. Adult females do not have
eyes, legs, or antennae. Males have
wings, well-developed legs, and
antennae. Some species reproduce
bisexually with fertilized eggs, while
others give birth without fertilization
to living young. Nymphs have short
legs and move about to feed. They later
become stationary, attached to the host
plant by their mouthparts.

49 Oyster Shell Scale
(*Lepidosaphes ulmi*)

Description: Female scale ⅛" (3 mm) long, less than
¹⁄₁₆" (1 mm) wide. Male scale smaller.
Both pale to dark brown, often
S-curved, tapering to a point. *Scale
suggests shape of an oyster shell* crossed by
several grooves.

Habitat: Woods, orchards, and lawns.

Range: Nova Scotia to Georgia mountains,
west to California, north to British
Columbia.

Food: Juices of woody vegetation, including
vines, and apple, pear, plum, ash, elm,
and maple trees.

Life Cycle: 40–100 oval white eggs overwinter
under female's scale, which still clings
to bark after female has died from frost.
Whitish nymphs disperse for 2–48
hours in spring. They pierce bark to
feed, develop scales, and mature by
mid-July. Usually 2 generations a year.

The Oyster Shell Scale, introduced
long ago on nursery stock from
Europe, has spread to more than 100
different American plants.

MEALYBUGS
(Family Pseudococcidae)

These soft-bodied, oval insects have well-developed legs and segmented bodies that are covered with powdery or mealy white wax. Many have distinctive points that project from the sides, or tail-like filaments that trail from the rear, or both. Adults and nymphs feed in compact groups, sucking juices from leaves, stems, and roots. Some species lay masses of eggs, which they surround with wax. Others give birth to living young and have no egg masses. Mealybugs are pests of citrus trees and greenhouse plants.

54 Long-tailed Mealybug
(*Pseudococcus adonidum*)

Description: 1/16–1/8" (2–3 mm). Pearly-white. Distinguished from other species by *1–5 long white filaments at rear*, often longer than body, and a series of shorter filamentous projections on each side, almost 1/2 body width.

Habitat: Outdoors on tropical and subtropical plants in the South; in greenhouses and homes in the North.

Range: Throughout North America.

Food: Plant juices of many kinds.

Life Cycle: Female gives birth to live nymphs. They gather in groups to feed and secrete a honeydew that attracts ants. Male nymphs prepare fragile silken cocoons, in which they shed the final nymphal skin, and emerge with 1 pair of wings. Females do not spin cocoons and grow gradually into adults. Several generations a year.

These tropical bugs are especially active on summer evenings, when winged males mate with wingless females. They attack a wide variety of ornamental and tropical plants, including dragon trees, birds of paradise, arum, poinsettia, avocado, lemon, mango, and oleander. Formerly called *P. longispinus*.

COCHINEAL BUGS
(Family Dactylopiidae)

These small, red insects are found on prickly pear cacti in the Southwest. Their soft, elongated bodies resemble those of mealybugs but are covered with waxy plates. Females are wingless. Males have 2 long, whitish wings but, unlike true flies, have only 1 segment in each tarsus rather than 5. Males lack mouthparts. The waxy scales on the female conceal minute eggs. Once eggs are extruded, the female dies, and hatchling nymphs crawl out from under the dead body. They disperse quickly, settling to feed and forming their own protective coverings. Native American Indians of the Southwest learned to make a crimson dye by drying female bugs and extracting red pigment. Even today old pharmacies often display a bottle of dried cochineal bugs on the shelf for use as a water-soluble dye.

52 Cochineal Bug
(*Dactylopius confusus*)

Description: Female ¹⁄₁₆–¹⁄₈″ (2–3 mm); male ½ length. Red with deep red to pink waxy scales under body. *Often concealed by dense tangled strands of white cottony wax.* Legs reduced. Male has 2 diverging filaments trailing from rear and long white wings.

Habitat: Deserts and arid areas.

Range: New Mexico to Mexico, northwest to California; also in Montana, Colorado, and Florida.

Food: Juices of cacti, especially prickly pear.

Life Cycle: Nymphs escape from beneath body of dead female and begin feeding. Females mature in place without moving after the 1st molt. They feed in all stages. Males do not feed in last nymphal stage, nor perhaps as adults.

Conspicuous clusters of Cochineal Bugs often feed side by side, covering large areas of cacti like a white furry rug.

Thrips
(Order Thysanoptera)

Thrips are such tiny insects that they often appear to be black specks moving over flowers. Most are minute (0.5 mm) or up to ⅛" (3 mm) long. There are 4,700 species worldwide, including 600 in North America. Thysanoptera means "fringed wing," alluding to the distinctive fringe of long hair on the wings. But many thrips have no wings at all. All have slender bodies, short antennae, and short legs that end in 1 or 2 claws. Some species have bladderlike enlargements at the ends of the legs. The heads are bluntly conical, bearing antennae and compound eyes towards the front. The unusual, asymmetrical mouthparts combine piercing and rasping elements, enabling thrips to saw through plant tissues and suck out juices. Most thrips curl the abdomen over the back when they crawl. Winged species are good fliers, staying airborne for hours.

Some species lay eggs, while others reproduce parthenogenetically without fertilization. Females with sawlike ovipositors cut slits into stems to lay their eggs. Females without ovipositors deposit eggs in crevices or under bark. Metamorphosis is simple: nymphs resemble adults but lack wings. Wing pads appear during the dormant, nonfeeding stage called the prepupa. In the pupal stage, most thrips are enclosed in cocoons, from which they emerge as true adults. There are usually several generations a year. Many thrips are serious pests, sawing into wheat, cotton, and citrus stems. Still others are beneficial predators, preying on mites and small insects and eating moth eggs. Some species transmit fungal and bacterial plant diseases.

BANDED THRIPS
(Family Aeolothripidae)

Banded thrips are tiny dark or bicolored insects, usually $\frac{1}{16}''$ (1.5 mm) or less long, with colored bands on their distinctive, fringed wings. Fore wings, when present, are broad and rounded at the end. The female's ovipositor is upcurved, and the antennae have 9 segments. When immature, most species prey upon small insects and mites. Adults are found on flowers.

75 Banded-wing Thrips
(*Aeolothrips fasciatus*)

Description: $\frac{1}{16}''$ (1.5 mm). Dark gray or brown to yellow. *Fore wings white to yellow, crossed by 2 dark brown bands.* Some wings short. Nymph is pale yellow, blending to orange at rear.

Habitat: Crop fields, grasslands, and gardens.

Range: Throughout North America.

Food: Smaller insects, especially other thrips.

Life Cycle: Eggs are inserted into stems and leaves. Nymphs complete feeding in early summer, then drop to ground. They overwinter in cells in the soil. 1 generation a year in the North, perhaps more in the South.

The Banded-wing Thrips is the most widely distributed species in this family. These insects are considered to be entirely beneficial because they eat many pests, such as aphids and mites. The Western Golden Banded-wing Thrips (*A. aureus*) is golden-yellow to dark brown, with dark brown at the tip of the abdomen. The Duval's Banded-wing Thrips (*A. duvali*) is yellowish with dark brown on its third antennal segment. Both have broader wings than the Banded-wing Thrips.

66 COMMON THRIPS
(Family Thripidae)

Common thrips, the largest family in the order, have narrower, more pointed wings than banded thrips. The female's ovipositor is downcurved and the antennae have 6–9 segments. These insects are usually $\frac{1}{16}$" (1.5 mm) or less long. Colors vary greatly among species, ranging from plain yellow, brown, or black to yellow and orange. Almost all species feed on cultivated plants and many are important economic pests of pears, gladioluses, onions, grasses, grains, and cereals.

Alderflies, Dobsonflies, and Fishflies (Order Megaloptera)

Fewer than 20 species of these distinctive, large insects have been identified in North America, and only 180 species worldwide. Because most have densely veined wings, they are sometimes included in Order Neuroptera with net-veined insects. But unlike neuropterans, megalopterans lack forking veins at the wing margins. Members of this order have large hind wings that are wider at the base than the fore wings. The order name means "ample wings." At rest wings are folded rooflike. These insects, ⅜–2¾ (9–70 mm) long, are clumsy fliers, often seen fluttering around lights in the evening. They live near ponds or streams, where their aquatic larvae feed on other insects. Adults eat little or nothing at all. Metamorphosis is complete: larvae transform to adults with wings during a distinct pupal stage. These insects are generally considered to be beneficial. They provide an important food source for freshwater fish, while their larvae help control the population of black fly larvae.

ALDERFLIES
(Family Sialidae)

Alderflies are black or dark brown insects, ⅜–¾″ (9–20 mm) long, with slender antennae about half as long as the body. They lack simple eyes. The broad, dark, or smoky wings are held rooflike over the body when at rest. Adults spend their brief lives of only 2–3 days running over and under foliage close to the shore. The aquatic larvae live 2–3 years in water under stones and prey upon small insects. Larvae have tapering, flattened bodies, 1⅛″ (30 mm) long, with 7 pairs of segmented gill filaments on each side (rather than 8 as in dobsonfly and fishfly larvae) and a long, tail-like filament at the rear. Unlike dobsonfly and fishfly larvae, they lack hooked anal prolegs. Larvae pupate in cells constructed in the soil above water level. There is 1 genus of alderfly in North America.

331 Alderflies
(*Sialis* spp.)

Description: ⅜″ (9–10 mm). Black to dark brown. *Wings partly clear to brownish black.* Antennae threadlike.

Habitat: Vegetation along rivers and streams.

Range: Northern North America, south in mountainous regions.

Food: Adult does not seem to eat. Larva preys on small aquatic insects and worms.

Life Cycle: Eggs are laid near water. Nymphs crawl into water and grow until autumn. Pupae overwinter concealed in underground cells near a stream or river. Adults emerge May–June, mating at night.

Larvae are voracious predators, crawling through mosses on river and stream bottoms in search of food. Although adults have biting mouthparts, they are rarely used, even in self-defense. Some species occur coast to coast. Other more local species differ in details evident mostly to specialists.

45 DOBSONFLIES AND FISHFLIES (Family Corydalidae)

Dobsonflies and fishflies are large, soft-bodied insects, ¾–2¾" (20–70 mm) long. All dobsonflies and some fishflies have milky or clear wings, but there are also fishflies with black or smoky wings. Dobsonflies are larger, with a wingspan of 2–4⅞" (50–125 mm) or more. Male dobsonflies have tusklike mandibles about 3 times longer than the head. Fishflies are smaller, having wingspans of less than 2" (50 mm) and short mandibles. Some have serrated or comblike antennae. Both dobsonfly and fishfly larvae live in streams under stones. Although dobsonfly larvae resemble alderfly larvae, they are larger, 1⅝–3½" (40–90 mm) long, and have 8 rather than 7 pairs of gill filaments, 1 pair of hooked anal prolegs, and no tail-like terminal filament on the abdomen. All are active predators. Larvae pupate in earthen cells under stones and debris near water.

329 Fishflies
(*Chauliodes* spp.)

Description: ¾–1" (20–25 mm). Body and wings ash-gray. Mandibles small, compound eyes dark and prominent, legs weak. *Antennae, ⅜" (9 mm) or more, have soft comblike projections along 1 side.*

Habitat: Vegetation along streams.

Range: Throughout North America; individual species more restricted.

Food: Adult probably eats nothing. Aquatic larva preys on small aquatic insects.

Life Cycle: Eggs are laid in masses near water, into which larvae crawl. They cling to vegetation with strong legs and seize prey in powerful jaws. Larvae pupate in cells prepared in mud along the shore. Adults are active at night May–August.

These large insects are easy to capture because they fly so slowly and are so poorly coordinated. They rest on foliage by day and fly to lights at night.

332, 333 Eastern Dobsonfly
(*Corydalus cornutus*)

Description: 2″ (50 mm), wingspan to 4⅞″ (125 mm). Head almost circular, prothorax squarish, slightly narrower. *Wings translucent, grayish* with dark veins. *Mandibles of male ½ as long as body,* curved and tapering to tips, held crossing one another; female's mandibles shorter, capable of biting forcefully.

Habitat: Close to fast-flowing water, on alders, willows, and other woody vegetation.

Range: East of the Rocky Mountains.

Food: Adult probably eats nothing. Larva preys on aquatic insects.

Life Cycle: Rounded masses containing 100–1,000 or more eggs are laid on rocks, branches, and objects close to water. Each mass is coated with a whitish secretion. Larvae drop into the water or crawl to reach feeding grounds. After 2 or 3 growing seasons, they crawl out of water and prepare pupal cells under stones or logs where they overwinter. Adults emerge in early summer.

Fishermen use dobsonfly larvae, called hellgrammites, as bait since trout seem to be attracted to them as a natural food. The small Western Dobsonfly (*C. cognata*) ranges from Texas to Arizona and Utah.

Snakeflies
(Order Raphidioptera)

Snakeflies have transparent, intricately
veined wings that closely resemble
those of net-veined insects. Many
entomologists formerly considered
snakeflies to be members of the Order
Neuroptera. Like the mantidfly, the
snakefly has a very long necklike
prothorax, but the fore legs are normal-
sized, resembling other legs, and arise
from the rear of the prothorax rather
than the front. The long head can be
raised high above the body. The narrow
fore wings are larger than the hind pair
and have a pigmented spot on the front
margin. Wings are folded rooflike at
rest and do not extend as far as the tip
of the female's long needlelike
ovipositor. Most snakeflies range from
½″ to 1″ (12–25 mm) long. Clusters of
eggs are deposited in bark crevices of
coniferous and deciduous trees. The
larvae's black bodies are difficult to
distinguish from those of beetle larvae.
Metamorphosis is complete: larvae do
not resemble adults. Both adults and
larvae are active predators, using their
strong legs to scurry after small insects.
Because they often feed on orchard
pests, such as aphids and caterpillars,
snakeflies are considered to be beneficial
insects. There are 50 species worldwide;
the 19 North American species belong
to 2 families, found only in the West.

OCELLATED SNAKEFLIES
(Family Raphidiidae)

This family includes 17 species. These dark insects have long, flat heads with very long antennae. Their compound eyes appear far forward on each side of the head, while 3 simple eyes, called ocelli, are arranged in a triangle between them, suggesting the common name "ocellated"—meaning having simple eyes.

334   Texan Snakefly
(*Agulla nixa*)

Description: ⅝–⅞" (15–22 mm) from the end of jaws to tip of folded fore wings. *Long slender antennae with more than 30 beadlike segments,* longer than head and elongated prothorax combined. Head dark reddish brown, body pale. Clear wings have many veins that fork just before the outer and hind margin. Female has long, upcurved, pointed ovipositor.

Habitat: Woods.

Range: Texas and adjacent areas of Mexico.

Food: Smaller insects.

Life Cycle: Clusters of eggs are inserted into crevices in bark, where larvae feed.

This species preys on the young of many detrimental insects. At least 16 other species are known from western states and British Columbia. All were formerly assigned to genus *Raphidia*.

Net-veined Insects
(Order Neuroptera)

Members of this order have many veins in their 4 transparent wings and an especially wide border of cross veins on the front margin of the fore wings. The net formed by these crisscrossing veins accounts for both the common and order name, which means nerve wings. There are about 4,500 species worldwide, with about 338 in North America. These insects include lacewings, mantidflies, antlions, and spongillaflies. All have 2 pairs of wings that are similar in size and have an elongate oval shape. At rest the wings are usually held rooflike over the body, and in use they beat in a poorly coordinated fashion. These insects have mouthparts adapted for chewing, and lack cerci. Their long, segmented antennae may be threadlike, clubbed, or toothed like a comb. Most species have large compound eyes.

Net-veined insects go through complete metamorphosis: larvae do not resemble adults. Most larvae are predators, although the larvae of mantidflies are parasites and those of the spongillaflies feed on freshwater sponges. All larvae have large, sickle-shaped mandibles that are used to seize and eat smaller insects or sponges. Pupation occurs in silken cocoons spun in sand or soil. Net-veined insects are economically beneficial because they help control destructive insects.

MANTIDFLIES
(Family Mantispidae)

Mantidflies resemble mantids, except
that they have wings spanning 1″
(25 mm) or more. Like the mantid, the
mantidfly has a long prothorax and
enlarged fore legs for grasping prey.
Eggs are laid at the ends of short stalks
attached to plants. Larvae have short
jaws and resemble larvae of some rove
beetles. They actively hunt for food,
capturing whatever small insects and
mites they can overpower. If they
discover a wasp's nest or a spider's egg
case, mantidfly larvae burrow inside.
Mantidflies are common in the South
and do not occur in the Northwest.

305, 306 **Brown Mantidfly**
(*Climaciella brunnea*)

Description: ⅞″ (23 mm) to tips of folded wings.
Wasplike. All brown, black, or banded
with yellow. *Prothorax long, necklike.
Fore legs greatly elongated.* Compound
eyes large; antennae short. Wings dark.

Habitat: Grasslands and forest edges.

Range: Eastern and southern states.

Food: Adult preys on small insects. Larva also
preys on small insects or is an internal
parasite in spider egg sac or wasp nest.

Life Cycle: White or reddish eggs, each on a short
stalk, are laid in clusters on vegetation.
Larvae are active predators, unless they
find suitable host to parasitize; then
they transform to grublike forms. They
spin silken cocoons and pupate in last
larval skins.

The similar Western Brown Mantidfly
(*C. b. occidentis*), 1″ (25 mm) long,
ranges west of the Rocky Mountains
from Mexico to Washington, east to
Colorado and Nevada. Formerly
assigned to genus *Mantispa*.

303, 304 Mantidflies
"False Mantids"
(*Mantispa* spp.)

Description: ¾–1" (20–25 mm) to tip of folded wings. *Prothorax long, necklike. Fore legs greatly elongated.* Brownish to yellow and black. Compound eyes large; antennae short, pale. Wings brownish.

Habitat: Grasslands and forest edges.

Range: New England to Georgia, west to California.

Food: Adult and larva prey on smaller and less aggressive insects, spiders, and spider egg masses. Larva also feeds in wasp nests when possible.

Life Cycle: Oval eggs on short stalks are laid in clusters on foliage. After finding a large food supply, such as spider egg mass or wasp nest, larvae become maggot-shaped. After feeding they spin cocoons and pupate in last larval skins. Adults active June–August in the East.

Mantidflies are often common in the same areas as their mantid look-alikes.

BROWN LACEWINGS
(Family Hemerobiidae)

Brown lacewings, ¼–⅝" (6–15 mm) long, have brown wings with a pattern of veins different than that of green lacewings. Many of the small cross veins between the first main vein and the fore wing margin are forked, whereas in green lacewings these veins do not fork. Females attach eggs directly to plant leaves; they are never suspended on individual stalks. Larvae, called aphid wolves, and adults are both predators. Larvae often carry debris around on their backs and pupate in cocoons spun in the soil. These insects are found in the woods.

328 Brown Lacewings
(*Hemerobius* spp.)

Description: ⅜–⅝″ (10–15 mm). Light to dark
brown. Compound eyes brown. Wings
often somewhat paler. Antennae and
wings with dense, short brown hair.

Habitat: Woods, forests, and fields.

Range: Throughout most of North America.

Food: Mostly aphids and mealybugs, nymphs
of scale insects, and similar soft prey.

Life Cycle: Eggs are attached to plants. Larvae
scavenge in soil debris and on plants.
They pupate in silken cocoons.

Only by observing fine details of these
insects can a specialist confidently
identify the many species of this genus.
There are at least 6 species in New
England, and 12 in areas in the West.

GREEN LACEWINGS
(Family Chrysopidae)

Green lacewings are common insects
often seen in weeds and grass or on the
leaves of trees and shrubs. They range
from ⅜″ to ¾″ (10–20 mm) in length.
These insects have 4 pale green, oval
wings. The very prominent eyes are
bright golden or coppery. Females lay
white eggs on foliage, each suspended
at the end of a long slender stalk. The
flat larvae are sometimes called aphid
lions because they prey on aphids and
other soft-bodied insects. Pupation
occurs in egg-shaped, silken cocoons
that are attached to plants.

336 Green Lacewings
(*Chrysopa* spp.)

Description: ⅜–⅝″ (10–15 mm). Body pale yellow
to pale green. *Compound eyes brilliant
golden or coppery,* hemispherical.

Threadlike antennae ⅔ body length.
Wings are clear with green veins and
are at least ¼ longer than body.

Habitat: Meadows, gardens, and forest edges.

Range: Throughout North America.

Food: Small insects, especially aphids and
nymphs of scale insects and their kin.

Life Cycle: Eggs on slender white silk stalks hang
from the underside of leaves. The
Eastern Green Lacewing (*C. ornata*) lays
eggs in single rows, whereas the
Western Green Lacewing (*C. majuscula*)
groups its eggs close together, each on
a separate stalk. All species pupate in
silken cocoons. Some fly in May and
June, others in late summer.

California Green Lacewings
(*C. californica*), found west of the
Rocky Mountains, are raised indoors by
the thousands for release in greenhouses
and vineyards, where they prey on
destructive mealybugs. Only a
specialist can distinguish the dozens of
American and Canadian *Chrysopa*
species.

GIANT LACEWINGS
(Family Polystoechotidae)

These slender, elongate insects, 1⅛"
(30 mm) long, have short, threadlike
antennae and long, oval wings,
spanning 1½–3" (38–75 mm), with
many cross veins. These large insects
are conspicuous when they fly feebly to
artificial lights at night. Little is known
of their larvae or behavior, because the
2 North American species are rare.

330 Giant Lacewings
(*Polystoechotes* spp.)

Description: 1⅛" (30 mm), wingspan 1⅝–3" (40–75 mm). Dark brown to black, body densely covered with short hair. Wings large and broad, dotted with black. *Antennae short.*

Habitat: Forest edges.

Range: Nova Scotia to Virginia, southwest to Texas, west to California, north to British Columbia.

Food: Unknown.

Life Cycle: Little known.

These giants are the largest neuropterans in North America. At night they sometimes fly to lights. Their wing venation is unlike that of members of other families.

403 ANTLIONS
(Family Myrmeleontidae)

Antlions resemble damselflies but have longer and clubbed antennae. Their soft, elongate bodies measure up to 1¾" (45 mm) with a wingspan up to 2½" (65 mm). The transparent wings have many cross veins. All are poor fliers. Adults drink nectar, nibble pollen, or do not eat at all. The common name comes from the voracious habits of the larvae. Often known as doodlebugs, the larvae have oversized heads with long spiny jaws, short legs, and bristles all over their bodies. Most hide at the bottom of small pit traps made in the sand and wait for ants and other small insects to tumble down the sloping sides. Larvae in some species do not build pits but lie buried in sand or hide among debris waiting for prey. Pupation occurs in a parchmentlike cocoon buried in the sand. The adult emerges from the cocoon through an opening cut in one end, leaving the conical door still in place as though hinged. Antlions are most common in the South and Southwest, where larval pits can be seen in most places with dry sandy soil.

402 OWLFLIES
(Family Ascalaphidae)

These large dragonflylike insects, 1⅜–1⅝" (35–40 mm) long, have a sturdy thorax, large compound eyes that meet on the midline above the head, and long, knobbed antennae. Their wings bear many cross veins along the front margin and span about 2½" (65 mm). The larvae resemble antlions but do not dig sand traps. They conceal themselves under loose debris and ambush small insects of many kinds.

Beetles
(Order Coleoptera)

The largest order in the animal kingdom, Coleoptera contains a third of all known insects—300,000 species worldwide and about 30,000 species in North America. Beetles range from large tropical insects, 5⅛" (130 mm) long, to small species less than 1/16" (1 mm). Some crawl on land; others fly or live in water. Beetles can be easily recognized by the tough, armorlike fore wings, called elytra, that cover the membranous hind wings used for flying. When the insect is at rest, the elytra usually meet in a neat line down the middle of the back. The order name Coleoptera means "sheath wings" and refers to the elytra. They are often brightly colored or patterned. There are many types of beetle antennae: threadlike or clubbed in various ways, having 10 or 11 segments. Most beetles have large, prominent compound eyes. Although almost all beetles fly, they do so apparently only to get to low vegetation or other natural features of their habitat. Chewing mouthparts with well-developed mandibles enable beetles to eat a broad range of materials. Many beetles are predators, others are scavengers, and a few are parasites. Beetles are known to eat leaves, bark, dung, wool, and other fabrics. Larvae, called grubs, can be either predacious or vegetarian and sometimes cover themselves in protective shelters. The larvae display various forms, but all have biting mouthparts. Metamorphosis is complete: larvae do not resemble adults and transform during a pupal stage. Most species produce 1 generation a year, mating in spring or summer, although some have as many as 4 generations. Some species attack plants and stored foods, while others pollinate flowers and eat plant pests.

TIGER BEETLES
(Family Cicindelidae)

Tiger beetles are usually shiny metallic bronze, blue, green, purple, or orange and range from ⅜″ to ⅞″ (10–21 mm) long. These fast runners have long legs and long antennae that arise from the top of the head. Most are diurnal, sun-loving species found on beaches and dry soil. Adults often burrow into sand at night or on hot days and are easy to see at night using a flashlight. All adults are ferocious predators that seize small insects with powerful sickle-shaped jaws. Their S-shaped larvae are also predators; they construct vertical burrows in dry soil and seize prey in strong jaws while anchoring themselves with hooks located on the 5th abdominal segment.

238 Beautiful Tiger Beetle
(*Cicindela formosa*)

Description: ⅝–¾″ (15–18 mm). Head, thorax, and underside are metallic blackish bronze with a few short white hairs. *Elytra have reddish-brown pattern with iridescent green on midline and edge,* white along sides.

Habitat: Sandy places with scattered vegetation.

Range: Throughout North America.

Food: Small insects and spiders.

Life Cycle: Eggs are left singly in the shade of a plant. Larvae make vertical burrows under plants, later use these hideaways as pupal shelters. 1 generation a year.

This tiger beetle is usually found alone, feeding on small insects. The western Gibson's Tiger Beetle (*C. f. gibsoni*), same size, has white elytra with only a narrow purplish line where they meet.

239 Dainty Tiger Beetle
(*Cicindela lepida*)

Description: ⅜–½" (10–12 mm). Head, thorax, and underside are mostly metallic dark green and coated with short white hair. *Elytra are ivory-white, each having a scrawled capital E and a few minute dots of bronze or dark green.*

Habitat: Sandy areas, especially along or near beaches.

Range: Eastern states to shores of the Great Lakes.

Food: Small insects and spiders.

Life Cycle: Eggs are left in pits under shade of plants. Larvae deepen pits to form vertical burrows where they later pupate.

This small tiger beetle is almost invisible as it runs over white sand.

207 Six-spotted Green Tiger Beetle
(*Cicindela sexguttata*)

Description: ½–⅝" (12–15 mm). Body, antennae, and legs *brilliant green to bluish green,* sometimes with purple sheen. *3–5 (commonly 4) white spots near rear of each elytron; spots occasionally absent.*

Habitat: Paths and adjacent open areas in deciduous or mixed woodland.

Range: Southern Nova Scotia to eastern South Dakota and south to South Carolina and Louisiana.

Food: Small insects and spiders.

Life Cycle: Female places eggs singly in small holes dug in bare ground, usually along shady paths. Larvae dig vertical burrows, where they later pupate. Adults emerge in summer. 1 generation a year.

The adult flies along forest paths, alights, and turns to face any intruders, as though judging when to fly again.

195 Dejean's Flightless Tiger Beetle
(*Omus dejeani*)

Description: ⅝–⅞" (15–21 mm). Black. Head and pronotum wrinkled; pronotum has a narrow shelflike flange along sides. *Elytra fused together along midline, conspicuously pitted,* sides narrowly turned under.

Habitat: Forest edges, especially at high elevations.

Range: Montana to Oregon, north to British Columbia.

Food: Small insects and spiders.

Life Cycle: Eggs are placed in pits. Larvae construct vertical burrows, then wait for prey. They also pupate in burrow. 1 generation a year.

Unlike most tiger beetles, adults cannot fly but scramble along the ground. Nocturnal, they hide by day and are seen occasionally as they run from under an overturned board or stone.

GROUND BEETLES
(Family Carabidae)

This large beetle family has over 3,000 species in North America. They are found under logs, rocks, and leaves in moist areas. These beetles measure ⅛–1⅜" (3–36 mm) long. Many are shiny black, but some are brightly colored. They have a conspicuous prothorax, narrow head, and long legs with spurs on the tibiae. The threadlike antennae arise from between large compound eyes. Most ground beetles rapidly pursue prey at night. A few eat pollen, berries, and seeds. Some species lay eggs in cells made of mud, twigs, and leaves. The larvae are predators, and take 1 year to grow from eggs to adults. Adults usually live 2–3 years, or rarely 4 years.

166 Bombardier Beetles
(*Brachinus* spp.)

Description: ¼–⅝" (5–15 mm). Head, prothorax, antennae, and legs brownish yellow. *Elytra grooved lengthwise, squarish at rear end;* powder-blue to darker blue, slightly metallic.

Habitat: Moist floodplains of rivers and near lakes, where temporary ponds form after storms.

Range: Throughout the United States and southern Canada.

Food: Larva feeds on pupating insects.

Life Cycle: Eggs are laid singly in mud cells on stones and plants. Larvae feed as external parasites until they kill the host, then scavenge on its remains. They pupate and overwinter in chamber of host. Adults emerge in spring after seasonal floods.

These beetles are named for their unusual defense mechanism—they emit from their anal glands toxic liquid that instantly vaporizes into puffs, making a protective screen that can stain people's skin.

209 Fiery Searcher
"Caterpillar Hunter"
(*Calosoma scrutator*)

Description: 1–1⅜" (25–36 mm). Black with dark greenish gold on sides of head and prothorax. Bluish luster on femora. *Elytra greenish or blackish blue, edged with gold,* fine lengthwise grooves and lines of elongate pits. Male has reddish hair inside curved middle tibiae and hind tibiae. *From above, prothorax appears broadly oval, constricted before base of the elytra.*

Habitat: Gardens, crop fields, and open woods.

Range: Throughout the United States and southern Canada.

Food: Caterpillars.

Life Cycle: Eggs are left singly on soil. Larvae
pupate in earthen cells. Adults
overwinter and live up to 3 years, active
May–November.

Adults and larvae sometimes climb
trees in search of prey, adults by day
and larvae at night. These insects are
highly beneficial because they prey
on caterpillars. The Fiery Hunter
(*C. calidum*), ¾–1″ (20–25 mm), has
6 rows of golden dots on its elytra. The
introduced European Caterpillar Hunter
(*C. sycophanta*), ¾–1⅛″ (20–30 mm),
has a dark blue thorax and bright,
golden-green elytra.

196   European Ground Beetle
(*Carabus nemoralis*)

Description: ¾–1″ (20–25 mm). Dull black with
glossy head; elytra violet or greenish
bronze. *Pronotum has iridescent violet
ridges around front and sides.* Each elytron
has 3 rows of pits.

Habitat: Gardens and open woods on sandy soil.

Range: Atlantic Coast to New Jersey, west to
Wisconsin; also Oregon to British
Columbia.

Food: Caterpillars and other soft insects,
sometimes earthworms.

Life Cycle: Eggs are dropped one at time under
debris in spring. Larvae may grow 2
years before pupating in late summer.
Adults emerge in fall, living 2 years.

A beneficial insect, this beetle destroys
cutworms. The Pinewoods Ground
Beetle (*C. serratus*), ¾–1″ (20–24
mm), is black with violet on the edges
of its pronotum and elytra, which have
3 rows of rectangular tubercles
separated by rows of small pits. Found
beneath logs and stones in wooded
areas, this species ranges from Georgia
through the Gulf states to New
Mexico, north to British Columbia.

206 Green Pubescent Ground Beetle
(*Chlaenius sericeus*)

Description: ⅜–⅝″ (10–16 mm). Elongate oval.
Bright green above, black below.
Antennae and legs pale brownish
yellow. *Elytra bluish or green with fine
lengthwise grooves and short silky hair.*

Habitat: Floodplains, along margins of lakes and
streams.

Range: Throughout North America.

Food: Small insects, slugs, snails, and other
animal matter.

Life Cycle: Eggs are deposited singly in purse-
shaped cells made of mud, twigs, and
leaves. Larvae pupate in fall and adults
emerge in spring. 1 generation a
year.

This beetle emits a leatherlike odor
when disturbed, then runs to a new
hiding place. Most species in this genus
are covered with silky hair. They are
distinguishable only by inconspicuous
details.

191 Common Black Ground Beetles
(*Pterostichus* spp.)

Description: ½–⅝″ (13–16 mm). Elongate. Shiny
black. Antennae, legs, and downturned sides of elytra reddish brown.
Head narrows behind eyes. *Elytra have
rounded forward corners,* lengthwise
grooves.

Habitat: Beneath stones, boards, and logs in
gardens, moist woods, and sometimes
fields planted with forage or grain.

Range: Throughout North America.

Food: Caterpillars and other soft insects.

Life Cycle: Eggs are left singly in upper soil mid-
to late summer. Larvae overwinter and
feed in early spring before pupating.
Adults emerge July–September.

This is one of the few genera of ground
beetles found coast to coast.

203 Boat-backed Ground Beetles
(*Scaphinotus* spp.)

Description: ⅝–⅞" (17–22 mm). Black with violet or coppery highlights. *Head and prothorax narrow.* Underlip deeply notched. Antennae long, slender, segment 5 and beyond with short hair. *Pronotum has ridge on both sides.*

Habitat: Moist woods.

Range: Massachusetts and New York to Florida, west to Texas.

Food: Slugs and snails.

Life Cycle: Eggs are laid under debris in May or June. Larvae overwinter and pupate when soil warms slightly in spring. Adults are active at night May–November in the South and June–September in the North.

This beetle's narrow head and prothorax allow it to reach inside a snail's curving shell using its long narrow mandibles to seize the animal inside.

34, 44 PREDACIOUS DIVING BEETLES
(Family Dytiscidae)

Diving beetles live in freshwater rivers, streams, lakes, pools, and even hot springs in northern regions. These brownish-black or dark green beetles have smooth, oval bodies, ¹⁄₁₆–1⅝" (1.5–40 mm) long. They swim by flexing their hind legs together simultaneously like oars, unlike other water beetles, which move their legs alternately. They can remain underwater for a long time, breathing air trapped in a chamber under the elytra. Periodically they come to the surface to renew this air. Both adults and larvae are predators. The larvae, called water tigers, are ¼–2¾" (5–70 mm) long. They have strong sickle-shaped jaws and can attack prey larger than themselves.

94 Small Flat Diving Beetles
(*Acilius* spp.)

Description: ½″ (12–14 mm). Oval, flattened.
Brown with brownish-yellow markings.
Flattened hind legs are fringed with golden-yellow hair. Leg color and markings on
upper body differ among species.

Habitat: Ponds, weedy margins of slow streams
and rivers.

Range: Southern Canada to northern United
States.

Food: Mosquito larvae, other small aquatic
insects, and water mites.

Life Cycle: Eggs are attached to underwater plants.
Larvae hunt prey, and when fully
grown pupate in a chamber in bottom
mud near the shore. Adults are active
April–May, October–November,
sometimes also beneath winter ice.
2 generations a year.

When disturbed, this beetle is quick to
dive, especially if it is refilling its air
reserve beneath the elytra. It sometimes
flies at night to other ponds and
artificial lights.

95 Large Diving Beetles
(*Dytiscus* spp.)

Description: 1–1⅝″ (25–40 mm). Oval. Dark
brown or black with yellow along sides
of prothorax and elytra, all often with
greenish tinge. Female has 10 parallel
grooves in each elytron. Legs are yellow
or brownish.

Habitat: Ponds, pools, streams, and rivers.

Range: Throughout North America.

Food: Less vigorous aquatic animals,
including small fishes, tadpoles, and
insect larvae.

Life Cycle: Eggs are thrust singly into underwater
plant stems. Larvae hunt prey and,
when fully grown, creep out of the
water to pupate in moist earth. Adults
overwinter and live 3 years or more.

The larvae are voracious predators.
The largest species are the Harris'
Diving Beetle (*D. harrisii*) and the
Pan-temperate Diving Beetle
(*D. marginalis*), both 1½–1⅝″ (38–40
mm), with yellow on the front and rear
margins of the pronotum. They occur
in the northern United States and
Canada. The Western Diving Beetle
(*D. marginicollis*), 1⅛″ (28 mm), has
yellow only on the sides of its
pronotum. The Understriped Diving
Beetle (*D. fasciventris*), 1–1⅛″ (25–
28 mm), has dark marks below its
brownish-red abdomen. The Vertical
Diving Beetle (*D. verticalis*), 1¼–1⅜″
(33–35 mm), has yellow stripes on its
elytra and is black underneath. The
Hybrid Diving Beetle (*D. hybridus*),
1–1⅛″ (25–28 mm), has yellow bars
across the tips of its elytra and is black
underneath.

96 Marbled Diving Beetle
(*Thermonectes marmoratus*)

Description: ⅜–⅝″ (10–15 mm). Broadly oval,
flattened, with smooth contours. Black.
Thorax and elytra have gold markings.
Abdomen has yellowish spots.

Habitat: Ponds.

Range: Western Texas, Arizona, California,
and Mexico.

Food: Smaller insects, water mites, and
aquatic worms.

Life Cycle: Eggs are attached to underwater plants.
Larvae emerge to pupate in mud near
pond. Adults return to pond. Larvae
and adults survive dry periods by
burrowing into mud.

Large swarms of these beetles
sometimes fly to artificial lights at
night. Other members of this genus are
widespread in northern states and
Canada. The Yellow-marked Diving
Beetle (*T. basilaris*), ⅜″ (9–10 mm), is

black above with a few dull yellow,
narrow edges and bars, reddish-brown
hind legs, and dull yellow fore and
middle legs.

WHIRLIGIG BEETLES
(Family Gyrinidae)

Whirligig beetles are oval, black
beetles, ⅛–⅝″ (3–15 mm) long, that
glide or skate across the surface of
ponds, lakes, and quiet streams in the
late summer and fall. Their compound
eyes are divided into upper and lower
parts that enable them to see above
and below the surface of the water
simultaneously. Their curiously short
antennae have 2 scooplike segments and
6 clubbed segments. The antennae
detect wavelets in the water surface
film, aiding in avoiding obstacles and
finding prey. These beetles have long,
slender fore legs and short, paddlelike
hind legs. They often cluster together
and swim rapidly in circles—
movements that have inspired their
name. Both adults and aquatic larvae
are predators.

92 Large Whirligig Beetles
(*Dineutus* spp.)

Description: ⅜–⅝″ (9–15 mm). Broadly oval,
flattened. Shiny to dull black, often
with a bronzy sheen. Scutellum
concealed. *Elytra have 9 impressed lines.*
Habitat: Surface film of ponds and slow streams.
Range: Throughout North America.
Food: Adult eats aquatic insects and insects
that fall into the water. Larva preys on
mites, snails, and small aquatic insects.
Life Cycle: Eggs are deposited in rows or masses on
submerged plants. Aquatic larvae hunt
prey, crawl out of water when fully
grown, and construct pupal cases of

sand and debris. Adults overwinter on plants or in mud. 1 generation a year.

Adults gather in restless swarms a few feet across. If threatened, a few may dive while the rest swim erratically away at the surface. The species within this genus can be identified only on the basis of male genitalia.

91 Small Whirligig Beetles
(*Gyrinus* spp.)

Description: ⅛–¼″ (3–7 mm). Broadly oval and flattened. *Shiny black above,* black to yellowish brown on sides and below. Distinct shield-shaped scutellum behind pronotum. *Each elytron has 11 rows of minute pits.*

Habitat: Surface film of ponds and slow streams.

Range: Throughout North America.

Food: Adult preys on aquatic insects and small insects that fall into the water. Larva eats aquatic insects and mites.

Life Cycle: Eggs are laid in masses or rows on submerged plants. Aquatic larvae hunt prey, then crawl out of the water onto rocks or vegetation to construct pupal shelters. Adults overwinter in leaf litter near ponds and streams. 1 generation a year.

Whirligig beetles in this genus have the scutellum more exposed than members of the genus *Dineutus*. The different species can be identified only on the basis of male genitalia.

WATER SCAVENGER BEETLES
(Family Hydrophilidae)

Water scavenger beetles are oval or elliptical insects, ⅟₁₆–1⅝″ (1–40 mm) long. Most are black or dull green and live in stagnant water along pond and lake edges, where they scavenge decaying plants. Terrestrial species feed on decayed animal matter, dung, and maggots. All have short, clubbed antennae and long maxillary palps. Unlike diving beetles, water scavengers kick their flattened hind legs alternately in swimming and surface for air headfirst. When they dive, a silvery envelope of air coats the underside of their bodies. The larvae are voracious predators and sometimes eat members of their own species.

93 Giant Water Scavenger Beetles
(*Hydrophilus* spp.)

Description: 1⅜–1½" (34–38 mm). *Elongate oval, strongly convex above, almost flat below.* Brown, gray or shiny black; legs reddish black. Short antennae clubbed, inconspicuous. *Long slender labial palps held forward,* easily mistaken for antennae. Spine under middle of thorax extends forward to fit a cup-shaped cavity.

Habitat: Ponds and slow streams.

Range: Throughout North America.

Food: Decaying remains of aquatic animals or small live animals.

Life Cycle: Female deposits 120–140 yellow eggs in silken cocoonlike egg case with a hornlike "mast." The case, ⅞–1" (22–24 mm) across and ⅝" (15 mm) deep, is left to float or attached to some underwater object. Larvae feed underwater. In late summer fully grown larvae leave water to prepare pupal cells in moist earth. Adults emerge in less than 2 weeks and return to water. 1–2 generations a year.

Some adults creep under litter on land to overwinter; others are active under ice all winter and live more than 1 year. On summer nights these insects often leave water and fly about, drawn to artificial lights. The widespread Giant Water Scavenger (*H. triangularis*), 1⅜–1½" (34–38 mm), is black with dark olive reflections.

HISTER BEETLES
(Family Histeridae)

Hister beetles have small, hard, shiny black, green, or bronze bodies, less than ⅜" (10 mm) long. The short, broad elytra leave 1 or 2 abdominal segments exposed, while the large pronotum conceals most of the head. The retractable, elbowed antennae end in a pronounced club. Most hister beetles live near dung, carrion, and other decayed organic matter, where they feed on small insects. Some flat species live under loose bark; other cylindrical species live in the open and attack leaf beetle larvae and caterpillars. A few inhabit termite and ant nests. The larvae of most are found in galleries of wood-boring beetles and feed on their larvae. The larvae of some species live in sand around roots of dune grasses and probably feed on larvae of weevils and flies.

199 Hister Beetles
(*Hister* spp.)

⅛–¼" (4–5 mm). *Shiny, jet-black.*
Description: Prothorax is concave in front, cupping rather small head and often concealing elbowed, clubbed antennae.
Habitat: Pastures and woods.
Range: Throughout North America.
Food: Fly eggs and larvae, small insects.
Life Cycle: Eggs are laid on decaying matter. Larvae grow rapidly if rich supply of food is available and later pupate in soil beneath food supply. Adults are active June–August. 1 generation a year.

These beetles use their sharp curved jaws to seize and cut up prey. If disturbed, they pull in their legs and antennae and remain completely still. The Twice-spotted Hister Beetle (*H. biplagiatus*), ¼" (5–6 mm), has 2 bright red, crescent-shaped spots. It is found in the eastern and central United States.

ROVE BEETLES
(Family Staphylinidae)

Rove beetles comprise a large family, having almost 2,900 species in North America. They are easily identified by the very short elytra, which cover only the first few abdominal segments. Most are black or brown, ranging ¹⁄₁₆–¾" (2–20 mm) long. These active beetles fly swiftly or run rapidly over the ground with the tip of the abdomen raised like a scorpion's stinger. Their sharp mandibles are usually crossed over the head but can inflict a painful stab if handled carelessly. Although a few are known to be parasitic, particularly in ants' nests, and some are scavengers, most rove beetles and their larvae prey upon mites, other insects, and small worms. They are usually found on mushrooms, flowers, or under bark.

227 Gold-and-brown Rove Beetle
(*Ontholestes cingulatus*)

Description: ½–¾" (13–20 mm). Dark brown with dense hair. Black hair forms spots on the head, thorax, and abdomen. *Glittering golden hair covers abdominal tip and forms a belt under thorax.* Large prominent eyes about midway along sides of head.

Habitat: Woods, and wherever carrion is found.

Range: Throughout North America.

Food: Adult eats small maggots, mites, and

beetle larvae. Larva feeds on carrion and fungi.

Life Cycle: Eggs are laid on or near carrion, where larvae feed. They pupate in chambers constructed in nearby soil. Adults are active spring–autumn.

These common beetles can raise their short elytra and spread their hind wings for flight in a fraction of a second. They run with the abdominal tips upturned.

---

240 Pictured Rove Beetle
(*Thinopinus pictus*)

Description: ⅝–¾" (15–20 mm). Elongate, *without wings*. Brownish yellow with *black dots, circles, and patterns*.
Habitat: Sandy beaches.
Range: Southern California to Alaska.
Food: Sand fleas and organic matter.
Life Cycle: Little known. Larvae live in beach drift and closely resemble adults but lack teeth on mandibles.

The Gray-and-black Rove Beetle (*Staphylinus maxillosus*), ½–¾" (13–18 mm), is shiny black with yellowish-gray bands across the elytra and abdomen. It is found near carrion throughout North America.

CARRION BEETLES
(Family Silphidae)

Carrion beetles are mostly large, black beetles with bright orange or reddish markings. They have flattened bodies with clubbed antennae and prominent coxae. The pronotum is bigger than the head and the elytra often leave exposed 1 or more abdominal segments. Most are over ⅜" (10 mm) long, but some measure ¹⁄₁₆–1⅝" (1.5–40 mm). Some species bury small carcasses and feed their larvae regurgitated food. Others feed on large carrion without burying it and let their larvae eat the same food as it decomposes. Some larvae prey on snails, and a few are plant feeders, destroying spinach, beets, and other vegetables.

177 Margined Burying Beetle
(*Nicrophorus marginatus*)

Description: ¾–1⅛" (20–27 mm). *Shiny black with 2 interrupted orange bands or series of patches across elytra.* Elytra leave exposed several abdominal segments. Antennae clubbed. Pronotum edges narrowly flattened.

Habitat: Deciduous and mixed woods, and adjacent fields.

Range: Throughout North America.

Food: Dying and dead animals, including mice or small birds; also fly larvae.

Life Cycle: Adults often drag a small carcass 16', bury it beneath loose dirt, then mate there. They remove fur or feathers, work body into a ball shape then lay eggs. Adults care for larvae until larvae pupate, sometimes in a side tunnel. Parents leave, perhaps to raise other families elsewhere. Adults live up to 15 months.

These nocturnal beetles hide during the day. Adults rub abdominal segments against the inner surface of elytra to make a rasping sound. This noise summons the newly hatched larvae to food or threatens competitors when more than two beetles of opposite sexes attempt to work on the same small carrion. Genitalia must be examined to differentiate the sexes.

176 Tomentose Burying Beetle
"Gold-necked Carrion Beetle"
(*Nicrophorus tomentosus*)

Description: ½–⅝" (12–17 mm). Glossy black. *Golden hair on pronotum and undersurface of thorax.* Elytra with yellowish-orange bands and yellow stripe under downcurved edges.

Habitat: Pastures and open woods.

Range: East of the Rocky Mountains.

Food: Adult and larva feed on small carcasses.

Life Cycle: Working together, adults bury small carcass or push out earth from under it; they then mate and lay eggs there. Larvae feed on carcass until they pupate.

If disturbed, this beetle may roll on its back and buzz with its wings like a bumble bee. Both the sound and beelike coloring help scare off predators. The widespread Large Black Burying Beetle (*N. pustulatus*), ⅞–1⅛" (22–30 mm), is black with reddish-orange spots near the tip of its elytra and red or orange on the antennal club. The Giant Carrion Beetle (*N. americanus*), 1⅛–1⅜" (30–35 mm), was widespread east of the Rocky Mountains, but now may be extinct.

174 American Carrion Beetle
(*Silpha americana*)

Description: ⅝–⅞" (17–22 mm). Broadly oval, gently convex above. *Pronotum is ivory to yellowish with black center. Elytra* brownish black *with 3 raised ridges connected by numerous dark cross-ridges.* Head, antennae, legs, and underside are black.

Habitat: Wherever carrion is found.

Range: East of the Rocky Mountains in the United States and southern Canada.

Food: Drying carrion, rat-sized and larger, fly larvae, and small larvae of other beetles.

Life Cycle: Eggs are laid singly on or near carrion. Larvae hatch in few days, feed under the carcass or in its cavities, and pupate in a cell excavated in soil nearby. Adults emerge May–July.

This beetle flies actively on warm days, locates carrion by scent, alights, and crawls quickly out of sight.

---

223 Northern Carrion Beetle
(*Silpha lapponica*)

Description: ½" (12–13 mm). Elongate, very flat, slightly constricted between prothorax and elytra. *Grayish black.* Head and thorax covered with yellowish hair.

Habitat: Wherever carrion is found.

Range: Pennsylvania to California, north to Canada and Alaska.

Food: Dry flesh and drying remains, particularly of mammals rat-sized and larger; large dead birds; also fly larvae and larvae of other beetles.

Life Cycle: Eggs are laid singly on or near carrion, hatching in a few days. Larvae feed under the carcass or in its cavities and pupate in a cell excavated in soil.

The Garden Carrion Beetle (*S. ramosa*), ½–¾" (12–18 mm), is velvety black with 3 raised lengthwise ridges on its elytra. It ranges from California to British Columbia, east to Montana, and south to New Mexico. It feeds on decaying plants and living weeds, grasses, and crop plants.

STAG BEETLES
(Family Lucanidae)

Stag beetles are named for the huge, branching, antlerlike jaws on males of the largest species. Most are black or reddish brown, ⅜–2⅜" (8–60 mm) long. Their elbowed antennae end in a club, but unlike those of scarab beetles, the plates of the club cannot be held tightly together. Males have large jaws, while females' jaws are smaller. These beetles are found in woods or along sandy beaches. Adults may feed on leaves and tree bark or aphid honeydew. The C-shaped larvae drink the juices of decaying wood. Because adults sometimes pinch with their mandibles, they are also called pinching bugs.

217   Elephant Stag Beetle
"Giant Stag Beetle"
(*Lucanus elephus*)

Description: Male 1¾–2⅜" (45–60 mm) including mandibles; female 1⅛–1⅜" (30–35 mm). Elongate, somewhat flat. Shiny reddish brown with blackish antennae and legs. Male's head is wider than prothorax and bears a crest above eyes. *Male's antlerlike jaws, as long as head and prothorax combined,* have small teeth along inner edge, and are forked at end. Female's head much narrower than thorax; jaws are barely longer than head.

Habitat: Woods.

Range: Virginia and North Carolina west to Oklahoma, northeast to Illinois.

Food: Adult may lap plant juices and aphid honeydew. Larva eats wet decaying wood.

Life Cycle: Eggs are laid in crevices of wet decaying wood. Larvae pupate in earthen cells near food source. Adults emerge July–August and live 2 or more years. 1 generation a year.

Males are formidable in self-defense, but these insects have difficulty righting themselves if overturned. They sometimes fly to lights at night.

187 Agassiz's Flat-horned Stag Beetle
(*Platycerus agassizi*)

Description: ⅜" (8–10 mm). Reddish black.
Oblong, flattened. Head much
narrower than prothorax or elytra.
Elytra rough between longitudinal
grooves and lines of pits. *Male's jaws as
long as head, slanting upward toward tips*,
female's smaller. *Antennae elbowed* with
end section clubbed. Legs long,
slender, fore tibiae expanded toward
outer end with fine teeth on outer side.

Habitat: Woods.

Range: California.

Food: Adult is not known to eat. Larva feeds
in dead wood of oak and madrone.

Life Cycle: Eggs are deposited on dead tree, often
on bark. Larvae gnaw their way into
decaying tree, particularly where moist.
They pupate in loose chambers among
wastes below bark, through which
adults emerge.

Except for the clubbed, elbowed
antennae, this beetle might easily be
mistaken for a ground beetle.
Specialists can identify more than 12
different species in this genus,
mostly in the West, with only 2 in
the East.

218 Reddish-brown Stag Beetle
(*Pseudolucanus capreolus*)

Description: ⅞–1⅜" (22–35 mm). Elongate
oblong, smooth. Dark reddish brown.
Fore legs have oval patch of golden hair.
Male's head about as wide as pronotum.
Male's jaws as long as head with a single
tooth on the inner side. Female's head
and jaws smaller.

Habitat: Deciduous woods, adjacent open spaces,
and city lots.

Range: Eastern United States and adjacent
Canada.

Food: Adult eats little. Larva feeds on juices

in decaying maple, apple, oak, and other deciduous trees.

Life Cycle: Spherical white eggs are deposited in rotting wood. Larvae take at least 2 years before pupating in nearby soil. Adults emerge June—July.

Adults fly all night. They are attracted to artificial lights or to fermenting sugar bait. They hide by day in moist soil or wet rotting logs and stumps. The Black Stag Beetle (*P. placidus*), ¾–1¼" (18–32 mm), has more than one tooth on its jaws. It lives in oak woods near the Great Lakes. The Western Stag Beetle (*P. mazama*), 1–1¼" (24–32 mm), eats cottonwood and ranges from Utah to northern Mexico.

189 Rugose Stag Beetle
(*Sinodendron rugosum*)

Description: ⅜–¾" (11–18 mm). Cylindrical. Jet-black. Head much narrower than thorax, both wrinkled. *Thorax has short curving horn,* more fully developed on male than female.

Habitat: Wet woods.

Range: California to British Columbia.

Food: Adult may lap plant juices or aphid honeydew. Larva feeds on wet wood of alder, willow, California laurel, and various oaks.

Life Cycle: Eggs are laid on bark or in crevices of dead wet trees. Larvae tunnel inward and later prepare pupal chambers from which adults emerge.

The Rugose Stag Beetle is considered beneficial because it helps recycle dead trees by transforming dead wood into excrement, which falls to the soil and provides nourishment for roots.

BESSBUGS
(Family Passalidae)

These large black beetles, 1⅛–1⅜"
(30–36 mm) long, have a short
forward-projecting horn on the head
and conspicuously grooved, parallel-
sided elytra. Their antennae are
elbowed and cannot be fully closed.
These beetles are related to stag beetles
but have a deeply notched lip. Adults
and larvae live in loosely organized
colonies in galleries inside tree stumps
and rotting logs. Adults chew decaying
wood and feed it to the larvae. When
disturbed, bessbugs make a squeaking
sound by rubbing roughened areas
under their wings across their back.
Their C-shaped larvae also make sounds
by scraping hind legs against the body.
These insects are also called patent-
leather beetles, betsy beetles, and
horned passaluses.

192 Patent-leather Beetle
"Horned Passalus"
(*Odontotaenius disjunctus*)

Description: 1¼–1⅜" (32–36 mm). Shiny black.
Elongate, flattened. Head has a *short
horn,* forward-directed. *Antennae clubbed
and elbowed.* Pronotum is squarish with
a deep middle groove, separated from
deeply grooved elytra by "waist."

Habitat: Deciduous forests.

Range: Eastern United States and adjacent areas
of Canada.

Food: Adult eats decaying wood. Larva feeds
on food prechewed by adults.

Life Cycle: Eggs are laid in galleries excavated by
both adults and larvae. Larvae feed and
pupate inside the same galleries. Adults
are active May–November.

Adults and larvae pass droplets of food
back and forth. This species was
formerly known as *Popilius disjunctus.*

SCARAB BEETLES
(Family Scarabaeidae)

Scarabs are stout beetles with large heads and pronotums. Many scarabs have beautiful metallic colors, and most measure ¼–2⅜" (5–60 mm) long. They have distinctive, clubbed antennae composed of leaflike plates, called lamellae, that can be drawn into a compact ball or fanned out to sense odors. The front tibiae are broad and adapted for digging. Tarsi have 5 segments, although a few species lack tarsi. The C-shaped larvae are yellowish or white. Both adults and larvae are nocturnal. Many are important scavengers that recycle dung, carrion, and decaying vegetable matter. Others are agricultural pests. They comprise a large family with about 1,300 North American species.

200 Tumblebugs
(*Canthon* spp.)

Description: ⅜–¾" (10–20 mm). *Dull black,* sometimes with bluish, greenish, or coppery tinge. Minute sculpturing or granulation on surface.

Habitat: Pastures and areas where dung of large mammals is available.

Range: Most of the United States.

Food: Dung.

Life Cycle: Male helps female form a ball of dung and roll it some distance, perhaps to compact it, before they dig a vertical tunnel and tumble the ball inside. A single egg is deposited on the dung ball, which is then covered with earth. Larvae feed alone and pupate in the soil close to dung. Adults are active August–September.

This beetle helps turn dry dung into fertilizer for plants and is considered beneficial. In ancient Egypt, the related

Sacred Scarab (*Scarabaeus sacer*) was associated with rebirth. Scarabs—both real and clay images—were placed on mummies in tombs.

234 Goldsmith Beetle
(*Cotalpa lanigera*)

Description: ¾–1″ (20–26 mm). Heavy, egg-shaped. Head, thorax, and scutellum are yellow to greenish with a metallic luster. *Elytra are yellow to beige.* Underside is covered with long, dense, woolly hair.

Habitat: Woods and adjacent fields.

Range: New England to Florida, west to Mississippi River basin, north to Ontario.

Food: Adult eats foliage, especially of poplar. Larva feeds on roots of poplar and other trees.

Life Cycle: Eggs are scattered on soil below trees. Larvae burrow to reach food, pupate in earthen cells at end of 1st or 2nd year. Adults emerge May–July.

This brilliantly colored beetle is the celebrated "gold bug" of Edgar Allan Poe's short story. The Western Mayate (*C. consobrina*), ¾–⅞″ (20–23 mm), has no hair. It feeds on cottonwood in Arizona. Adults may strip a tree of its foliage in a single night.

208 Green June Beetle
(*Cotinus nitida*)

Description: ¾–⅞″ (20–23 mm). Robust, elongate, somewhat flattened. *Head dark, with a horn.* Pronotum and elytra are metallic green, with brownish yellow on the sides; *the underside is glittery green and brownish yellow.* Tibiae green; femora are brownish yellow. Larva, to 2″ (50 mm), is yellowish white with brown head.

Habitat: Gardens, orchards, open woods, and crop fields, particularly above sandy soil.

Range: New York to Florida and Gulf states, north to Missouri.

Food: Adult drinks pollen from open flowers, such as hollyhock, and devours ripening fruits, especially peaches, and the foliage and fruits of many trees and shrubs. Larva eats roots of grasses, alfalfa, vegetables, tobacco, ornamental plants, and many other plants.

Life Cycle: Grayish, spherical eggs are laid in soil with high organic content. Larvae often emerge after a prolonged rain, crawl on their backs over soil or through sod, and overwinter deep in soil. Larvae develop in earthen cells near soil surface and pupate in the late spring of the 2nd year after hatching. Adults emerge June–July. 1 generation a year.

Adults fly noisily at night in search of food. This beetle is often an agricultural pest because its larvae destroy the roots of valuable plants, especially tobacco.

228 Brown Fruit Chafer
(*Euphoria inda*)

Description: ½–⅝″ (13–16 mm). Somewhat flattened. Head and pronotum are bronzy black. *Elytra bronzy yellowish brown speckled with black.* Underside and legs are blackish with yellow-brown hair. All black variants are sometimes seen.

Habitat: Meadows and thickets.

Range: Atlantic Coast to eastern Texas, north to Manitoba.

Food: Adult eats rotting fruit, nectar, and pollen. Larva feeds on dung, decaying plants, and rotting wood.

Life Cycle: Eggs are left on decaying vegetation, rotting wood, or dung, where larvae feed and overwinter. Adults emerge in early summer.

When flying, this beetle buzzes like a
bumble bee. The less common Green
Fruit Chafer (*E. fulgida*), ½–¾″ (13–
18 mm), is shiny green with white on
each side of its abdomen and has a
brownish pronotum and elytra.

204 Glossy Pillbug
(*Geotrupes splendidus*)

Description: ½–¾″ (13–18 mm). Stout,
hemispherical. *Bright metallic green,
purple, or bronze.* Elytra deeply pitted in
lengthwise rows. *Stout black legs adapted
for digging.*

Habitat: Pastures occupied by cattle.

Range: Connecticut to Florida, west to
Louisiana, north to Indiana.

Food: Adult and larva eat dung.

Life Cycle: In winter adults dig tunnels beneath
droppings of cattle, transfer large
pellets of dung to the bottom of each
burrow, and deposit a single egg on the
dung. Larvae are usually ready to
pupate close to remaining dung by
early summer, emerging and dispersing
as adults by autumn. This species may
be found throughout the year beneath
dung and carrion.

This beetle often appears on warm
winter days, when its handsome
metallic body glistens in the sunshine.
Since adults bury far more dung than
their larvae need, the residue serves as
fertilizer close to plant roots, enriching
the soil and making these insects
especially beneficial.

233 Grapevine Beetle
(*Pelidnota punctata*)

Description: ¾–1″ (18–25 mm). Almost
egg-shaped. Dull reddish brown to
brownish yellow above with *2 black dots*

*on sides of pronotum and 3 black dots on
side of each elytron.* Top of head,
scutellum, and underside blackish,
tinged with green.

Habitat: Thickets, vineyards, and woods.

Range: Eastern half of the United States.

Food: Adult feeds on leaves and fruits of wild
and cultivated grapes. Larva eats
decaying wood in tree stumps.

Life Cycle: Eggs are laid on wet stumps and
rotting fallen trees, where larvae feed.
They pupate in adjacent soil. Adults
emerge May–September.

Grapevine Beetles fly rapidly, usually in
curves.

219 May Beetles
"June Beetles"
(*Phyllophaga* spp.)

Description: ¾–1⅜" (18–35 mm). Bulky. Shiny
reddish brown to almost black.
Antennae with 3 platelike segments
that form a club at right angles to other
segments. Head, pronotum, and elytra
usually without markings or grooves.
Hind wings well developed. Larva is white
with brown head and has 6 prominent
legs.

Habitat: Deciduous forests and grasslands.

Range: Throughout North America.

Food: Adult eats foliage. Larva feeds on roots
of trees, shrubs, and herbs.

Life Cycle: Female constructs earthen cell for
50–100 cylindrical white eggs, which
hatch in 2–3 weeks. Larvae feed slowly,
taking up to 3 years before pupating in
cell below the soil surface. Adults
emerge in late spring.

The familiar buzzing flight of these
beetles was celebrated by William
Shakespeare as "the shard-borne chafer
with his drowsy hum."

242 Ten-lined June Beetle
(*Polyphylla decimlineata*)

Description: 1–1⅜" (25–35 mm). Brown with 1 white mark on each side of head, 1 broader white stripe down middle of pronotum, *1 short and 4 long white stripes on each elytron.* Long brownish hair under thorax. Antennal club has long and broad plates.

Habitat: Forests and woods.

Range: Rocky Mountain states and provinces; also the Southwest.

Food: Larva feeds on roots of woody plants, including fruit trees.

Life Cycle: Eggs are laid on soil near host plants, where larvae work their way into the soil and feed on roots. Pupal chambers are built shallowly underground. Adults emerge in July.

These beetles fly low over fields on warm evenings and sometimes gather around artificial lights at night. The similar Lined June Beetle (*P. crinita*), ⅞–1⅛" (23–28 mm), has no white scaly patches on its head, a single broken white stripe down the pronotum, and 7 silvery stripes on the elytra. It is found from Kansas and New Mexico to California, north to British Columbia.

212 Japanese Beetle
(*Popilla japonica*)

Description: ⅜–½" (8–12 mm). Oval, sturdy. *Body bright metallic green; elytra mostly brownish or reddish orange.* Grayish hair on underside and 5 *patches of white hair along each side of abdomen with 2 white tufts at tip.* Male has pointed tibial spurs; female's are rounded.

Habitat: Open woods and meadows.

Range: Maine to South Carolina.

Food: Adult damages leaf tissues and ripening fruit of more than 200 plants, including vines, flowers, shrubs, and trees. Larva feeds on roots, especially those of grasses, vegetables, and nursery plants.

Life Cycle: Elongate, yellowish-white eggs are deposited on soil, 1–4 at a time. Fully grown larvae overwinter in soil and pupate in the spring. 1 generation of adults emerges in summer when blackberries ripen. In the North, cycle takes 2 years.

The Japanese Beetle was introduced accidentally in 1916 on iris roots imported from Japan and has been a major pest for years. Its numbers have been reduced by the controlled use of parasitic tachnid flies and tiphiid wasps that prey on beetle larvae.

METALLIC WOOD-BORING BEETLES
(Family Buprestidae)

This large family gets its name from the metallic copper, green, blue, or black sheen of adults. They have hard, cylindrical to flattened oval bodies, ⅛–1⅛″ (3–30 mm) long. Their short antennae may be threadlike, saw-toothed, or comblike. Adults fly quickly, sometimes visiting flowers, more often alighting on tree bark of dying or dead trees. Eggs are laid on twigs and branches. The larvae, known as flat-headed wood borers, excavate galleries inside trees to obtain food. Some produce galls on alder, roses, blue beech, ironwood, and hazelnut; a few larvae live in pine cones or herbaceous plants. Almost 700 species are found in the United States.

210 Golden Buprestid
(*Buprestis aurulenta*)

Description: ½–¾″ (14–19 mm). Elongate oval, flattened. *Iridescent bluish green to green* with red or copper along elytral margins and where elytra meet. Head rounded in front of eyes; antennae threadlike, extending backward beyond pronotum. Larva is yellowish white, lacks legs, and has a small head and spoon-shaped mandibles.

Habitat: Coniferous forests.

Range: Rocky Mountains to the Pacific Coast.

Food: Adult eats pollen, nectar, and some foliage. Larva feeds on the wood of conifers, such as Douglas fir, spruce, and pine.

Life Cycle: Cylindrical white eggs are laid on bark of twigs, often on those already attacked by previous generations of the same species. Larvae pupate in a tunnel. Adults chew through bark to escape in June–July.

Adults are wary and fly quickly, often to bask in the sun on the bark of trees. Although these beetles are beautiful, adults and larvae are very destructive. The eastern Metallic Green Wood Borer (*B. fasciata*), slightly larger, is brilliant green with 3 yellow areas on each elytron in male or 2 bordered with black in female. Larvae live under the bark of maple, beech, poplar, and pine.

179 Western Pine Borer
"Sculptured Pine Borer"
(*Chalcophora angulicollis*)

Description: 1–1⅛″ (25–30 mm). Elongate oval. Black or dark shiny brown with bronze, irregular, shallowly sculptured areas on pronotum and elytra. *Underneath iridescent bronze. Body narrows slightly halfway to rear.*

Habitat: Coniferous forests.

Range: Idaho to New Mexico, west to
California, north to southernmost
Alaska; probably also western Alberta.

Food: Adult feeds on foliage. Larva invades
firs and western yellow pine.

Life Cycle: Elongate white eggs are laid on bark of
large branches and tree trunks. Larvae
eat through to inner bark, filling the
tunnels with feces. Fully grown adults
cut through bark in June–July.

The related Virginia Pine Borer
(*C. virginiensis*), ⅞–1⅛″ (23–30 mm),
is dull black with shiny bronze
sculptured areas on its pronotum and
elytra. Its larvae attack pines and range
from Canada to northern Florida.

237 Divergent Metallic Wood Borer
"Flatheaded Cherry Tree Borer"
(*Dicerca divaricata*)

Description: ⅝–⅞″ (16–21 mm). Cylindrical. Pale
brown or gray with bronze highlights
above, shiny coppery below. Elytra
have scattered, raised smooth areas; *tips
of elytra usually diverge.*

Habitat: Forests and orchards.

Range: Newfoundland to Georgia, west to
California, north to Alaska.

Food: Adult eats little. Larva feeds on dying
and dead trees of many kinds.

Life Cycle: Eggs are laid on bark. Larvae feed on
inner wood and pupate in tunnels close
to bark. Adults active May–August.

These larvae damage coniferous timber
trees and many orchard trees. They are
generally unnoticed on ash, birch, elm,
ironwood, and maple trees.

CLICK BEETLES
(Family Elateridae)

Click beetles get their name from the sharp clicking sound overturned beetles make when they flip themselves into the air, often landing upright. They accomplish this amazing feat by snapping a fingerlike spine on the underside of the thorax into a groove below the mesothorax. These minute to medium-sized beetles have elongate, flattened, and parallel-sided bodies, ¹⁄₁₆–2″ (2–50 mm) long, with a large flexible prothorax that has pointed rear corners. Adults eat plants and live on leaves in decayed wood or on the ground. The larvae, called wireworms, inhabit the soil or rotting wood, where they eat roots, seeds, or other insects, often seriously damaging crops of potatoes, wheat, corn, cabbage, cauliflower, radishes, and cotton.

---

186 Eastern Eyed Click Beetle
"Big Eyed Click Beetle"
(*Alaus oculatus*)

Description: 1–1¾″ (25–45 mm). Elongate. Shiny black with small, white dotlike scales on back. *Pronotum has 2 large, velvety black eyespots* surrounded by a dense ring of white scales. Elytra have thin lengthwise ridges with speckled depressions in between.

Habitat: Deciduous and mixed woods, especially around rotting timber.

Range: East of the Rocky Mountains, except in the Southeast.

Food: Adult seems to eat little. Larva attacks roots of many kinds of plants and preys on small animals in soil.

Life Cycle: Eggs are laid in soil. Larvae grow slowly and pupate in unlined cells below the ground or within rotting wood. Adults are numerous spring–September.

Adults sometimes rest on trunks of
orchard trees, particularly on pruned
trees. The related Blind Click Beetle
(*A. myops*), 1–1½" (25–38 mm), occurs
in the Southeast and is reddish brown
to black. It is called "blind" because its
whitish-gray eyespots are vaguely
outlined and hard to see. The Arizona
Eyed Click Beetle (*A. lusciosus*), 1¾"
(45 mm), has circular eyespots and
yellowish-white scales on its body.

222 Fire Beetles
(*Pyrophorus* spp.)

Description: ¾–1" (20–25 mm). Elongate, slender,
somewhat flattened. Pale to dark
brown. *2 luminous organs at rear corners
of flaring pronotum and 1 below abdomen.*

Habitat: Subtropical lowlands and areas with
thornscrub.

Range: Mexico to southwestern United States;
also southern Florida.

Food: Adult eats pollen and small insects,
such as aphids and scale insects. Larva
feeds on a mixture of plant materials,
small invertebrate animals, and larvae
of other beetles.

Life Cycle: Luminous eggs are deposited in or on
soil. Luminous larvae scavenge, grow
slowly, and pupate in soil, perhaps
several years later. Adults are active for
a few weeks in May in Florida and late
June in the Southwest.

Tropical members of this genus, up to
2" (50 mm) long, are known as *cucujos*
and often caught and worn as a
luminous decoration by partygoers. If
one beetle is held in the fingers and
placed on a newspaper in a darkened
room, it can supply enough light to
read the print near the insect.

FIREFLIES OR LIGHTNING BUGS
(Family Lampyridae)

Because of the luminous organ on the
tip of their abdomen, fireflies are
among the most familiar insects. All
are soft-bodied brownish or blackish
beetles, ¼–¾" (5–20 mm) long. The
pronotum is prolonged so far forward
that the head is concealed from above.
On summer nights they blink their
green or yellow light organs to attract
mates. Adults evidently do not feed,
yet they are attracted to moth baits.
The larvae, which are also often
luminous, prey on small animals,
including snails. Nocturnal, some live
in moist places under debris on the
ground, others beneath bark and
decaying vegetation.

171, 173 **Pyralis Firefly**
(*Photinus pyralis*)

Description: ⅜–½" (10–14 mm). Head concealed
from above by rounded front of
pronotum. *Pronotum is rosy pink with
dull yellow edges and black spot in center.*
Elytra are mostly blackish brown with
dull yellow on sides and down middle.
Both sexes have *flashing yellow light,*
smaller in female, which does not fly.

Habitat: Meadows.

Range: East of the Rocky Mountains.

Food: Adult eats nothing. Larva feeds on
insect larvae, slugs, and snails.

Life Cycle: Eggs are left on damp soil. Larvae
overwinter at end of 1st and 2nd year,
then pupate in chambers in moist soil.

This species is named after a fly fabled
for rising from the fire. The smaller
Scintillating Firefly (*P. scintillans*), ¼–
⅜" (5–10 mm), is yellow and pink and
has a large black spot on its pronotum.
It is found from New England to
Kansas and Texas.

172 Pennsylvania Firefly
(*Photuris pennsylvanicus*)

Description: ⅜–⅝" (9–15 mm). Elongate, flattened. *Head visible from above,* eyes large, widely separated. Antennae threadlike. Head and pronotum are dull yellowish, latter with a black spot surrounded by reddish ring. *Elytra* are brown or gray and *have yellow bands along sides near midline* and a narrow pale stripe down middle. Both sexes have *flashing green light.* Larva is spindle-shaped with light organ below abdomen at rear.

Habitat: Meadows and open woods.

Range: Atlantic Coast to Texas, north to Manitoba.

Food: Soft-bodied insects, snails, slugs, mites; also their own species.

Life Cycle: Eggs are concealed singly among rotting wood and humid debris on ground. Larvae hatch in spring. Fully grown larvae overwinter in pupal chambers just below soil surface and pupate the following spring. Adults emerge early summer–late August.

Eggs, larvae, and pupae are all luminous. This firefly flashes its light every 2 or 3 seconds while in flight.

SOLDIER BEETLES
(Family Cantharidae)

Soldier beetles resemble fireflies but lack light-producing organs and have protruding heads. These small beetles are less than ⅝" (15 mm) long. Their soft, thin elytra are usually yellow, orange, or red and covered with short, downy hair. Many species visit flowers, taking nectar and pollen, but most prey on other insects as adults and larvae. A few are omnivorous, feeding on wheat, potatoes, celery, and other vegetables. The adults are common on grasses and foliage of goldenrod, milkweed, hydrangea, and other plants. The larvae overwinter and pupate in cells in the soil. Some larvae are also found beneath bark and debris.

162 Pennsylvania Leather-wing
(*Chauliognathus pennsylvanicus*)

Description: ⅜–½" (9–14 mm). Elongate, parallel-sided. Elytra and pronotum brownish yellow with a *broad, black lengthwise mark on rear of each elytron* and short black crossband on pronotum. Head, antennae, legs, and undersurface black. *Antennae more than ½ as long as body.*

Habitat: Meadows, fields, and gardens.

Range: East of the Rocky Mountains.

Food: Adult eats pollen, nectar, and small insects. Larva devours grasshopper eggs, small caterpillars, and beetles.

Life Cycle: Eggs are deposited in soil or among ground litter. Fully grown larvae pupate in cells in the soil in spring, emerging as adults after warm weather.

Several species native to the Great Plains are used for biological control of Corn Earworm (*Heliothis zea*) caterpillars.

168 Downy Leather-wing
(*Podabrus tomentosus*)

Description: ⅜–½" (9–14 mm). Elongate. Elytra black; whitish hair makes them appear grayish blue. *Head, 1st antennal segment, prothorax, and legs reddish yellow;* rest of antennae darker. Prominent compound eyes. Larva is pink with velvety hair and 2 lines on thorax.

Habitat: Meadows, fields, and gardens.

Range: United States and southern Canada.

Food: Aphids, spider mites, and other small soft-bodied prey.

Life Cycle: Pale yellow eggs are laid in masses on
or in soil. Larvae pupate in cells below
the soil. Adults emerge in May in the
South and in June in the North; they
are active until frost.

This beetle is considered beneficial in
gardens because it devours aphids and
other pests.

NET-WINGED BEETLES
(Family Lycidae)

These small soft-bodied beetles, ¼–¾"
(5–19 mm), slightly resemble fireflies
but are easily recognized by the
network of ridges on their elytra. They
have bright yellow or red flattened
bodies with a large pronotum that
covers the head, bulging eyes, and large
long antennae. Many western species
have a snout. Adults are commonly
seen on flowers and plants, feeding on
plant juices and small insects. The
larvae are found under loose bark,
where they prey on numerous other
insects.

163 Banded Net-wing
(*Calopteron reticulatum*)

Description: ⅜–¾" (10–19 mm). Flat, shaped like
an inverted heart. Head is almost
hidden beneath pronotum. Brownish or
orange-yellow with black band across
elytra and at tip. *Elytra soft, flares to 3–
4 times as wide as base with network of
ridges* and numerous cross-ridges.
Habitat: Moist woods and meadows.
Range: Eastern United States and southern
Canada.
Food: Adult feeds on juices of decaying plant
matter. Larva preys on small insects and
mites under bark.
Life Cycle: Eggs are deposited on bark of dying

and dead trees. Larvae feed and later pupate in cells under bark. Adults active in August.

Certain day-flying moths mimic this beetle's colors, presumably because it is distasteful to birds. The End-band Net-wing (*C. terminale*), ⅜–⅝″ (11–17 mm), is black only across the bottom third of the elytra and is found throughout North America.

160 Golden Net-wing
(*Lycostomus loripes*)

Description: ¼–⅜″ (5–10 mm). Brownish yellow; *antennae are blackish* beyond the 1st segment. Larva is yellowish brown.

Habitat: Meadows, fields, and gardens, on flowers.

Range: Arizona at moderate elevations.

Food: Adult feeds on pollen and nectar, leaf tissues, and small soft insects, such as aphids. Larva probably preys on small insects under bark.

Life Cycle: Eggs are deposited singly on loose bark. Larvae work their way to inner bark, then pupate beneath bark in late summer of 2nd year. Adults emerge 3rd summer.

Adults are often seen mating on clustered flowers. Sometimes these insects gather in pairs side by side. This species was formerly assigned to genus *Lycus.*

SPIDER BEETLES
(Family Ptinidae)

These minute, pear-shaped beetles superficially resemble spiders because of their long, thin legs and long, threadlike antennae. Most are brownish and less than ¼″ (6 mm) long. Their hind wings have a long fringe of hair that often reaches beyond the elytra. Both adults and their C-shaped larvae are scavengers, feeding on dried organic matter, including wool, museum specimens, desiccated animals, dung, plants, stored seeds, and dried fruits. Some spider beetle species inhabit ant nests, where they scavenge the stored foods.

684 Spider Beetles
(*Mezium* spp.)

Description: ⅟₁₆″ (2 mm). *Pear-shaped*, strongly convex; swollen elytra. Dark reddish brown, yellow, or black, with hair and scales. *Antennae and legs long, slender*, pale brown to yellow.

Habitat: Wherever grain is stored.

Range: Worldwide.

Food: Grain and other vegetable matter, decaying or dried remains of animals, sometimes woolens and other nonsynthetic textiles.

Life Cycle: Eggs are scattered where food is available. Larvae feed and leave droppings among dry food. Some pupate in loose silken cocoons. Continuous generations as long as food remains available.

The American Spider Beetle (*M. americanum*), ⅛″ (3 mm), is dull yellow with glossy black elytra and a blunt projection on each side of its nearly cylindrical thorax. It feeds on such exotic substances as cayenne pepper, tobacco seed, and opium. The European Spider Beetle (*M. affine*), same size, is almost black.

685 Texan Spider Beetle
(*Niptus abstrusus*)

Description: ⅛″ (3–4 mm). Brown. Long legs are covered with brown hair. *Prothorax almost spherical.* Head projects only slightly. Long antennae. *Globular elytra* are nearly twice as wide as prothorax.
Habitat: Pastures and barnyards.
Range: Texas.
Food: Decaying organic matter, including animal droppings.
Life Cycle: Little known.

The related Golden Spider Beetle (*N. hololeucus*), same size, is reddish brown and covered by dense brownish-yellow hair and bristles. It is a worldwide pest that attacks seeds, dry plants, and stored cotton.

BRANCH AND TWIG BORERS
(Family Bostrichidae)

Branch and twig borers are mostly cylindrical beetles, ¹⁄₁₆–¾″ (2–20 mm) long. The large pronotum has either many tubercles or sharp teeth, and the small head is carried downward so it is seldom visible from above. The 8- to 10-segmented antennae end in a 3- to 4-segmented club. These beetles bore into living trees, dead twigs, and seasoned lumber, where they and their C-shaped larvae feed. A western species known as the Short Circuit Beetle can gnaw through lead-sheathed cables. A few species attack stored grain.

226 Apple Twig Borer
(*Amphicerus bicaudatus*)

Description: ¼–⅜″ (6–9 mm). Cylindrical. Brownish black to reddish brown. *Pronotum has tubercles near head.* Elytra are coarsely pitted in lengthwise rows.

Habitat: Orchards and woods.

Range: Nova Scotia to Virginia, west to New Mexico and Colorado.

Food: Twigs of apple, pear, cherry, and other trees, and grapevines.

Life Cycle: Adult uses mandibles to cut a deep pit in host plant, where its eggs are deposited. Larvae tunnel deeper, often killing twigs. Adults emerge in June in the North.

The Western Twig Borer (*A. cornutus*), ⅜–½" (9–12 mm), has 2 tubercles on the rear of each elytron. Its larvae are found in almond, apricot, fig, orange, and pear trees and grapevines from Mexico and Texas to southern California.

CHECKERED BEETLES
(Family Cleridae)

Checkered beetles are brightly patterned and sometimes checkered with red, orange, yellow, and blue. Their small, narrow, elongate bodies, ⅛–½" (3–12 mm) long, bristle with erect hair. Their bulging eyes protrude beyond the cylindrical pronotum. The 5-segmented tarsi have membranous lobes. Most antennae end in a club. Adults visit flowers and rest on foliage and trunks of dying or dead trees, where they prey on the larvae of wood-boring insects. Most larvae eat bark beetle larvae, although some complete their development on grasshopper egg pods, or as parasites of wasp and bee larvae. Some species may also feed on stored animal and vegetable food products.

144 California Checkered Beetle
(*Aulicus terrestris*)

Description: ¼-⅜" (6—9 mm). Elongate. *Blue-black with red marks,* covered with upright, fine, stiff hair. Antennae hairy and clubbed.

Habitat: Dry, grass-covered hills.

Range: California.

Food: Adult preys on moth caterpillars, especially of owlet moths. Larva feeds on lubber grasshopper eggs.

Life Cycle: In spring newly emerged female searches for caterpillar, kills it by piercing its skin, and eats it. While female is still feeding, male arrives and begins to mate. Later female lays eggs singly under stones or in soil near grasshopper eggs. Larvae feed and pupate in soil. Adults overwinter and are active late April—early June. 1 generation a year.

After only 10 minutes of feeding, the female's abdomen grows many times its normal size. Sometimes many males arrive to mate and try to dislodge the first male—all while the female continues to eat.

141 Elegant Checkered Beetle
(*Chariessa elegans*)

Description: ⅜" (9—10 mm). *Head, thorax, legs, and underside orange-red.* Elytra are dark metallic blue-black. Tarsi and comblike antennae black. Body hairy.

Habitat: Meadows, fields, and woods.

Range: Texas to California, and north to Oregon.

Food: Adult eats nectar, perhaps pollen. Larva preys on larvae of wood-boring insects.

Life Cycle: Eggs are laid in crevices and holes of bark. Larvae seek prey below the bark and inside tunnels in wood. They pupate in soil or in galleries under

bark. Larvae, pupae, or adults
overwinter.

A more widely distributed relative, the
Pilose Checkered Beetle (*C. pilosa*), ⅜–
½" (8–12 mm), is black with reddish-
yellow sides and has a middle stripe on
its prothorax. It ranges from southern
New England to Florida, west to Texas
north to British Columbia.

230   Slender Checkered Beetles
(*Cymatodera* spp.)

Description: ¼–⅜" (5–11 mm). *Slender* with long
slender legs and threadlike antennae.
Earth-brown or slightly paler, *covered
with spiny yellow hair.* Generally
unmarked, but elytra pitted shallowly
in lengthwise rows.

Habitat: Woods with oaks.

Range: Throughout North America; individual
species more localized.

Food: Adult preys on gall wasps. Larva
parasitizes gall wasp larvae.

Life Cycle: Female actively hunts on foliage and
twigs of oaks for gall wasps, then lays
eggs close to the site of gall wasp eggs,
sometimes attacking gall wasps. After a
few weeks, beetle larvae hatch, invade
developing galls, feed on wasp larvae as
parasitoids, eventually killing hosts and
ending development of galls. Beetle
larvae usually pupate in galls. Larvae,
pupae, or adults overwinter. Adults
emerge in summer.

These beetles help reduce the number
of gall-formers on oak trees and are
considered to be highly beneficial.

142 Red-blue Checkered Beetle
(*Trichodes nutalli*)

Description: ⅜" (8–11 mm). Elongate, almost
cylindrical, with many upright fine stiff
bristles. Mostly *dark blue-black,*
sometimes purplish or greenish; *with 3
orange to reddish crossbands.* Antennae
pale to dark brown with 3-segmented
club.

Habitat: Meadows, fields, and gardens; adult on
flowers and foliage; larva in nests of
wasps and bees.

Range: East of the Rocky Mountains; also
straying into Idaho and British
Columbia.

Food: Adult preys on thrips and other small
insects and eats pollen. Larva feeds on
the larvae of wasps and bees.

Life Cycle: Eggs are laid on flowers. Larvae attach
themselves to bees and wasps, ride to
nests, and prey on or parasitize their
larvae. Beetle larvae or pupae
overwinter. Adults emerge in
midsummer.

This handsome little insect is easily
noticed on daisy heads and other open
flowers, where it often rests or feeds
voraciously.

FLAT BARK BEETLES
(Family Cucujidae)

Flat bark beetles have very flat,
elongate bodies, less than ½" (14 mm)
long, and threadlike antennae that
usually end in a club. Most are yellow,
brown, or reddish. These beetles are
common under loose bark, where they
prey upon smaller insects and mites. A
few species feed on stored grain and dry
or semidry plant materials. The
elongate larvae have prominent heads
and bristly bodies; they are often
considered beneficial because they
destroy bark beetle adults and larvae.

The larvae of some species parasitize the pupae of long-horned beetles and the larvae of braconid wasps.

140   Red Flat Bark Beetle
(*Cucujus clavipes*)

Description: ⅜–½″ (10–14 mm). Elongate, sides almost parallel, constricted between prothorax and elytral bases. Bright yellowish red with black antennae and eyes; elytra dull red below. *Elytra extremely flat, sometimes slightly concave.* Larva is brownish yellow.

Habitat: Woods under bark of ash and poplar, especially of recently felled trees.

Range: Throughout North America.

Food: Adult apparently eats nothing. Larva preys on larvae of bark beetles and wood borers.

Life Cycle: Eggs are laid under bark, where larvae seek prey. Fully grown larvae construct circular pupal cells from small particles of the decaying bark and wood, where they overwinter. Adults are active March–December.

This species is much larger and more brightly colored than most others.

PLEASING FUNGUS BEETLES
(Family Erotylidae)

Pleasing fungus beetles are minute to medium-sized beetles, less than ⅛″ to ¾″ (3–20 mm) long, that feed on fungus under decaying logs and bark. Most have oval, shiny black bodies with attractive red, orange, or yellow patterns—inspiring their common name "pleasing." Their clubbed antennae arise in front of or between the eyes, and their feet end in broad, hairy pads. Adults overwinter under bark, often close together. The larvae, which have sensory appendages at the end of their cylindrical bodies, are found wherever fungus grows.

139 Rough Fungus Beetle
(*Cypherotylus asperus*)

Description: ½–⅝" (14–17 mm). Elongate, oval, rough. Shiny black head and thorax; *elytra dull brownish yellow or lavender, with small black dots.*

Habitat: Meadows, fields, woods, and gardens.

Range: Colorado, New Mexico, and Arizona.

Food: Adult feeds on pollen, nectar, and some fungus. Larva eats fungus.

Life Cycle: Eggs are deposited in bark crevices on dead trees and fallen logs. Larvae creep in spaces beneath bark or through channels within decaying wood, pupating when humidity is high. Adults emerge in summer.

This rough-textured beetle is unusual, because most pleasing fungus beetles are smooth.

178 LADYBUG BEETLES
(Family Coccinellidae)

Ladybugs are among the most familiar beetles, easily recognized by their round, often spotted bodies, less than $\frac{1}{16}$–⅜" (1–10 mm) long. Most are shiny red, orange, or yellow with black markings, or black with red or yellow markings. Some resemble leaf beetles but have 3 rather than 4 tarsal segments. Both adults and larvae are predators, mostly of aphids. They are common on plants and often overwinter as adults in large swarms under fallen leaves or bark. The active spindle-shaped larvae are usually covered with spines, bright spots, and bands. In warm climates there are many generations a year. During the Middle Ages, these beetles rid grapevines of insect pests, and in appreciation were dedicated to "Our Lady," hence their common name. In Britain they are called ladybird beetles, in the United States, usually ladybugs.

146   Two-spotted Ladybug Beetle
        (*Adalia bipunctata*)

Description: ⅛–¼" (4–5 mm). Almost hemispherical, slightly longer than broad. Head and thorax black marked with yellow. *Elytra orange with 2 large black spots.* Underside black to reddish brown. Larva is velvety black spotted with yellow and white.

Habitat: Meadows, fields, and gardens; also in houses.

Range: Throughout North America.

Food: Adult and larva feed on aphids and other small insects.

Life Cycle: Clusters of bright yellow eggs are attached to foliage and other supports near a food supply, where larvae later feed. Pupae are black with yellow spots and hang by back tip from leaf surfaces. Adults overwinter in the North. Many generations a year.

This little beetle is one of the most familiar ladybugs. The Western Two-spotted Ladybug Beetle (*A. frigida*), ⅛–¼" (3–5 mm), has a white head and white pronotum with an M-shaped black mark. It also has orange-red elytra with 2 faint dark spots, a broken black band, and a transverse spot near its tip. It is found across northern Canada to New York State and New England, often at higher elevations.

147 Nine-spotted Ladybug Beetle
(*Coccinella novemnotata*)

Description: ¼" (5–7 mm). Almost hemispherical.
Head and thorax black with yellowish
or white markings on margin; legs
and underside black. *Elytra* are
yellowish red or orange *with 9 black
spots* (4 on each elytron and 1 on the
scutellum).

Habitat: Meadows, crop fields, gardens, and
marshes.

Range: North America, except the Southwest.

Food: Aphids, small soft insects, and mites.

Life Cycle: Lemon-yellow egg clusters are attached
to leaves near aphids. Larvae feed, then
pupate without cocoons, attached to
leaves. Adults overwinter in large
groups and emerge May–September.

The pattern of black spots on this
ladybug varies geographically. The
Seven-spotted Ladybug Beetle
(*C. septapunctata*), same size, has 7
black spots. Recently introduced into
New Jersey, it occurs throughout most
of the Northeast. The Three-banded
Ladybug Beetle (*C. trifasciata*), ⅛–¼"
(4–6 mm), has 3 black bands across its
orange or yellow elytra and is found in
the North.

145 Spotless "Nine-spotted" Ladybug
(*Coccinella novemnotata franciscana*)

Description: ¼" (5–7 mm). Almost hemispherical.
Head and thorax black with yellow or
white markings; legs and underside
black. *Elytra yellowish red* with or
without 1 black spot at scutellum.

Habitat: Forest edges.

Range: California.

Food: Aphids and similar soft small insects.

Life Cycle: Clusters of eggs are attached to
underleaf surfaces. Larvae prey on
aphids, then pupate without cocoons
attached to leaves. Adults overwinter.

Despite its nearly spotless elytra, structural details reveal that this beetle is a subspecies of the widespread Nine-spotted Ladybug Beetle.

148 Convergent Ladybug Beetle
(*Hippodamia convergens*)

Description: ¼–⅜″ (6–8 mm). Oval, convex above. *Pronotum black with white border and 2 converging white stripes. Elytra* are red or orange *with 13 black spots* (1 spot at scutellum and 6 on each elytron); sometimes spots are enlarged to form 3 transverse bands. Larva is velvety black with 8 orange spots. Pupa is black with red spots.

Habitat: Woods, meadows, and gardens.

Range: Throughout North America.

Food: Aphids and other small insects.

Life Cycle: Female may lay up to 500 eggs during a lifespan of a few months; clusters of 5–30 eggs are attached on leaves and twigs. Larvae feed, then pupate attached by the back end to some support. Many generations a year, if food supply is good.

Large numbers of ladybugs occasionally find their way into houses in autumn looking for places to spend the winter. In the West huge swarms fly into mountain canyons, overwinter under leaves, and return to valleys in the spring. Overwintering beetles are sometimes purchased by mail and freed near crops that are vulnerable to aphids.

183 Ashy Gray Ladybug Beetle
(*Olla abdominalis*)

Description: ⅛–¼″ (4–6 mm). Almost completely hemispherical. Pale yellow to ashy gray. *Pronotum and elytra have many small black dots and streaks in a recognizable pattern:* 2 dots between eyes, 2 dots and an M-shaped mark on pronotum, 6–8 dots in a row across base of elytra, and 8 spots toward rear.

Habitat: Deciduous forests, and areas with scattered trees.

Range: Southwestern states.

Food: Aphids, especially on walnut trees.

Life Cycle: Eggs are attached to leaves near aphids. Larvae feed, then when fully grown attach themselves to leaves and pupate.

This ladybug is one of the principal aphid eaters in California.

DARKLING BEETLES
(Family Tenebrionidae)

Darkling beetles are slow-moving, small to medium-sized insects, $\frac{1}{16}$–$1\frac{3}{8}''$ (2–35 mm) long. Most are dull brown or black with heavily striated or warty elytra. Their antennae are often clubbed. Fore and middle tarsi have 5 segments, while the hind tarsi have 4. The larvae resemble click beetle larvae but do not have a conspicuous upper lip. Most adults and larvae are nocturnal and scavenge on rotting wood. Some species attack stored foods, rugs, clothing, skins, insect collections, and dried plants. Others live in termite and ant nests, and some are seen in desert areas. There are over 1,400 species in North America, most of which are in the West.

225 Horned Fungus Beetle
(Bolitotherus cornutus)

Description: $\frac{3}{8}$–$\frac{1}{2}''$ (8–12 mm). Oblong, rough. Dull black to dark brown. Clubbed antennae arise far apart. *Male has 2 hornlike projections on pronotum* that extend forward beyond head *and*

1 forked horn on middle of head; female has only tubercles on pronotum.

Habitat: Woods.

Range: Eastern United States and adjacent Canada.

Food: Fungus tissue, particularly woody bracket type on dying and dead tree trunks.

Life Cycle: Eggs are deposited in or on fungi. Larvae pupate within woody fungi or in nearby soil if larvae have fed on less firm fungal tissues. Adults are active June–August.

If disturbed, adults feign death and remain motionless, resembling fragments of rotted wood.

197 Broad-necked Darkling Beetle
(*Coelocnemis californicus*)

Description: ⅞–1" (22–25 mm). *Prothorax wider than long, almost flat above.* Elytra convex, rounded. Uniformly dull black with minute to large pits in surface. Dull tan stripes on inner surface of tibiae and tarsi.

Habitat: Deserts and semidesert regions.

Range: Southern California into southern Oregon.

Food: Fungus and windblown organic matter.

Life Cycle: Incompletely known.

This beetle moves with its head down, abdominal end raised, and hind legs extended. It stands stiffly if disturbed and during the heat of day burrows into cooler soil or hides beneath opaque objects. This species was formerly known as *C. dilaticollis.*

224 Plicate Beetle
(*Noserus plicatus*)

Description: ½" (12–14 mm). Dull grayish black.
Pronotum and elytra are extremely
hard, wide, *covered with rows of wartlike
projections.*

Habitat: Under bark, chiefly of oak trees.

Range: California.

Food: Decaying plant tissues and fungi.

Life Cycle: Little known.

This beetle has thick, warty elytra. It
resembles the Horned Fungus Beetle
but lack its distinctive horns.

190 Yellow Mealworm Beetle
(*Tenebrio molitor*)

Description: ½–⅝" (13–16 mm). Elongate oval,
somewhat flattened. Shiny reddish
black to black. *Pronotum has prominent
corners.* Head rounded, wider than long.
Elytra have many fine lengthwise
grooves. Larva, to 1¼" (32 mm), is
yellow to pale reddish brown.

Habitat: Storage warehouses, granaries, and
barns.

Range: Worldwide.

Food: Stored grain, bran, and cereals,
occasionally packaged foods, and other
stored dry foods eaten by people.

Life Cycle: White bean-shaped eggs are coated
with a sticky secretion and left singly or
in small groups on or in food materials.
Larvae often overwinter, pupate in
spring, and emerge as adults in early
summer. 1 generation a year.

Introduced accidentally from Europe by
colonists, this beetle is often reared
deliberately as food for pet fish,
amphibians, and reptiles. The Dark
Mealworm Beetle (*T. obscurus*), ½–⅝"
(14–17 mm), is dull black or reddish
brown and has a less prominent head.
Its pronotum and elytra are roughly

granular instead of shiny and the larva is darker. Mealworm beetles are the largest beetles that infest grain, milled flour, bran and many other cereal products.

185   Ironclad Beetle
(*Zopherus haldemani*)

Description: ¾–1″ (20–26 mm). Extremely hard-bodied. Head, body, and elytra are *dull ivory-yellow marked with black* on top of head, on sides of pronotum, and over elytra. Yellow markings along femora, tibiae, and as spots on tarsal segments.
Habitat: Deserts.
Range: Southwest.
Food: Fungi.
Life Cycle: Incompletely known.

This nocturnal beetle hides in dark crevices by day and often feigns death when picked up or disturbed. It is highly resistant to water loss.

FIRE-COLORED BEETLES
(Family Pyrochroidae)

Fire-colored beetles have contrasting bright red or orange pronotums and black bodies. These small beetles are usually ¼–¾″ (6–20 mm) long, and have soft, flattened bodies with bulging eyes and a distinct, necklike constriction behind the head. Both the pronotum and head are narrower than the elytra. Most have comblike antennae, but in a few species the male's antennae have spectacular feathers along one side. Adults often visit flowers or rest on leaves in the sun, feeding on nectar and pollen. The larvae prey on smaller insects under bark and in crevices of decaying wood.

170 Fire Beetles
(*Dendroides* spp.)

Description: ⅜–½″ (9–13 mm). Head and pronotum are reddish orange to brownish yellow; elytra same color or black. Eyes are separated by less than their diameter. Antennae rise in front of eyes; male's antennae are comblike.

Habitat: Forests, woodlands, and clearings.

Range: Throughout North America.

Food: Adult eats pollen. Larva preys on smaller insects.

Life Cycle: Eggs are laid on the bark of dead and dying trees. Larvae hunt prey beneath bark. Adults are active June–July.

The widespread Canadian Fire Beetle (*D. canadensis*), ⅜–½″ (9–13 mm), has black antennae, head, and elytra and reddish-yellow pronotum and legs. The Eastern Fire Beetle (*D. concolor*), same size or slightly smaller, is entirely brownish yellow and ranges throughout eastern Canada and the United States. It is often found beneath the bark of dead pines.

BLISTER BEETLES
(Family Meloidae)

These beetles contain the chemical cantharidin, which causes blisters on human skin. Blister beetles have soft, elongate bodies, ⅜–1⅛″ (9–28 mm) long. Their broad head is usually wider than the prothorax and connected by a narrow "neck." Plant eaters, adults are commonly found on flowers and leaves; some are important crop pests. The larvae undergo a complex development, hypermetamorphosis, and appear in several forms: they are first slender and long-legged, later grublike. Some parasitize grasshopper eggs, others ride bees from flowers to the nest, where they attack bee larvae.

161 Striped Blister Beetle
(*Epicauta vittata*)

Description: ½–⅝" (13–17 mm). Dull yellow above, black below with yellow hair. Thorax and *each elytron with 2 black stripes.* Bulbous head loosely linked to nearly cylindrical thorax, somewhat narrower than closed elytra.

Habitat: Fields, including pastures, and croplands.

Range: Nova Scotia to North Carolina, west to Louisiana, north to Saskatchewan.

Food: Adult eats crops and weeds. Larva feeds on grasshopper eggs that have been deposited in soil.

Life Cycle: Egg clusters of 100 are deposited in holes made in soil and hatch in 10–21 days. Larvae burrow in search of grasshopper eggs, pupate in 2 weeks, and overwinter in soil. Adults emerge in great numbers in early summer. Usually 1 generation a year.

The population of these beetles increases or decreases according to the availability of grasshopper egg masses. Larvae are beneficial—one larva can destroy 30 or more eggs, which is a full pod for some grasshoppers. But adult beetles are detrimental when they attack field crops. The Black-striped Blister Beetle (*E. lemniscata*), ½–⅝" (13–15 mm), has 3 narrow black stripes on each elytron and ranges from Pennsylvania to Georgia, west to Louisiana, north to Missouri.

167 Arizona Blister Beetle
(*Lytta magister*)

Description: ⅝–1⅛" (15–28 mm). Blue-black with *dull orange head and pronotum.* Legs are brownish red, tarsi blackish.

Habitat: Deserts.

Range: Arizona.

Food: Adult eats plant tissues of desert

shrubs. Larva attacks grasshopper eggs in soil.

Life Cycle: Incompletely known.

Nuttall's Blister Beetle (*L. nuttalli*), ⅝–1″ (15–25 mm), is metallic purple or green with pale purplish elytra. It is widespread in east Manitoba and in the Rocky Mountains from New Mexico to Alberta.

---

202 Short-winged Blister Beetle
"Oil Beetle"
(*Meloe angusticollis*)

Description: ½–⅝″ (13–15 mm). Stout. Almost entirely dark metallic blue to violet-black. Head is wider than prothorax. *Antennae short, beadlike* with 11 segments; male's are dilated and often bent about halfway between base and tip. *Elytra short,* pointed, overlapping at base, spreading to expose broad abdomen. No hind wings.

Habitat: Crop fields and meadows.

Range: Southern Canada and northern United States.

Food: Adult eats herbaceous foliage, especially potatoes. Larva is parasitic on wild bees.

Life Cycle: Batches of eggs are laid in the ground near bee nests. Larvae climb up plants and ride bees to nest. Larvae transform from long-legged to grublike forms before pupating. Adults are active in spring and fall.

If disturbed, this beetle feigns death by falling on its side. Leg joints exude droplets of liquid that cause blisters.

LONG-HORNED BEETLES
(Family Cerambycidae)

Most long-horned beetles have elongate cylindrical bodies and back-sweeping antennae that are often 3 times longer than the body. Some measure only ¼" (6 mm), although a few giants reach 3" (75 mm) in length. Many are admired for their beautiful colors and long antennae. Usually the femora are large, and tarsi have 5 segments. Adults feed on wood, roots, leaves, and pollen and are rarely carnivorous. Some, especially brightly colored species, feed on flowers during the day. Many others come out from under logs and loose bark only at night. Most larvae, called round-headed borers, are white. They tunnel into wood and often ruin cut logs in forests, although a few species invade living trees. Sawdust around a hole in bark is a common sign of the presence of long-horned larvae inside. This large family has over 1,200 species in North America.

---

188 Black-horned Pine Borer
(*Callidium antennatum*)

Description: ⅜–½" (11–14 mm). *Flattened. Bright bluish black or purplish.* Antennae shorter than elytra, which have low lengthwise ridges and fine pits. Femora swollen. Larva is yellowish white, legless.

Habitat: Coniferous forests and lumber yards.

Range: Throughout North America.

Food: Larva feeds on sapwood of pines, spruce, Douglas fir, and other conifers.

Life Cycle: Eggs are laid on bark of dying or felled trees that have not been stripped of bark. Larvae bore between the bark and wood, eating out extensive passageways. They pupate beneath bark, emerging as adults May–July. 1 generation a year.

This borer appears to favor felled trees that have been seasoned for one winter.

232 Yellow Douglas Fir Borer
(*Centrodera spurca*)

Description: ¾–1" (18–26 mm). Cylindrical. Yellowish brown. Large round head same width as *prothorax,* which *has short conical spines on each side.* Elytra rounded at front and rear, with lengthwise ridges and pits. Antennae long.

Habitat: Forests, coniferous and mixed.

Range: California to British Columbia.

Food: Larva eats deadwood of Douglas firs.

Life Cycle: Eggs are laid on bark. Larvae tunnel in, pupate close to the surface, and overwinter. Adults emerge in spring.

The Yellow Douglas Fir Borer was formerly included in the genus *Parapachyta.* The Brown Douglas Fir Borer (*Pachyta liturata*), ¾–⅞" (18–23 mm), has blunt spines on its prothorax. Its larvae feed on Douglas firs at high elevations in the West.

158 Long-jawed Longhorn
(*Dendrobius mandibularis*)

Description: ¾–1¼" (18–33 mm). Elongate oval, narrowed at rear. Black with 2 *incomplete yellow crossbands* interrupted by black along midline. Antennae black and yellow; male's are much longer than body. *Narrow clawlike mandibles longer than head.*

Habitat: Hardwood forests.

Range: Mexico to Texas and New Mexico, west to California.

Food: Larva feeds on dead branches of various trees, including citrus and paloverde.

Life Cycle: Eggs are laid on bark. Larvae tunnel in to feed and pupate near surface. Adults emerge July–August.

Nocturnal, these beetles are found under the bark of fallen trees during the day. They are hunted by woodpeckers and other birds that eat their larvae.

164 Elder Borer
"Cloaked Knotty-horn"
(*Desmocerus palliatus*)

Description: ⅝–1″ (17–24 mm). Shiny metallic blue with *yellow or reddish yellow across elytral front.* Each elytron has 3 narrow lengthwise ridges. *Antennal segments have enlarged knotlike tips.*

Habitat: Low grounds and stream edges.

Range: Central and eastern United States and Canada.

Food: Adult eats pollen. Larva feeds on elderberry roots.

Life Cycle: Eggs are laid close to soil on elderberry stems. Larvae burrow into stems, progress into roots, and pupate in soil. Adults appear June–September.

This handsome beetle is easily recognized by its contrasting color pattern.

229 Twig Pruners
(*Elaphidionoides* spp.)

Description: ⅜–⅝″ (9–17 mm). Slender, almost cylindrical. Brown except *thick shiny base of antennae dark reddish brown.* Body covered by patches of fine, short grayish-yellow hair. Male's antennae are longer than body, female's slightly shorter.

Habitat: Deciduous forests.

Range: Eastern United States and Canada.

Food: Larva eats tender wood of twigs.

Life Cycle: Eggs are laid on twigs, into which larvae tunnel. Later larvae cut through twigs, causing them to break off. Pupation occurs in twigs or in soil.

Species differ noticeably in the coarseness of spines at the tips of elytra and in the shape of the pronotum. The Spiny Twig Pruner (*E. mucronatus*) has a cylindrical pronotum and coarse spines. Its larvae feed on dead hardwoods in the East. The southeastern Gray Twig Pruner (*E. villosus*), ½–¾" (12–18 mm), has a barrel-shaped prothorax, and its larvae eat twigs of living hardwoods, pruning off new growth.

221 Pine Sawyer
"Spined-neck Longhorn"
(*Ergates spiculatus*)

Description: 1⅜–2¼" (40–58 mm). Slender, somewhat flattened. Head and pronotum rough above, dark reddish brown. Elytra lighter reddish brown. *Sides of pronotum have a few large and many small sharp teeth.* Elytra have a few lengthwise ridges.

Habitat: Forests, mostly above 3,937' (1,200 m).

Range: Montana to Arizona, west to California, north to British Columbia.

Food: Larva eats sapwood and heartwood of pines and Douglas firs.

Life Cycle: Female lays eggs in crevices of bark on dead trees and stumps. Larvae pack tunnels with wood fibers and pupate just below the bark. Fully grown in 2–3 years. Adults emerge July–August.

Pine Sawyer larvae destroy fallen trees, logs, and poles.

231 Golden-haired Flower Longhorn
(*Leptura chrysocoma*)

Description: ⅜–⅝" (9–17 mm). Body slender, head and prothorax wide; elytra even wider, shorter than abdomen. Head black; *thorax and elytra covered with velvety golden or yellow hair.* Antennae are

½ as long as body.

Habitat: Forests and forest edges.

Range: Coast to coast in northern North America; also Rocky Mountains to California, and the Appalachians to Georgia.

Food: Adult drinks nectar and eats pollen. Larva eats wood of pine, alder, and other trees.

Life Cycle: Eggs are laid in bark crevices. Larvae tunnel inward, develop for a year or more before pupating below bark. Adults chew their way to freedom.

This handsome insect is quick to fly from any disturbance.

157 Locust Borer
(*Megacyllene robiniae*)

Description: ½–¾" (12–20 mm). Elongate, stout. Velvety black with *golden-yellow bars on head, pronotum, and elytra,* including "W" in middle of body. Antennae are dark brown, ⅔ male's body length; female's ½ as long. Legs are reddish brown.

Habitat: Woods with black locust trees.

Range: Eastern and southern United States and eastern Canada.

Food: Adult eats goldenrod pollen and nectar. Larva eats sapwood of black locust.

Life Cycle: Female cuts deep pits into bark and deposits eggs one at a time inside. Larvae bore inward and pupate under bark. Adults emerge in late summer. 1 generation a year.

The similar-looking Painted Hickory Borer (*M. caryae*), ½–¾" (12–20 mm), attacks hickory, black walnut, butternut, osage orange, and mulberry.

193 Black Pine Sawyer
"White Spotted Sawyer"
(*Monochamus scutellatus*)

Description: ⅝−1" (15−25 mm). Pronotum has prominent spine in middle of each side. Elytra elongate, rounded at tip. Black but *appearing bronze* because of gray and brown hair. Antennae have scarlike area on outer end of 1st segment; male's are bare and nearly twice length of body; female's hairy, shorter. Larva is white with brown head.

Habitat: Coniferous forests.

Range: Eastern North America.

Food: Adult gnaws on bark of small twigs. Larva eats inner wood of pine.

Life Cycle: Female cuts deep pits in bark and lays 1−6 eggs inside each. Larvae bore U-shaped tunnels, packing wastes behind them, and almost reach their starting point where they pupate. Adults chew their way to freedom. Cycle takes 1−4 years depending on conditions.

The sound of gnawing larvae can be heard by pressing an ear against the bark of infested trees. This beetle does considerable damage to recently felled trees and those scorched by fire. The Spotted Sawyer (*M. maculosus*), ⅝−1" (15−25 mm), is dark brown and has irregular bluish spots on its elytra. It attacks pines from Colorado and New Mexico to California, north to British Columbia. The larger Northeastern Sawyer (*M. notatus*), ½−1⅛" (12−28 mm), is brown with black spots, white flecks, and silver-gray hair on its elytra. It attacks pines, spruces, and firs. The Eastern Sawyer (*M. titillator*), ¾−1⅛" (20−28 mm), is brown and mottled gray with brown and black hair. It feeds on pines from Quebec to Georgia.

245 Cylindrical Hardwood Borer
(*Neoclytus acuminatus*)

Description: ¼–¾" (6–18 mm). Elongate,
cylindrical. Reddish brown. *Elytra
parallel-sided, crossed by 4 narrow yellow
bands* interspersed with dark brown.
Antennae are longer than head and
prothorax combined.

Habitat: Forests, woods, and adjacent clearings.

Range: Eastern United States and Canada.

Food: Larva eats oak, hickory, ash, other
hardwoods, and unseasoned lumber.

Life Cycle: Eggs are laid on bark of living or dead
trees. Larvae tunnel inward, later
pupate close to bark. Adults emerge
during first warm days of spring.

Adults are common on clustered
flowers. If approached, they scramble
quickly, drop off, or fly away.

169 Cottonwood Twig Borer
(*Oberea quadricallosa*)

Description: ⅜–½" (8–14 mm). Slender,
cylindrical. Elytra, head, antennae, and
undersurface of thorax black to grayish.
*Pronotum yellow to orange with 4 black
spots.* Legs and abdomen orange. Larva
is whitish.

Habitat: Woods and river margins.

Range: California to British Columbia.

Food: Adult probably does not eat. Larva
feeds in twigs of cottonwood, poplar,
and willow.

Life Cycle: Female cuts 2 rows of pits in young
stem about 6" or so from tip and lays
eggs between rows. Larvae bore into
inner wood, work down toward soil,
and pupate in tunnel. Adults emerge
late May–early June.

The similar Raspberry Cane Borer
(*O. bimaculata*), same size, has a yellow
pronotum with 2 black spots. Its larvae
bore into canes of live blackberry and

raspberry. The Dogwood Twig Borer (*O. tripunctata*), ⅜–⅝″ (8–16 mm), has some red on its abdomen and 3 black spots on the pronotum. It attacks viburnum, elm, and fruit trees, as well as dogwoods. Both range throughout North America.

181 Cottonwood Borer
(*Plectrodera scalator*)

Description: ⅞–1⅝″ (22–40 mm). Shiny black with *pattern of thick white hair on pronotum and irregular crossbands on elytra.* Black antennae almost ½ again as long as body. Black legs with white hair. Pronotum has sharp projection on each side.

Habitat: River banks and adjacent higher ground, with cottonwoods or poplars.

Range: Atlantic Coast to the Mississippi River basin, north to Manitoba.

Food: Larva eats wood of cottonwood and poplar.

Life Cycle: Eggs are laid on bark in fall. Larvae tunnel into wood, overwinter, then pupate in chambers below bark. Development into adults takes 2–3 years. Adults usually emerge July–September.

These beetles attack tree bases, then move up the bark to the leaves.

194, 220 Giant Root Borers
(*Prionus* spp.)

Description: ⅞–3″ (22–75 mm). Shiny, reddish brown, yellowish brown, or black. Flattened; sides of pronotum usually have coarse spines. *Long thick antennae.* Antennal segments conical; overlapping in male, cylindrical in female. Larva has dark brown head, white to yellowish-gray body.

Habitat: Coniferous forests, mixed or deciduous.
Range: Southern California north to Alaska,
           east across Canada and the northern
           United States, south to northern
           Florida, west to New Mexico.
Food: Larva eats live, dying, and
           decomposing trees, shrubs, and woody
           vines.
Life Cycle: Eggs are laid singly in soil close to food
           supply. Larvae eat to inner bark. After
           3 or more years, larvae prepare egg-
           shaped pupal cells inside wood. Adults
           emerge July–August.

Primarily nocturnal, adults buzz loudly
in flight. They are attracted to light,
sometimes crashing violently against
windowpanes. They are the largest
North American long-horned beetles,
rivaling in weight the Rhinoceros
Scarab.

180   Ribbed Pine Borer
(*Rhagium inquisitor*)

Description: ½–¾″ (13–18 mm). Cylindrical. Dull
           gray with reddish-brown, black, and
           whitish hair. Head and thorax about
           equal width; tooth on each side of
           thorax. Antennal segments 5–11 thick.
           Elytra have prominent lengthwise ridges.
Habitat: Coniferous forests.
Range: Throughout North America.
Food: Larva eats inner bark and sapwood of
           various pines.
Life Cycle: Eggs are laid on bark. Larvae tunnel
           into sapwood and later construct
           circular pupal cells beneath the bark,
           fencing themselves in with long
           threadlike pieces of sapwood resembling
           birds' nests. Adults emerge in early
           spring and are active until October.

These beetles destroy pines and other
conifers, often boring into freshly cut
logs. Formerly assigned to the genus
Stenocorus.

184 California Laurel Borer
"Banded Alder Borer"
(*Rosalia funebris*)

Description: 1–1½" (25–38 mm). Prothorax
whitish with large bluish-gray spot at
center, sharp points at sides. *Elytra have
3 bluish-gray bands, 3 whitish bands, and
2 small whitish dots.* Head black.
Antennae long, bluish gray and white
with black tips.

Habitat: Forests and woods.

Range: California to Alaska, also New Mexico.

Food: Larva feeds on California laurel, Oregon
ash, and New Mexico willow.

Life Cycle: Eggs are laid on bark. Larvae tunnel
inward, later prepare pupal chambers.

This striking beetle is recognized
easily, but its life cycle is still
not fully known.

241 Ivory-marked Beetle
(*Eburia quadrigeminata*)

Description: ½–1" (13–25 mm). Elongate.
Brownish-yellow with *2 pairs of ivory-
white spots on each elytron.* Antennae
slightly longer than body (in male),
slightly shorter than body (in female).
Prothorax with sharp spine on each
side. Paired spines at ends of elytra.
Larva to 1½" (39 mm); prefers dry,
solid hardwood.

Habitat: Forests, particularly with hardwoods,
and lumberyards.

Range: East of Great Plains.

Food: Adult eats foliage and new twig
growth. Larva bores in wood.

Life Cycle: Eggs are laid singly in breaks through
bark. Larvae eat tunnels inward,
especially in oak, ash, maple, hickory,
chestnut, cypress, honey locust, beech,
and elm.

Larvae of the Ivory-marked Beetle are
heartwood-borers, making their way

deep within a tree to the heartwood
found near the center. For this reason,
the species is of some economic
importance, often turning up in
lumberyards where wood has been
stacked for seasoning. The Ivory-
marked Beetle is quite similar to the
Spotted Apple Tree Borer (*Saperda
cretata*), another of the long-horned
beetles that occur in much the same
range. Both the Spotted Apple Tree
Borer and the Round-Headed Apple
Tree Borer (*S. candida*) are also
destructive to trees.

---

143   Red Milkweed Beetle
"Eastern Milkweed Longhorn"
(*Tetraopes tetraophthalmus*)

Description: ⅜–½" (9–14 mm). Elongate, almost
cylindrical. Red on top with *4 black
dots on pronotum. Elytra have* varying
pattern of *black streaks and dots.*
Triangular scutellum and underbody
dark. Legs mostly grayish black tinged
with red. *Antennae gray.* Thorax has
tubercles on sides.

Habitat: Meadows and roadsides with milkweed.

Range: Eastern United States and Canada.

Food: Larva bores in stems and roots of
milkweed.

Life Cycle: Eggs are left on milkweed stems near
ground or slightly below soil. Larvae
bore into stems, overwinter in roots,
and pupate in spring. Adults emerge in
early summer, complete life cycle by
autumn.

These beetles make squeaking sounds
by rubbing together rough areas on the
thorax. Adults are immune to poison
in milkweeds, but their larvae are
poisonous to birds. The Western
Milkweed Longhorn (*T. femoratus*), ½–
⅝" (12–15 mm), has gray rings around
each antennal segment and lacks long
black marks on its elytra.

159 Notch-tipped Flower Longhorn
(*Typocerus sinuatus*)

Description: ⅜–½″ (9–14 mm). Head and bell-shaped prothorax black. Elytra are yellow to tan, with 3–4 pairs of black or dark brown spots on sides and small spots at tip, often joined across midline. *Elytral tips notched,* outer projection longer than inner.

Habitat: Forests and adjacent clearings.

Range: Eastern United States and Canada.

Food: Adult eats pollen and nectar, especially sumac. Larva feeds on decayed wood.

Life Cycle: Eggs are laid in decayed wood. Larvae tunnel inside, overwinter, and pupate in spring. Adults emerge in early summer.

The Banded Longhorn (*T. velutinus*), ⅜–½″ (9–14 mm), has reddish-brown elytra crossed by 4 narrow black bands, a black pronotum with short yellow hair, and black head and underside. It is common on sumac and goldenrod, while its larvae feed on decaying birch, hardwoods, and some conifers.

244 Willow Borer
(*Xylotrechus insignis*)

Description: ½–⅝″ (14–15 mm). Almost cylindrical, narrowed between prothorax and elytra. *Male* velvety reddish brown *with yellow lines on face and prothorax and with 3 marks across elytra.* Female is black with reddish antennae and legs and with variable yellow crossbands on prothorax and elytra. *Antennae longer than head and prothorax.* Elytra do not reach tip of abdomen.

Habitat: Rivers and floodplains.

Range: California, Oregon, and Arizona.

Food: Adult nibbles pollen. Larva bores into willow.

Life Cycle: Eggs are laid on bark of living or dead trees. Larvae tunnel inward, later pupate close to bark surface. 1 generation a year.

These beetles are often seen on flower clusters; they run with an antlike gait. Other species of this genus are dark brown with yellow markings. All have a Y-shaped ridge on the front of the head, often marked with yellow.

SEED BEETLES
(Family Bruchidae)

These small, stout beetles have broad, oval bodies, less than $\frac{1}{16}-\frac{3}{8}''$ (1–10 mm) long, with elytra that do not cover the abdominal tip. The small downturned head ends in a short snout but is rarely visible beneath the roughly triangular pronotum. Most seed beetles are black or brown, although some are reddish or yellowish or have mottled patterns. Larvae bore into seeds, where they feed and pupate, sometimes even before the plant releases its seeds. Many are major pests, feeding on stored grain, dry peas, and beans. Some species have a single generation a year, while others are continuous breeders.

135 Bean Weevil
(*Acanthoscelides obtectus*)

Description: $\frac{1}{16}-\frac{1}{8}''$ (2–4 mm). Flattened. Body black but appears velvety brownish gray or beige because of fine hair. Elytra have a pattern of brownish-black spots. Legs reddish. Antennae swollen slightly to form a slender club.

Habitat: Warehouses, and wherever leguminous seeds are stored

Range: Worldwide.

Food: Most dry beans, cowpeas, and lentils.

Life Cycle: White eggs are laid on beans, hatching in 3–9 days. Larvae carve twisting tunnels within beans, where they feed for 12 days–6 months and later pupate for 8–25 days. Adults emerge from the seed by cutting little round holes. 6 or more generations a year.

As long as the food supply lasts, the Bean Weevil continues to breed. Unlike the Pea Weevil, it does not need moisture inside seeds.

LEAF BEETLES
(Family Chrysomelidae)

Leaf beetles can be recognized by short antennae, less than ½ the length of the body. Most are convex, oval, and very small, measuring less than ¹⁄₁₆–½″ (1–13 mm). Many have bright, metallic colors that glisten in the sun. These beetles feed on leaves and flowers. The larvae attack roots, devour leaves, or tunnel within them. Many are important agricultural pests. This family has nearly 1,400 species in North America.

182 Dogwood Calligrapha
(*Calligrapha philadelphica*)

Description: ⅜″ (8–10 mm). Oval, hemispherical. Head and thorax dark metallic olive-green; *elytra ivory-white marked with dark green stripes and spots.* Antennae, legs, and mouthparts dark reddish brown.
Habitat: Woods, parks, and forest edges.
Range: Atlantic Coast in Canada down the Appalachians to Georgia, west to the Mississippi River basin, north to Nebraska.
Food: Foliage of dogwood, basswood, elm, and other trees.
Life Cycle: Eggs are attached to leaves and

branches. Larvae feed on leaves, drop to soil to pupate, overwinter in soil as pupae or young adults. Beetles fly actively May–August.

At the slightest disturbance these beetles drop to the ground and become almost impossible to see. The pattern of spots and lines varies considerably within each species.

215   Milkweed Tortoise Beetle
"Argus Tortoise Beetle"
(*Chelymorpha cassidea*)

Description: ⅜–½" (9–12 mm). Oval, convex. Brick-red or yellow, transparent around edges. Pronotum has 6 black dots; *each elytron has 6 black dots, 1 on midline;* resembling a ladybug beetle. Larva is yellowish green or orange-yellow.

Habitat: Meadows and roadsides.

Range: Throughout North America.

Food: Foliage of wild morning glory and related plants; sometimes milkweed, raspberry, maize, and sweet potato.

Life Cycle: Eggs are laid in clusters of 15–30 on leaves. Larvae feed there until fully grown, then drop to soil, pupate, and may overwinter before emerging as adults in midsummer.

This is one of the largest North American leaf beetles. It is called "Argus" after the mythical 100-eyed Greek monster, because it stretches out its bright red head beyond the pronotum, as though it were a single red eye. It is sometimes a pest. The Golden Tortoise Beetle (*Metriona bicolor*), ¼" (5–6 mm), is brilliant brassy gold. This eastern garden pest attacks morning glories, eating pinlike holes in the leaves.

211 Dogbane Leaf Beetle
(*Chrysochus auratus*)

Description: ⅜″ (9–11 mm). Oblong, strongly convex. *Head and thorax bright shiny green. Elytra often coppery, brassy, or bluish.* Threadlike antennae, widely separated at base. Larva is white with brown head.

Habitat: Meadows and along roadsides.

Range: Atlantic Coast to Texas, north to Saskatchewan.

Food: Dogbane and other members of the milkweed family.

Life Cycle: Yellow eggs are laid on the ground or on host plant. Larvae tunnel through soil to roots, feed, and pupate in soil.

When disturbed, these beetles drop to the ground and hide among leaf litter. If caught, they exude a foul-smelling secretion.

205 Milkweed Leaf Beetle
(*Chrysochus cobaltinus*)

Description: ⅜″ (9–10 mm). Cylindrical. *Pronotum rounded, narrower than elytra,* which are evenly rounded to meet at rear. Body and elytra glossy dark blue to metallic green. Antennae and undersurfaces bluish black. Threadlike antennae, widely separated at base.

Habitat: Meadows and roadsides.

Range: New Mexico to California, north to British Columbia.

Food: Adult eats foliage of milkweed and occasionally oleander or peach trees. Larva feeds on roots of host plant.

Life Cycle: Yellow eggs are laid on host plant near or at ground level. Larvae tunnel through soil to roots and pupate in soil. Adults emerge end of summer.

When approached, these beetles drop to the ground and hide. They give off a foul-smelling secretion when handled.

150   Willow Leaf Beetle
(*Chrysomela lapponica*)

Description: ¼–⅜″ (7–9 mm). Oval, flattened. *Elytra either reddish above with black spots* resembling those of ladybug beetles *or unspotted dark metallic purple.* Both forms black below. Larva, to ⅜″ (9 mm), is black, rough.

Habitat: River edges and floodplains.

Range: United States and southern Canada.

Food: Willow foliage.

Life Cycle: Eggs are laid in spring on willow leaves, where larvae feed. Full-grown larvae pupate while hanging head downward from a leaf or twig.

The Unspotted Poplar Leaf Beetle (*Lina tremulae*), ⅜–⅝″ (9–15 mm), is bronze-black with red elytra. It was introduced accidentally from Europe and is now common in both eastern and western states and provinces. The Spotted Poplar Leaf Beetle (*L. scripta*), ¼–⅜″ (7–9 mm), has a dark pronotum with pale yellow edges and yellowish-green to dull red elytra with variable purplish spots and lines.

149   Spotted Asparagus Beetle
(*Crioceris duodecimpunctata*)

Description: ¼–⅜″ (7–9 mm). Elongate, somewhat flattened; head slightly constricted behind eyes. *Head and pronotum reddish brown. Elytra tan to reddish brown with 6 black spots on each* (or 5 if 2 spots are joined to make broader mark). Larva, to ⅜″ (9 mm), is orange with black head and legs.

Habitat: Vegetable gardens.

Range: Most of North America.

Food: Green parts of asparagus.

Life Cycle: Female attaches greenish eggs singly to leaves. Orange larvae bore into plant and eat out pulp. After 15 days or less, they drop to soil and prepare thinly lined pupal cells close to the soil surface. 1st generation of adults appears in late July, the last usually in early September. Depending on the climate, last of 2-5 generations overwinter as adults in stems of fallen plants or other debris.

This pest was introduced accidentally from Europe to Maryland in 1881. Adults may completely defoliate asparagus plants; the larvae attack the green fruits.

214 Clavate Tortoise Beetle
(*Deloyala clavata*)

Description: ¼″ (7 mm). Almost circular, slightly convex resembling *flat turtlelike shell*. Brown above, straw-yellow underneath. *Pronotum and elytra transparent with brown patches near elytral edges.*

Habitat: Fields and wastelands.

Range: Widespread from New England to South Carolina.

Food: Foliage of plants in the morning glory family.

Life Cycle: Eggs are deposited on leaves, which larvae later attack, eating all but veins. Larvae drop to soil and pupate in fallen leaves. Adults overwinter and are active May–September.

The Clavate Tortoise Beetle pulls in its antennae when threatened, just as a turtle might withdraw its head and legs.

153 Spotted Cucumber Beetle
(*Diabrotica undecimpunctata*)

Description: ¼″ (5–7 mm). Oval, flattened, widest at middle. Head, antennae, underside, and legs black; pronotum is dark. *Elytra are yellowish green with 12 black*

spots (2 nearest head join when wings
are closed). Larva is yellow with brown
head.

Habitat: Meadows, wastelands, and gardens.

Range: West of the Rocky Mountains.

Food: Many wild plants and crops, including
citrus.

Life Cycle: Female deposits yellow eggs in soil near
food plants. Larvae feed on roots and
stems until fully grown, then construct
pupal cells below soil.

One of the most destructive beetles is
the Eastern Spotted Cucumber Beetle
(*D. u. howardi*), ¼″ (6–7 mm), which
is greenish yellow with black spots and
has brown and yellow antennae. It
damages foliage, flowers, and pollen of
cucumbers, melons, corn, potatoes, and
peanuts. The Striped Cucumber Beetle
(*D. acalymma vittata*), ¼″ (5–7 mm), is
yellow with 3 black stripes, black head
and legs, and black and yellow
antennae. It is common from Mexico to
Colorado, and northeast to Canada.
Adults attack beans, peas, corn, and
flowers; larvae feed on plants in the
cucumber family.

198 Waterlily Leaf Beetles
(*Donacia* spp.)

Description: ¼–½″ (6–12 mm). Elongate, slender,
flattened. *Bronze, with green or yellow
gloss. Antennae about ½ as long as body;
legs long; tarsi broad.* Elytra are pitted in
lengthwise rows and conceal tip
of abdomen.

Habitat: Water lilies, pickerel weed, and other
floating plants on ponds, slow streams,
and protected shores of large lakes.

Range: Throughout North America.

Food: Adult feeds on foliage and pollen of
water lily, skunk cabbage, and similar
plants. Larva eats submerged parts of
plants.

Life Cycle: Female cuts holes through water lily

pads, then with tip of abdomen
cements eggs to the underwater surface
of leaves. Larvae spin silken pupal
shelters below leaves. Adults emerge in
10 months. Adults fly June—September.

Active by day, these beetles are fast
fliers. Only a specialist using a
microscope can distinguish the 36
different species in this genus.

151 Swamp Milkweed Leaf Beetle
(*Labidomera clivicollis*)

Description: ⅜–½" (8–12 mm). Oval, strongly
convex. Bluish or greenish black; *elytra*
orange to yellow *with 2 black marks in
an X across midline.* Male has 1 large
projecting tooth on outside of each
front femur and 1 on inside. Larva is
orange.

Habitat: Marsh and stream edges where swamp
milkweed grows.

Range: Most of North America.

Food: Foliage and sometimes flowers of
milkweed.

Life Cycle: Elongate yellow eggs are cemented to
the underside of milkweed leaves.
Larvae grow rapidly, drop to the
ground, and pupate there. Adults
emerge by late summer.

These beetles often overwinter as adults
in mullein plants and creep deep among
the woolly leaves, which shrivel but do
not fall off until the dead plant
collapses in spring.

154 Three-lined Potato Beetle
"Old-fashioned Potato Bug"
(*Lema trilineata*)

Description: ¼" (6–7 mm). Reddish yellow. Black
eyes. Pronotum has 2 black dots
(sometimes absent). *Elytra have 3 black*

lengthwise stripes. Thorax is constricted in front of elytra. Larva is grayish yellow.

Habitat: Meadows and potato fields.

Range: Throughout North America.

Food: Members of the nightshade family, including potatoes.

Life Cycle: Yellow eggs are laid singly or clustered on leaves. Larvae pupate in silk-lined cells in the soil. Adults overwinter. 2 generations a year.

Voracious larvae gather in clusters on potato leaves, nibbling lacy holes and eventually consuming all but the midvein. Unlike the larvae of other potato-feeders, they are blanketed in a wet froth of their own secretions. Adults can be distinguished from the Striped Cucumber Beetle by the constriction behind the thorax.

243 Colorado Potato Beetle
(*Leptinotarsa decimlineata*)

Description: ¼–⅜" (6–11 mm). Oval, convex. Pronotum yellow-brown to orange-yellow with 2 black spots, often joined to produce a V-shaped mark, and a number of smaller black spots on each side. *Each elytron has 5 black lines on whitish background.* Larva is dark red, maturing to orange with black head and black spots on sides.

Habitat: Mountain meadows and potato fields.

Range: Most of the United States and Canada.

Food: Potato foliage and wild members of the nightshade family.

Life Cycle: Small clusters of orange elongate eggs are attached to the underside of leaves. Larvae reach full size in 15–20 days, then drop to soil, where they pupate. Adults emerge in 10–15 days. 1–2 generations a year.

One native species of tachinid fly and at least 3 kinds of predacious bugs attack adults or prey upon their larvae. The larger False Potato Beetle (*L. juncta*), ⅜–½″ (11–12 mm), has a black dot on each femur and lives on nightshades in the South.

155 Girdled Leaf Beetle
(*Pachybrachis circumcinctus*)

Description: ¹⁄₁₆–¼″ (2–6 mm). Broad. Yellowish; pronotum reddish-brown. Elytra have black or brown stripes. Antennae long. Fore femora thicker than hind femora.
Habitat: Meadows and woods.
Range: California to Oregon.
Food: Adult and larva eat leaves; adult also eats pollen.
Life Cycle: Eggs are laid on grasses, weed foliage, and trees, including elms. Larvae are often gregarious. Pupae are attached to leaves and stems without cocoons. Adults emerge in late summer, overwinter, and disperse in spring or early summer.

Adults are noticed more often on flowers than on foliage. Their small bodies resemble the fecal pellets of large caterpillars, which may protect them from insectivorous birds. This genus has more than 35 species in North America, distinguished by color patterns and inconspicuous details.

PRIMITIVE WEEVILS
(Family Brentidae)

Primitive weevils are easily recognized by their narrow, elongate bodies and the female's long, straight beak. These beetles, ¼–1⅛″ (6–30 mm) long, have beadlike antennae, a pear-shaped prothorax, and swollen femora. Primarily tropical insects, they live under loose bark and in the galleries of rotting logs, where they feed on wood-eating insects, fungi, and sometimes sap. The larvae bore in rotting wood and sometimes living trees.

246 Oak Timberworm Beetle
"Primitive Weevil"
(*Arrhenodes minutus*)

Description: ¼–1⅛" (7–30 mm). Cylindrical, elongate. *Prothorax pear-shaped.* Dark brownish red; elytra with yellowish streaks or spots and with lengthwise grooves. *Female has long slender beak* with minute jaws at tip; male has pincerlike flattened jaws, toothed on inner surface. Antennae with 10–11 segments arise in front of eyes. *Femora stout, particularly fore pair.*

Habitat: Deciduous forests, beneath bark of dying or dead beech, oak, poplar, and maple trees.

Range: Eastern North America.

Food: Adult feeds on fungi, other insects, and liquid exuded by wood. Larva eats fungi and wood.

Life Cycle: Female uses beak and jaws to cut a cylindrical hole through bark into wood, while male stands nearby to help free female if beak gets stuck. Eggs are laid in hole, then female uses hairy antennae to brush off beak. Larvae bore galleries in solid wood and pupate under bark, where adults overwinter. Adults active May–September.

Adults can be found under loose bark at any season.

SNOUT BEETLES AND WEEVILS
(Family Curculionidae)

These hard-bodied beetles comprise the largest family of insects, with 40,000 species worldwide and 2,500 in North America. The head is elongated into a slender, downcurved beak or snout with elbowed or clubbed antennae partway down and mandibles at the tip. They range from less than $\frac{1}{16}''$ (1 mm) to $1\frac{5}{8}''$ (40 mm) long. All feed on plants. Females bore into fruits, seeds, and stems to lay eggs, and their C-shaped larvae are equally destructive. Many are major agricultural pests, although a few species have been used for biological control of weeds.

---

131 Boll Weevil
(*Anthonomus grandis*)

Description: $\frac{1}{8}-\frac{1}{4}''$ (4–7 mm). Egg-shaped. Dark grayish black to brown, *covered with yellowish hairlike scales.* Elytra pitted and grooved. Female's beak $\frac{1}{2}$ as long as body. Antennae elbowed with club at tip. Each fore femur has 2 spurs or teeth (the inner spur is much longer than the other); each middle femur has 1 tooth.

Habitat: Wherever cotton grows.

Range: Southeastern United States to California.

Food: Adult eats cotton seedpods, called bolls, and cotton flower buds; also okra and hollyhocks. Larva feeds inside bolls.

Life Cycle: Female drills holes into bolls, laying 1 egg inside each boll and eventually depositing 100-300 eggs. Larvae feed and later pupate inside boll. Many generations a year, the last overwintering as adults in trash and ground litter.

The infamous Boll Weevil is the bane of cotton farmers throughout the United States. Accidentally introduced from Mexico in the 1800s, it causes cotton to yellow and drop off plants.

134 Black Oak Acorn Weevil
(*Curculio rectus*)

Description: ⅜" (9 mm). Egg-shaped, constricted between prothorax and elytra. Brown covered by pale brown hairlike scales and sometimes small brown or yellow spots. *Beak slender;* female's longer than body and almost straight near head, male's shorter than body and uniformly curved. 2nd antennal segment is at least as long as 3rd.

Habitat: Deciduous forests.

Range: Eastern United States west to Arizona.

Food: Acorns of black, white, and red oaks.

Life Cycle: Female bores small circular hole through acorns, lays 1 egg or more in separate pockets in hole, and seals opening with a fecal pellet that looks like a white dot on the outside of dry acorn. Larvae feed and pupate inside. Adults are active June—August. 1 generation a year, corresponding to acorn crop.

This weevil can be easily recognized by its long beak. The Eastern Black Oak Acorn Weevil (*C. baculi*), ¼" (6–7 mm), is grayish to reddish brown with short pale gray scales; its curved beak is ⅗ the body length in female, or ½ body length in male. It is found from New York to Alabama, northwest to Nebraska.

136 Lesser Cloverleaf Weevil
(*Hypera nigrirostris*)

Description: ⅛" (3–4 mm). *Reddish brown to black in East or green with brown legs and head and black beak in West.* Both covered with short stiff hair that make 2 pale stripes on pronotum and greenish, gray, or yellowish markings elsewhere.

Habitat: Clover fields.

Range: Throughout North America.

Food: Leaf and flower buds of various clovers,

especially red clover.

Life Cycle: Oval greenish eggs are thrust into buds, where they hatch in about 7 days. The larvae attain full size in 17–20 days and pupate in oval cocoons. Adults emerge in 7 days, feed, and overwinter. They are active April–July.

Accidentally introduced from Europe, this beetle is now widespread. Formerly it was placed in genus *Phytonomus*.

133 Alfalfa Weevil
(*Hypera postica*)

Description: 1/8–1/4" (3–6 mm). Elongate, egg-shaped. Pale brown to black, covered with *hair and scales* that *form a darker stripe down back and a pattern of irregular dots*. Antennae and legs are pale brown.
Habitat: Alfalfa fields.
Range: Chiefly western states.
Food: Buds and foliage of alfalfa, occasionally clovers and vetch.
Life Cycle: Lemon-yellow eggs are laid in clusters in plant stems, overwinter, and hatch in April. Larvae feed inside buds and then on leaves. They pupate in cocoons attached to the host plant or in the curl of fallen leaves. Adults emerge in summer and overwinter in the crowns of plants among debris. 1 generation a year.

The Alfalfa Weevil was introduced from Europe in 1904. This beetle was formerly placed in genus *Phytonomus*.

132 Pine Weevils
(*Lixus* spp.)

Description: 1/8–1/4" (4–6 mm). Light to dark brown or black with pale spots of sparse hair. Body elongate, cylindrical, tapering at each end; slightly constricted between prothorax and elytra. Elytra are wider than prothorax, grooved lengthwise, and pitted. Clubbed antennae arise from sides of slender *cylindrical beak,* which is *as long as prothorax.*

Habitat: Coniferous forests.

Range: Northern states and Canada, and at higher elevations to Georgia and California.

Food: Adult eats soft plant tissues. Larva feeds on young stems of pines, firs, spruces.

Life Cycle: Female cuts pits in bark of leader stem, laying a single egg in each pit. Larvae gather in ring around the base of the leader, or bore in separately. They pupate in cocoons made of wood fibers and chips and overwinter as young adults.

Glistening drops of resin on tree bark often indicate where adults feed or where females have cut egg pits. Because larvae cause trees to grow crooked, they greatly diminish their value for lumber. The pine weevil formerly was placed in the genus *Pissodes* to differentiate it from the larger Rhubarb Weevil (*L. concavus*), ⅜–½" (10–13 mm), and its close relatives, which feed in the stems of curly dock, sunflower, thistle, rhubarb, and other plants.

138 Rose Weevil
(*Rhynchites bicolor*)

Description: ¼" (5–7 mm). *Stout, beak as long as head.* Back of head, pronotum, and elytra red; rest of body black. Elytra are wider than pronotum.

Habitat: Open fields with wild roses, blackberry and raspberry; also in gardens.

Range: Throughout North America.

Food: Pith of canes, also rose hips and buds.

Life Cycle: Eggs are deposited in rose buds and hips after female removes pits, which are also a food source. Larvae bore into plant tissue, where they remain for up to 2 years while developing to full size. They then pupate and overwinter in soil. Adults emerge in June. 1 generation a year.

Several western subspecies differ in color and size and feed on different plants. The related Juniper Weevil (*R. nano*), $\frac{1}{16}-\frac{1}{4}''$ (2–5 mm), is blue instead of red and attacks juniper in California.

---

130 Agave Billbug
(*Scyphophorus acupunctatus*)

Description: $\frac{3}{8}-\frac{3}{4}''$ (10–19 mm). Elongate oval, hard. Black. *Beak is thickest at base, downturned. Elytra are deeply grooved,* shorter than abdomen.

Habitat: Deserts.

Range: Mexico north into Texas, New Mexico, and Arizona.

Food: Adult eats the sap of agave plants. Larva feeds on the flower stalks of agave plants.

Life Cycle: Eggs are laid in pits cut at the base of flower stalk. Larvae tunnel inward and pupate inside the plant or in soil.

This beetle's elytra are extremely hard, protecting it from the drying sun of its habitat. The related Yucca Billbug (*S. yuccae*), $\frac{5}{8}''$ (15–17 mm), feeds on yucca plants in California.

---

137 Stored-grain Billbug
"Granary Weevil" "Elephant Bug"
(*Sitophilus granarius*)

Description: $\frac{1}{8}''$ (3–4 mm). Hard. Shiny blackish or chestnut-brown. Pronotum has elongate pits, elytra deep grooves. *Beak stout,* $\frac{1}{2}$ length of pronotum, which is almost as long as elytra. *Last abdominal segment often exposed beyond elytra.* Antennal club oval. No flying wings.

Habitat: Grain fields and storage centers, and grasslands.

Range: Worldwide.

Food: Dry seeds and seedlings of grasses, grains on farms.

Life Cycle: Females cut through seed coverings, lay eggs in pits, and seal the seeds with a gelatinous secretion. Larvae feed in seeds (or seedlings, if sprouting) and pupate inside kernels, emerging as adults 4 weeks after hatching. Adults live 7–8 months. A single pair can produce 6,000 young a year in 4–6 generations.

This pest often attacks harvested wheat, using its prominent beak, which accounts for its alternate common name, "Elephant Bug." The related Rice Billbug (*S. oryzae*), $\frac{1}{16}$–$\frac{1}{8}$" (2–3 mm), often has 4 red or yellow spots on its elytra, round pits on the pronotum, and hind flying wings. It infests stored grains, especially rice, and is capable of producing 6,000 young in 6 generations a year.

BARK AND AMBROSIA BEETLES
(Family Scolytidae)

These small cylindrical beetles carve elaborate tunnels in the inner bark or wood of trees. Most are black or brown and are usually less than $\frac{1}{4}$" (5–7 mm) long. They have a short beak (scolytid means "cut short" in Greek) and widely separated, short, elbowed, and clubbed antennae. This family contains 2 distinct groups: the bark, or engraver, beetles and the ambrosia, or timber, beetles. Both have the same form, but bark beetles have a large spine on the tip of each front tibia. Bark beetles and their larvae live under bark, where they make branching galleries and eat wood.

Ambrosia beetles tunnel into hardwood and make holes, which are filled by fungi, their "ambrosia," which they eat.

213 Red Turpentine Beetle
(*Dendroctonus valens*)

Description: ¼–⅜" (6–9 mm). Elongate, *egg-shaped;* widest halfway along elytra. Pale to dark red above, medium red to black below. Pronotum shiny, slightly hairy, and irregularly pitted. Elytra have short erect hair and indistinct pits in lengthwise rows. Legs are hairy. Antennae clubbed.

Habitat: Coniferous forests.

Range: Northeastern and western United States, and southern Canada.

Food: Inner bark of pines and sometimes of spruce, larch, or fir trees.

Life Cycle: Adults excavate branching galleries in the inner bark, slightly scoring sap wood. Eggs are packed among wood dust. Larvae feed close together, mining large chambers.

Most infestations occur within 20 feet of the ground, severely injuring trees. This is the largest species in the genus *Dendroctonus*.

216 Pine and Spruce Engraver Beetles
(*Ips* spp.)

Description: ⅛–¼" (4–6 mm). *Cylindrical.* Black, brown, or reddish brown, sometimes with fine yellow hair. Head concealed from above by pronotum but flattened club of antennae project. Front tibiae widen toward tip, with several fine teeth on outer surface. Tip of *elytra* deeply excavated and *edged with coarse teeth.*

Habitat: Coniferous forests.

Range: Throughout North America; individual
species more localized.

Food: Cambium and phloem tissues of inner
bark.

Life Cycle: Adults cut cylindrical tunnels through
bark to feeding area, then expand brood
galleries, where eggs are laid. Larvae
excavate galleries further in a pattern
resembling an engraving when bark is
peeled away. Larvae pupate inside the
galleries. Adults emerge in summer.

Spending most of their time in
galleries, adults come into the open on
warm sunny days, when they fly off and
disperse widely. Their tunnels provide
openings for fungus, which often
hastens the death of a tree already in
decline.

201 Bark Beetles
(*Scolytus* spp.)

Description: ⅛–¼″ (3–5 mm). Shiny reddish
brown, dark brown, or black.
Cylindrical, rounded sharply at front of
thorax and behind elytra. Concave
below at rear of abdomen. *Male's
abdomen usually has blunt spines below*,
pointing to rear. Elytra grooved,
pitted. Antennae clubbed.

Habitat: Deciduous and mixed forests.

Range: Throughout North America.

Food: Inner bark of various trees, including
hickory, pecan, walnut, Douglas fir.

Life Cycle: In early spring female cuts holes for
eggs in bark of dying or dead trees and
deposits each egg in separate side
tunnel of inner bark, hollowed parallel
to grain. Larvae bore deeper,
overwinter, and pupate in 2nd summer.
Adults emerge late summer or fall.

These beetles do major damage to trees.
The European Elm Bark Beetle
(*S. multistriatus*), ⅛″ (3 mm), is shiny
reddish brown and the male has 1 blunt

spine below its abdomen. It carries the
fungus that causes Dutch elm disease,
and ranges east of the Mississippi and
also from Colorado and Nevada to
California, north to Washington. The
Hickory Bark Beetle (*S. quadrispinosus*),
⅛–¼" (3–5 mm), is black or dark
brown and the male has 4 side spines
beneath its abdomen. It attacks
hickory, pecan, and butternut trees
throughout the eastern United States.
The Fir Engraver (*S. ventralis*), same
size, is black and has no spines. It feeds
on fir trees along the Pacific Coast to
British Columbia.

Twisted-winged Parasites
(Order Strepsiptera)

Only 300 species of this small order are known worldwide, including about 60 in North America. Minute insects, seldom more than ⅛" (4 mm) long, strepsipterans differ markedly according to sex, and exhibit a complicated form of parasitism on other insects. Males are dark brown to black and have bulging compound eyes, reduced chewing mouthparts, and unusually thick, forked or comblike antennae. The male's fore wings are small club-shaped organs that twist when the insect flies, accounting for both its common and order names. The male's hind wings resemble large fans with very few veins. All adult males are nonparasitic, as are females in the Family Mengeidae. They are found under stones and debris, where they attack bristletails. Most female strepsipterans and all larvae, however, are parasites. All females lack wings, legs, antennae, and eyes, so they somewhat resemble beetle larvae. Females never leave their pupal skins. They use their mouthparts to burrow into hosts, leaving the rear part of the body protruding between the host's abdominal segments. These numerous hosts include wasps, bees, and homopterans.

To mate, the male strepsipteran deposits sperm between the female's adult and pupal skins. These sperm flow down to the genitals at the tip of the female's body. Metamorphosis is complete. At first the active larvae have slender bristly bodies and well-developed legs. They then molt to become legless, maggotlike forms with reduced mouthparts. Larvae feed on tissues of the host's testis or ovary, usually sterilizing the host and often causing it to change color.

Scorpionflies and Kin
(Order Mecoptera)

These unusual-looking insects include about 85 species in North America and 400 worldwide. Their slender, soft reddish-brown to gray-black bodies, $\frac{1}{16}$–1″ (2–26 mm) in length, have long legs and extremely elongated, snoutlike heads that end in biting mouthparts. Antennae are long and threadlike, and the compound eyes are large and well developed. Winged species have 4 oval to elongate wings with many veins and cross veins. Despite the order's common name, only males of the Family Panorpidae have large conspicuous genitalia that curl upward over the abdomen like a scorpion's stinger. But mecopterans do not sting or bite. Metamorphosis is complete. The caterpillarlike larvae live in loose soil and moss, where they scavenge on dead insects or other organic material. Adults eat ripe fruit, dead insects, or even bird droppings. Mecopterans have changed little from fossils 250 million years old. Some entomologists believe that now-extinct members of the group are the ancestors of fleas, flies, butterflies, and moths. Classification of the group is mostly based on detailed differences in wings and legs. The 4 families in this order include scorpionflies, snow scorpionflies, hangingflies, and earwigflies.

SNOW SCORPIONFLIES
(Family Boreidae)

These dark-colored insects creep over the snow in early spring but spend most of the year in wet mosses, where they feed. Most are $1/16-1/4''$ (2–5 mm) in length and have long antennae, fairly long slender legs, and a prolonged face that ends in biting jaws—features that distinguish them from insects with similar habits. The wings are greatly reduced or absent. Males have bristlelike wings used to grasp females during mating; females have small scalelike wings. Unlike other mecopterans, the female has a long pointed ovipositor. There are 15 species of snow scorpionflies in North America; most occur in the West.

74 Snow Scorpionflies
(*Boreus* spp.)

Description: $1/8''$ (3 mm). Spindle-shaped. Black or brown. Compound eyes large. Antennae threadlike and $1/2$ as long as body. Legs long, slender. Female has long ovipositor. Mostly *wingless*.

Habitat: Woods on moss clumps.

Range: Colder parts of North America; at lower elevations in the North, higher in the South.

Food: Small insects and other animals found among mosses, possibly also moss itself.

Life Cycle: Eggs are laid among mosses. Larvae pupate in moss clumps or adjacent soil probably emerging before soil freezes and overwintering among mosses.

On warm winter days conspicuous numbers of adults hop across the snow. They are often mistaken for Snow Fleas which are much smaller. Only a specialist can distinguish the different species of snow scorpionflies.

COMMON SCORPIONFLIES
(Family Panorpidae)

Common scorpionflies get their name
from the male's bulbous, upward-
curving genitalia, which are similar to a
scorpion's stinger. These slender-bodied
brownish insects, ½–¾" (12–20 mm)
long, have dark spots or bands on their
4 membranous wings. They have long,
threadlike antennae and elongated faces
with projecting jaws. Adults feed on
dead and dying insects, nectar, and
rotting fruit. The spiny, caterpillarlike
or grublike larvae live in burrows, but
hunt for insects on the ground.

409 Scorpionflies
(*Panorpa* spp.)

Description: ½–¾" (12–20 mm), wingspan to 1⅛"
(30 mm). Slender. Reddish to yellowish
brown with dark spots or brown
crossbands on wings. Antennae
threadlike, more than ½ as long as
body. *Head prolonged into snout.* Distinct
neck. *Male genitalia large, pear-shaped,
held forward above abdomen like a
scorpion's stinger.* Wings long, narrow,
with many cross veins.

Habitat: Moist deciduous woods and adjacent
open areas.

Range: East of the Rocky Mountains.

Food: Adult feeds on nectar, fruits, dying and
dead insects. Larva eats organic matter
and preys on insects.

Life Cycle: Eggs are deposited in small masses
on soil. Larvae live in short burrows in
the soil, emerging to scavenge for food
on the surface. They overwinter and
pupate in underground cells. Adults
appear in early summer.

Scorpionflies serve as food for spiders,
dragonflies, robber flies, insectivorous
birds, and other predators. More than
40 species are found in North America.

HANGINGFLIES
(Family Bittacidae)

These yellowish-brown, long-legged insects, ¾–1" (20–26 mm) in length, resemble crane flies. They are often seen dangling by the fore legs from twigs and leaves. They use their hind legs to catch prey. Hangingflies have 2 pairs of membranous wings that narrow at the base. At rest wings are usually held over the abdomen. Adults actively prey upon small insects, such as aphids, flies, and caterpillars; the caterpillarlike larvae feed on dead insects, organic matter found among ground litter, or occasionally small live insects.

405 Green Stigma Hangingfly
(*Bittacus chlorostigma*)

Description: 1" (25–26 mm), wingspan 2–2¼" (51–57 mm). *Slender, resembling crane fly.* Brown and yellow. Eyes large. Head downturned, ending in prominent mandibles. Each tarsus has 1 claw. *Wings have conspicuous pale green patch* (stigma) on front margin.

Habitat: Moist meadows and grasslands.

Range: Southern California.

Food: Adult feeds on smaller insects. Larva preys on small insects, also scavenges on dead animals and other organic matter.

Life Cycle: Eggs are often left in moss clumps, where caterpillarlike larvae hunt for prey.

Appearing harmless, this insect hangs by the tarsi of its fore legs at rest. But its hind legs will suddenly dart out to seize flying prey. The Eastern Woodland Hangingfly (*B. stigmaterus*) same size, is pale brown, lacks colored stigma, and occurs in the eastern United States, west to Kansas.

Fleas
(Order Siphonaptera)

Fleas are small, wingless insects that live as external parasites on mammals and birds. Most are less than ¼" (5 mm) long. About 16,000 species are known worldwide, including 250 species in North America. Fleas have distinctive, laterally flattened abdomens with many spines and bristles. They can easily slip between the feathers or hair of their hosts. Their bodies have tough skins, making them extremely difficult to kill. The legs have enlarged coxae, which enable fleas to make spectacular leaps of a foot or more. Fleas have only minute, compound eyes or none at all. The short, 3-segmented antennae are often concealed in grooves on the head. Fleas use mouthparts with 3 piercing stylets to suck blood. Unlike lice, they do not necessarily confine themselves to one type of host.

Most fleas lay eggs in dirt or lint, or in their host's nest. Some species attach eggs directly to the host's hair or feathers. All undergo complete metamorphosis. Flea larvae are whitish, legless, wormlike creatures with hooks on the end of the abdomen that are used to cling to the host's feathers or hair. Unlike the parasitic adults, the larvae scavenge on organic matter. They pupate in silken cocoons and may remain dormant for months until stimulated by vibrations indicating that a host may be approaching. A few species are carriers of diseases, including bubonic plague, which was spread throughout Europe in the Middle Ages by fleas that had fed on infected rats.

COMMON FLEAS
(Family Pulicidae)

This family contains most of the fleas that attack people and domestic animals. Many are named after their principal host, although they may attack other animals as well.

---

76 Cat Flea
(*Ctenocephalides felis*)

Description: $\frac{1}{16}$" (2 mm). Female slightly larger than male. *Strongly flattened side to side.* Pale to dark reddish brown. Head gently curved front to back. Whiskerlike spines along lower edge of cheek. Hind femora have 7–10 bristles along each inner surface.

Habitat: Among hair on live cats.

Range: Worldwide.

Food: Adult feeds on cat blood. Larva eats organic debris in cat's bedding, under rugs, or in dusty corners.

Life Cycle: Female drops 50–100 eggs, usually among cat's bedding. Ivory-colored larvae pupate in bedding or cracks.

This parasite is often confused with the Dog Flea (*C. canis*), $\frac{1}{16}$–$\frac{1}{8}$" (2–4 mm) which has a small first spine along its cheek, 10–13 bristles on the inner surface of the hind femora, and a more strongly curved head.

---

68 Human Flea
(*Pulex irritans*)

Description: $\frac{1}{16}$" (1–2 mm). *Strongly flattened side to side.* Pale, changing to dark brown. Compound eyes incurved along lower border toward front. 1 large bristle below each eye; 1 bristle at base of lower mandibles. Male's abdominal tip turns upward, female's downward. No comblike spines on head or pronotum.

Habitat: Adult lives in dense hair and clothing. Larvae are found under rugs and among accumulated lint incorporating organic matter, including the contents of carpet sweepers and vacuum cleaners.

Range: Worldwide.

Food: Adult feeds on blood from humans, pigs, rodents, dogs, coyotes, cats, mules, and deer. Larva eats organic debris, scavenged from hosts' nests.

Life Cycle: Female scatters about 500 eggs during lifetime, which may be 18 months if well fed, 4 months if starved. Larvae emerge in 1–9 days and feed for 8–32 days. They pupate in cracks for 3–35 days but may wait months to emerge, aroused by vibrations from a passing potential host. Generations are produced at intervals of 4–6 weeks in the East, 9–11 weeks in the West.

These fleas can transmit bubonic plague after taking blood from an infected person. The Common Coyote Flea (*P. simulans*) of the Southwest differs in small details. It probably accompanied coyotes into the Northeast and may also attack dogs.

69 Oriental Rat Flea
"Common Flea"
(*Xenopsylla cheopis*)

Description: ¹⁄₁₆–¹⁄₈″ (1.5–3 mm). *Strongly flattened side to side.* Pale to dark reddish brown. Large bristle in front of almost circular compound eyes. No comblike spines below head or behind pronotum.

Habitat: Nests of rats and other rodents, and human habitations. Often common around wharves and ships.

Range: Worldwide.

Food: Adult sucks blood from rodents, humans, and occasionally other mammals. Larva scavenges for organic matter among litter.

Life Cycle: Eggs are scattered on host, hatching in 9–12 days. Larvae mature in 4–5 weeks. Well-fed adults live up to 1 year, but can survive without food for 1 month or more.

Probably a native of the Nile Valley, this flea spread with rats. In Europe during the 14th to 17th centuries, rat fleas carried the bubonic plague, known as the Black Death, which killed nearly half the European population. The disease disappeared when thatched roofs were replaced by tile, slate, or shingle —eliminating the rats' habitat.

Flies
(Order Diptera)

This order has over 86,000 known species, with about 16,300 in North America, and many more awaiting discovery. Prevalent in all habitats, flies are easily distinguished from other insects because they have only 1 pair of normal wings. The second pair, just behind the first, is represented by 2 knobbed organs, the halteres, thought to stabilize the body during flight. Many flies have a membranous lobe, or calypter, at the base of each wing overlying the haltere; its presence is a useful aid in recognizing certain families. Most flies have large compound eyes and mouthparts that are modified for piercing, lapping, or sucking fluids. The antennae range from short, 3-segmented organs to long, threadlike structures; they are feathery in midges and mosquitoes, clubbed in mydas flies. Flies exhibit complete metamorphosis. The larvae of most species are soft, legless, and headless. They are called maggots and live in soil, decaying material, or as parasites of vertebrates, snails, or other insects. The aquatic, mobile larvae of mosquitoes, midges, and certain other groups are more slender and have an obvious head. Some blood-sucking flies are carriers of diseases, such as malaria and yellow fever. Others, which feed on unsanitary substances, carry bacteria that cause such diseases as typhoid and dysentery. Some flies are agricultural pests. On the positive side, flies are valuable as pollinators of flowers, scavengers, and a source of food for wildlife; parasitic and predatory flies help control other insect pests. Identification of many flies requires attention to such technical details as wing venation, bristles, the presence or absence of certain segments, and the structure of the genitalia.

CRANE FLIES
(Family Tipulidae)

These gray or brown flies resemble giant mosquitoes, ⅜–2½″ (8–65 mm) long. They can be recognized by the long legs, which break off easily. They have 3 simple eyes and a V-shaped groove on the top of the thorax. Unlike mosquitoes, crane flies do not have a long proboscis and cannot bite. They differ from phantom crane flies mostly in wing venation. Adults are very common in a wide variety of habitats and frequently enter houses. The larvae called leatherjackets because of their tough skin, live in shallow water or moist soil and feed mainly on decaying plant matter.

408 Wood-boring Tipulid
(*Ctenophora vittata*)

Description: ⅝–1″ (17–25 mm). *Elongated.* Head and comblike antennae black. Thorax and abdomen reddish yellow to reddish orange with long black stripe down back. *Legs long,* reddish yellow with dark bands. *Wings transparent amber.*

Habitat: Wet woodlands and orchards.

Range: California to British Columbia.

Food: Adult eats little. Larva scavenges in wet decaying wood.

Life Cycle: Black eggs are deposited singly on decaying wet wood. Larvae tunnel into wood and, fully grown, pupate near bark surface.

This wood-borer increases the damage caused by rot in neglected or poorly pruned trees by enlarging the area of decay.

382 Giant Western Crane Fly
(*Holorusia rubiginosa*)

Description: 1–1⅜″ (25–35 mm), wingspan to 2¾″ (70 mm). *Large.* Reddish brown or olive-green to orange with some white markings on sides and under thorax. *Legs long.* Wings clear. Larva, to 2⅛″ (55 mm), is dull brown with tough skin.

Habitat: Humid areas, edges of forests, and streams.

Range: Southern California to British Columbia and Idaho.

Food: Adult does not eat. Larva scavenges on decaying plant matter.

Life Cycle: Eggs are laid on wet mud near open water, into which larvae creep. Larvae pupate in unlined cells in mud near waterline.

This is the largest fly west of the Rocky Mountains. Adults easily lose legs and cannot regenerate them. The Giant Eastern Crane Fly (*Pedicia albivitta*), ¾–1⅛″ (20–30 mm) with a wingspan to 3⅛″ (80 mm), has distinctly patterned broad brown bands on its wings.

406, 407 Crane Flies
(*Tipula* spp.)

Description: ⅜–2½″ (8–65 mm), wingspan to 3″ (75 mm). *Slender;* abdomen longer than thorax and head combined. Grayish brown to golden, depending on species. Antennae with many segments, threadlike or narrowly feathery in males. Thorax has deep V-shaped crease above and acute or round point between wing bases. Females of some species wingless. *Legs very slender,* usually twice as long as body. Females have sharp ovipositor. Larva, ½–1½″ (13–38 mm), is grayish to pale brown, depending on species.

Habitat: Humid areas and wet ground, often near streams or lakes in mud or wet moss.

Range: Worldwide.

Food: Adult does not eat. Larva feeds on decaying vegetation, fungi, roots, leaves of emergent and terrestrial plants, and, less often, animal matter.

Life Cycle: Slender eggs are usually laid in or on moist soil. Fully grown larvae pupate in soil or mud, where pupae usually overwinter. Adults emerge in spring. 1 or more generations a year.

Mating swarms of males "dance" above a bush or treetop waiting to seize females. Then each pair settles on foliage to mate. The larvae are often eaten by skunks and moles; adults are devoured by birds and bats. The Central Crane Fly (*T. cunctans*), ½–¾" (12–18 mm), has a gray head and thorax, and yellowish abdomen with a brown middorsal line. It ranges from New Brunswick south to Alabama, west to Colorado, north to Manitoba. The female California Range Crane Fly (*T. simplex*), ⅜–½" (9–13 mm), is grayish-brown and wingless with short legs.

PHANTOM CRANE FLIES
(Family Ptychopteridae)

Phantom crane flies are delicate, long-legged flies, about ⅜" (9 mm) long, found in shady woods and other damp places. They differ from true crane flies in wing venation and details of tarsal structure. Adults are often seen in small swarms drifting slowly across a forest clearing on a breeze. The larvae live in water or wet soil and breathe air through a long snorkel-like tube.

404 Phantom Crane Fly
(*Bittacomorpha clavipes*)

Description: ⅜–½" (9–13 mm). Long, thin. Black. *Legs delicate, long with white bands,* especially on swollen tarsi, which are concave below. Wings clear, black veins.

Habitat: Moist woods and stream margins.

Range: Eastern United States and Canada.

Food: Adult eats little or nothing. Larva feeds on organic matter.

Life Cycle: Eggs are deposited singly or in small clusters at the edge of fresh water. Larvae scavenge in shallow water, breathing air through a snorkel-like tube at tip of abdomen. Fully grown larvae pupate in moist soil. Adults are found May–September.

The Phantom Crane Fly soars slowly through the air with its legs extended. The swollen tarsi catch air currents, helping it stay aloft. When it flies into shade, only the white leg bands are visible, and the insect seems to appear and disappear like a phantom.

NET-WINGED MIDGES
(Family Blephariceridae)

Net-winged midges are slender flies, ¼–½" (6–13 mm) long. They have long, delicate legs and a network of fine lines between the wing veins, caused by the way the wings were folded in the puparium. Net-winged midges resemble mosquitoes but have threadlike, not feathery, antennae and lack a proboscis. They are found along swift-flowing mountain streams. The aquatic larvae and pupae live attached to submerged rocks. At the end of the pupal stage, the pupae break loose and float to the surface. Their casings then burst open, instantly freeing the adults, which fly off.

410, 411 Comstock's Net-winged Midge
(*Agathon comstocki*)

Description: ⅜" (10 mm). Spindlelike; thorax domed. *Dull yellowish gray with dark gray pattern.* Eyes large, black, not quite meeting. Antennae slender with short hair. Legs long, especially hind pair. Wings broad, clear; knoblike halteres long with flat, black triangular tips pointing toward body.

Habitat: Near fast-flowing streams.

Range: California to British Columbia.

Food: Adult does not eat. Larva eats algae.

Life Cycle: Eggs are laid on wet rocks beside fast streams. Larvae creep into the water, using special ventral suckers to move and cling to rocks. After pupal stage is completed, pupae float to the surface, burst open, and adults fly off.

Midges are most often seen resting on evergreen needles. This species was formerly included in the genus *Bibiocephala*.

33 PHANTOM MIDGES
(Family Chaoboridae)

These small, delicate flies are ⅛–½" (3–13 mm) long. They closely resemble mosquitoes, except they have a very short proboscis that is not used for biting and have wing scales only on the margins. Phantom midges thrive near ponds, lakes, and other bodies of water. They sometimes gather in great swarms that resemble columns of smoke. The larvae, to ¾" (18 mm) long, are aquatic predators of small swimming insects and crustaceans. They use their antennae as grasping organs. The name phantom midge comes from the fact that the larvae are transparent and difficult to see in the water unless light hits them just right. *Chaoborus* and *Corethra* are the principal genera.

31, 32 MOSQUITOES
(Family Culicidae)

Mosquitoes are slender, delicate flies, usually less than ¼" (6 mm) long, with slender legs. They are easily recognized by the long, sharp proboscis, which immediately distinguishes mosquitoes from midges, fungus gnats, and other similar insects. The male has very feathery antennae unlike the female's sparsely hairy antennae. The wings of both sexes are coated with delicate scales. At rest the wings are neatly folded over the abdomen. These common insects are best known for the biting habits of females, which must have a blood meal before they can produce eggs. Males do not bite and generally feed on plant juices. Mosquito larvae, called wrigglers, grow to ¾" (18 mm) long and the pupae to ¼" (6 mm). They live in quiet bodies of water, ranging from fluid-filled leaves of pitcher plants to stagnant ponds and lakes. Of all insects, mosquitoes are probably the most harmful to people, because they help transmit such serious epidemic diseases as malaria and yellow fever.

386 Summer Mosquito
(*Aedes atlanticus*)

Description: ⅛–¼" (3–6 mm). Thorax brown to dark brown with light stripe down center. Abdomen dark brown. Legs dark brown with femora lighter below. Wings have dark brown scales.

Habitat: Near shallow temporary pools.

Range: New Jersey to Florida, west to Texas, north to Kansas.

Food: Male sips plant juices; female takes blood of mammals, including humans. Larva feeds on microscopic aquatic plants.

Life Cycle: Eggs are dropped in temporary pools.

Larvae, active March–November in the South, feed, pupate, and emerge over an extended period. More than 1 generation a year.

The mosquito population increases after summer rains, because the number of places where they can breed multiplies.

389   Snow Mosquito
(*Aedes communis*)

Description: ¼″ (5–6 mm). Brown with dark brown scales and some golden-yellow to gray scales. Abdomen has dark and light bands above. Wings have brown scales.
Habitat: Forested regions with shade.
Range: Northern United States and Canada.
Food: Male drinks plant juices; female takes blood of mammals and birds. Larva feeds on microscopic algae in water.
Life Cycle: Eggs are dropped on pools of melting snow, hatching early in spring. Larvae develop slowly in cold water. Adults remain near sites of emergence. 1 generation a year.

Unlike most mosquitoes, female Snow Mosquitoes are active even in cold weather, when the snow is still on the ground.

380, 381   Golden Saltmarsh Mosquito
(*Aedes solicitans*)

Description: ⅛–¼″ (4–6 mm). Golden-brown; male unmarked, *female's abdomen with middorsal silvery-white stripe and silvery-white or golden-yellow bands.* Female's proboscis and tarsi have white and black bands. Thorax feathery. Male's antennae feathery, female's threadlike. Wings smoky, female's spotted.
Habitat: Near brackish and salt water; larvae also in swimming pools.

Range: New Brunswick to Florida, west to
Texas, north to Nebraska.

Food: Male feeds on plant juices; female sucks
blood from wild and domestic animals
and humans. Larva feeds on algae and
single-celled animals.

Life Cycle: Eggs are laid singly on vegetation or
other surface at waterline. Larvae hatch
and pupate in stagnant water.

The Black Saltmarsh Mosquito
(*A. taeniorhynchus*), ⅛" (4 mm), is black
with broad white bands, no middorsal
stripe, and a different pattern of white
on the proboscis and tarsi. It is found
in Massachusetts, south along the
Atlantic Coast, and west along the Gulf
of Mexico into southern California.

387 Tree-hole Mosquito
(*Aedes triseriatus*)

Description: ¼–⅜" (5–8 mm). Brown to black with
many silvery-white and dark brown
scales in streaks and patches on head
and thorax. *Abdomen appears to have dark
and light bands;* blue-black scales above,
patches of white scales along sides.
Hind femora yellowish near body, dark
toward tips; white spots on knee joints.
Wings have brown scales.

Habitat: Mature forests, in tree holes.

Range: Quebec to Florida, west to Texas, north
to British Columbia.

Food: Male drinks plant juices; female takes
blood of mammals and birds. Larva
feeds on microscopic aquatic plants
and algae.

Life Cycle: Eggs are deposited in water,
accumulated in tree holes, tubs, cans,
and other containers. Larvae are present
in most months in the South. Many
overlapping generations a year.

Of all mosquitoes that breed in tree
holes, this is the most widely
distributed.

384 Malaria-carrying Mosquitoes
(*Anopheles* spp.)

Description: ⅛″ (4 mm). Dark brown. Head bears patch of pale hair. Antennae narrowly feathery. Proboscis, palps, and thorax have dark scales. *Abdomen lacks scales* but may be hairy. Legs slender. Each wing has several dark patches of scales. *Upon landing, it does headstand with hind legs up in the air* and head pointing downward.

Habitat: Deciduous and mixed forests, and around human habitations. Larvae are found in small pools of rainwater.

Range: Throughout North America; individual species more local.

Food: Male feeds on plant juices; female takes blood from warm-blooded animals and humans. Larva feeds on microscopic algae.

Life Cycle: Eggs are laid singly on water surface film, hatching in 2–3 days. Larvae float parallel to surface, swimming to greater depths if disturbed. They pupate in about 2 weeks. Adults emerge after 2–3 days. Many survive winter weather.

These mosquitoes transmit infectious malaria, obtained from taking blood of infected people. The Eastern Malaria Mosquito (*A. quadrimaculatus*), same size, has 4 dark patches of scales on each fore wing and ranges east of the Rocky Mountains from Quebec and Nova Scotia to Florida, west to Mexico, and north to North Dakota. The Western Malaria Mosquito (*A. freeborni,* formerly believed to be identical with the European *A. maculipennis*), same size, has a bronze patch at the tip of each black-spotted wing. It is common from Montana to Texas, south to northern Mexico, north to British Columbia.

419 House Mosquito
(Culex pipiens)

Description: ⅛–¼" (4–5 mm). *Thorax light brown to brownish gray.* Abdomen banded white and brown above. Proboscis brown. Wings brown. Male's antennae more feathery than female's. Abdomen kept parallel to support although hind legs normally do not rest but are held raised at an angle.

Habitat: Near swamps, ponds, and other bodies of stagnant water.

Range: Throughout North America; subspecies more limited in distribution.

Food: Male drinks plant juices; female takes blood from birds and mammals, including humans. Larva feeds on microscopic algae.

Life Cycle: Eggs are deposited in raftlike masses of 100–300 on water surface film. They hatch in 1–5 days. Larvae feed head down in water. They pupate after 1–2 weeks. Adults emerge after a few days. Many generations are possible, the last overwintering as adults.

The Northern House Mosquito (*C. p. pipiens*), found in the northern United States and Canada, is the most common night-flying mosquito. The Southern House Mosquito (*C. p. quinquefasciatus*) is common in the Southeast, ranging west to California.

388 Unmarked Slender Mosquito
(Culiseta inornata)

Description: ¼–½" (5–13 mm). *Slender.* Male golden-brown with tip of abdomen exceptionally enlarged. Female dark with many yellowish and white scales. Abdominal tip slender, pointed. Wings dark.

Habitat: Near irrigation ditches, stagnant water, and seepage pools.

Range: New Hampshire to Florida, west to

California, north to Canadian Northwest Territories.

Food: Male drinks plant juices; female takes blood from birds and mammals, especially cattle and rabbits. Larva feeds on microscopic algae and bacteria.

Life Cycle: Eggs are deposited on surface film of water in sunny locations in spring. Larvae hatch in a few days, feed, pupate, and emerge in 3–5 weeks. Adults are abundant in spring and fall, seeking shelter in dry weather.

Females sometimes bite people at dusk or dawn, but usually are much less aggressive than other mosquitoes.

PUNKIES
(Family Ceratopogonidae)

Punkies are minute flies, 1/16–1/4″ (1–5 mm) long, that resemble tiny, short-legged mosquitoes. They differ from midges in folding their wings over the abdomen when at rest, rather than holding them out to the side. Females of a few species bite people and are intensely annoying despite their small size, but most species either attack other insects or feed on flower juices or pollen. They are most numerous around ponds and streams, where the eggs are laid in gelatinous masses on the water. The larvae are scavengers or predators in water or wet soil.

415 Bodega Black Gnat
(*Leptoconops kerteszi*)

Description: 1/16–1/8″ (1–3 mm). *Glossy black.* Antennae have 13 segments. Wings clear to slightly milky.

Habitat: Near tidal marshes and saline lakes with humus-rich soil just above waterline.

Range: Nebraska to California, north to
Washington.

Food: Adult drinks nectar. Female seeks
blood from birds and mammals,
including humans. Larva feeds on
organic matter and small insect larvae.

Life Cycle: Eggs are deposited on wet, humus-rich
soil. Orange larvae feed in wet soil,
tolerating moderate salinity. They
pupate close to surface. Adults are seen
April—June.

The larvae of this gnat were first
discovered along streams that empty
into the Bodega Bay, California,
inspiring their common name. Females
bite fiercely, and the painful bites often
swell. They are found in large numbers
in some places and are so annoying that
they frequently drive people away.

MIDGES
(Family Chironomidae)

A large and abundant group of flies,
midges are very similar to mosquitoes
and are often mistaken for them. They
are $\frac{1}{16}$–$\frac{3}{8}''$ (1–10 mm) long, and seem
more delicate than mosquitoes. Unlike
mosquitoes, they do not bite and have
no long proboscis. When at rest, they
hold their wings out to the side, rather
than folding them neatly over the back.
The antennae, especially the males', are
usually more feathery than those of
mosquitoes, and the wings have no
delicate scales along the wing veins.
Adults are numerous around water,
where many species congregate in large
and conspicuous swarms during the
mating season. The larvae live in water
or wet soil and are scavengers or
predators. In some species the larvae
have red hemoglobin in their body
fluids and thus are called
"bloodworms."

379 Green Midges
(*Tanytarsus* spp.)

Description: ¼–⅜″ (5–10 mm). Cylindrical; abdomen tapering to a point. Head small, projecting beyond cylindrical prothorax. *Pale green;* prothorax faintly marked with brown above. Antennae feathery with long hair on segments. Legs long, pale brown; front tarsi long. Wings milky, transparent.

Habitat: Meadows and woods, near wetlands.

Range: Most of North America.

Food: Wet organic matter, mostly in silt at bottom of quiet ponds, streams, or marshes.

Life Cycle: Males perform aerial dances in large swarms, often above a shrub or treetop, usually in the evening. Females enter swarm and are seized by males. Pairs drop below swarm to mate. Mated females thrust eggs through water surface film. Larvae wriggle through bottom silt to reach decomposing plant matter. At end of pupal stage, pupae float to surface and explode suddenly to release adults.

Midges do not bite but are often mistaken for mosquitoes, particularly when swarming. On cooler days they usually swarm in the late afternoon when they are warmed by the sun.

48 BLACK FLIES
(Family Simuliidae)

Black flies are small, rather squat, grayish or blackish flies with large, rounded wings and a distinctively humpbacked shape. They are ¹⁄₁₆–¼″ (2–5 mm) long. Adults are most common in spring and early summer. The females bite and in forested regions can be extremely annoying. The larvae, to ¼″ (5–7 mm) long, are found in streams with a high oxygen content.

They attach themselves to rocks by means of a suction disk at the end of the abdomen. Pupae are shaped like cornucopias and, like the larvae, are attached firmly to rocks. North American black flies do not transmit any serious diseases, but in tropical areas some species carry a roundworm parasite that can cause blindness in people.

416 Black Flies
(*Simulium* spp.)

Description: 1/16–1/8" (2–4 mm). *Humpbacked,* head pointing downward. Grayish brown to shiny black. Antennae thick often with many segments. Wings smoky to clear; veins near front margin heavy, others delicate.

Habitat: Near running water in forests, mountains, and tundra.

Range: Labrador south to Georgia, west to California and Mexico, north to Alaska.

Food: Male and female feed on nectar. Female sucks blood from birds and mammals. Larva is filter feeder, eating particles such as diatoms and bacteria.

Life Cycle: Eggs are laid on stones or leaves at the edge of rapidly flowing streams, or on the water surface itself. Larvae tumble into water. Fully grown larvae pupate in cocoons that coat rocks in water, resembling moss. Adults burst out, rise on a bubble of trapped air, and fly away in late spring and early summer.

Biting adults are the bane of the North Country and mountain resorts, particularly early in the season. Some species transmit waterfowl malaria, which accounts for up to half of the deaths of ducks, geese, swans, and turkeys.

MARCH FLIES
(Family Bibionidae)

March flies are rather hairy, blackish flies, often marked with yellow or red. They have short, thick antennae attached below the eyes, and relatively long legs with large femora and usually large "thorns," or "combs," on the undersides of the front tibiae. They measure ⅜–½" (9–13 mm) long. Adults characteristically fly slowly and seemingly laboriously. They are often seen in large numbers over fields and other open places, especially in spring and early summer. Because they fly within a few feet of the ground, they frequently foul the windshields of fast-moving cars. The larvae live in the soil, feeding on roots and decaying plant matter.

---

412 March Flies
(*Bibio* spp.)

Description: ⅜–½" (9–13 mm). Thorax bulging; abdomen flattened. *Male's eyes huge, divided into upper and lower part,* which has smaller facets; female's eyes smaller, head flatter, narrower. Antennae short, downcurved. Body shiny black to dark brown. *Wings clear or whitish with yellowish-brown veins and yellowish-brown patch* (stigma).

Habitat: Meadows, suburban gardens, and city parks.

Range: Throughout North America, some species more limited.

Food: Larva feeds on decaying plant matter around the roots of herbaceous vegetation, occasionally harming tubers.

Life Cycle: Female deposits egg mass of 200–300 eggs in loose soil, then dies. Larvae scavenge on roots, later pupate in soil. Adults emerge in early spring, but not necessarily in March.

March flies are clumsy fliers, often bumping into plants and even people. Because they swarm in mating dances in early spring, they are called March flies.

FUNGUS GNATS
(Family Mycetophilidae)

These slender, delicate flies, ⅛–⅜″ (3–9 mm) long, are similar to mosquitoes but lack the long proboscis and have simple, threadlike antennae and a humped thorax that partly conceals the head from above. The coxae are greatly lengthened, giving these flies the appearance of having an extra joint in each leg. The short-lived adults are common in damp places. The larvae feed on fungus and sometimes damage commercial mushrooms.

385 Fungus Gnats
(*Mycetophila* spp.)

Description: ⅛–¼″ (3–6 mm). Slender, *mosquitolike*. Brownish to grayish yellow, sometimes streaked and ringed with dark brown. Thorax hairy. Legs blackish, long, slender; *coxae greatly elongated*. Wings smoky.

Habitat: Moist dark woodlands and shaded valleys; indoors in potted houseplants.

Range: Throughout North America.

Food: Some adults feed on flowers. All larvae eat fungi, decaying wood, and other wet plant matter.

Life Cycle: Eggs are laid on or in food materials, where larvae feed. They pupate near surface of food. Adults emerge in summer outdoors, or in any season indoors.

Mosquitolike adults flit close to wet areas. The many species can only be distinguished by a specialist on the basis of details of wing venation, body bristles, and genitalia.

SOLDIER FLIES
(Family Stratiomyidae)

These rather stocky flies, ⅜–½" (9–12 mm) long, are often metallic or banded with yellow and black. They closely resemble hover flies or wasps but do not hover and almost entirely lack bristles on the body, which is covered with short, fine hair. These sluggish, inactive flies spend most of their time resting on flowers or foliage. Some larvae are aquatic; others live in dung or under bark. They feed on algae, decaying plant matter, or small insects.

519   Soldier Flies
(*Odontomyia* spp.)

Description: ⅜–½" (9–12 mm). *Head and thorax about equally wide;* abdomen broader, somewhat flattened. Coloration varies among species. Thorax is usually black above, yellow to yellowish green on sides. *Abdomen has black and yellow bands. Antennae L-shaped.* Wings clear.

Habitat: Woods and grassy fields near ponds and sluggish streams.

Range: Most of North America.

Food: Adult drinks nectar. Larva feeds on algae.

Life Cycle: Eggs are deposited at edge of water. Aquatic larvae stick tip of abdomen through surface film to take in air. Fully grown larvae pupate in mud on bottom or shore. Adults are active in late summer.

These handsome flies have sharply demarcated colors. Only a specialist can identify the different species, mostly from genitalia and details of color markings.

HORSE AND DEER FLIES
(Family Tabanidae)

These stout, broad-headed flies, ⅜–1⅛″ (9–28 mm) long, have bulging, often brightly colored eyes. Unlike most other flies, their flight can be silent, and they are best known for landing stealthily on exposed skin and delivering a painful bite. Yet only the females bite; the males feed mainly on nectar and pollen at flowers. Adults are especially common around ponds, streams, and marshes, where the predacious larvae live in shallow water or moist soil.

424 Deer Flies
(*Chrysops* spp.)

Description: ⅜–⅝″ (9–15 mm). Body somewhat flattened, head smaller than that of horse fly. Black with yellow-green markings on thorax and most of abdomen. Antennae cylindrical. *Eyes bright green or gold with zigzag or other patterns.* Hind tibiae have 2 spurs at tip. Wings have distinctive brownish-black pattern. Larva is yellowish white or greenish with brown rings.

Habitat: Deciduous and mixed forests, meadows, roadsides, and suburbs near water.

Range: Throughout North America.

Food: Male drinks plant juices; female sucks blood from mammals. Larva feeds on small aquatic insects.

Life Cycle: Shiny black eggs are laid in clusters on leaves of emergent plants just above water. Fully grown larvae pupate in mud at edge of water. Adults emerge May–August.

A deer fly circles over its intended victim before settling, then immediately bites. Some transmit bacteria that cause tularemia in rabbits, hares, and occasionally people. The most common species, the Callidus Deer Fly (*C. callidus*), has black on its thorax and V-shaped black marks on abdominal segments 2, 3, and 4. It pesters animals and people during June and July, from Maine to Florida, west to Texas, north to British Columbia.

429   American Horse Fly
(*Tabanus americanus*)

Description: ¾–1⅛″ (20–28 mm). Large, broad. Head tan to ash-gray between large green eyes and on rear surface. Antennae reddish brown. Thorax brownish to blackish with gray hair. Abdomen is blackish red-brown with short gray hair across rear margin. Hind tibiae do not have spurs. Wings smoky; brown to black near base.

Habitat: Near swamps, marshes, and ponds.

Range: Newfoundland to Florida, west to Texas and northern Mexico, north to Canadian Northwest Territories.

Food: Male eats pollen and nectar; female takes blood of large mammals. Larva preys on aquatic insects and other small animals.

Life Cycle: Egg masses are attached to plants overhanging fresh water, into which larvae drop. Larvae overwinter in muddy bottom 2 winters, then pupate in spring. Males are short-lived, but females may survive until fall.

When the female bites, the wound inflicted often continues to bleed for several minutes because the fly's saliva contains an anticoagulant that prevents clotting. A single animal may suffer a debilitating loss of blood if many of these insects attack it.

427 Black Horse Fly
(*Tabanus atratus*)

Description: ¾–1⅛″ (20–28 mm). Jet black.
Thorax has fine whitish, yellowish, or
black hair. Abdomen has bluish luster.
Hind tibiae do not have spurs. Wings
are brownish to black, unpatterned.
Larva is white with black bands.

Habitat: Meadows and open grasslands, near
marshy areas or slow streams.

Range: Quebec south to Florida and Gulf
states, west to New Mexico, north to
Pacific Northwest.

Food: Male drinks nectar; female sucks blood
from large mammals, especially cattle,
horses, mules, and hogs. Larva preys on
small aquatic insects.

Life Cycle: Female attaches egg masses to plants
overhanging fresh water. Larvae drop
into water, feed, and then overwinter in
mud for 2 winters, pupating in spring.
Males have very short life-spans;
females survive until fall.

This horse fly lands on its victim's
neck, head, or back, quickly slices the
skin with its bladelike mouthparts, and
sucks out blood. Some animals become
seriously weakened if they suffer
repeated attacks and loss of blood.

428 Three-spot Horse Fly
(*Tabanus trimaculatus*)

Description: ½–⅝″ (12–16 mm). Mostly brownish
or blackish. Thorax is brownish gray
with 4 longitudinal black to dark
brown stripes above, often incomplete.
*Abdomen brownish with gray area and
3 small beige triangular spots.* Eyes
brownish gray to black. Wings smoky
brown.

Habitat: Near ponds, marshes, and streams.

Range: Massachusetts to Florida, west to
Texas, north to Minnesota.

Food: Male drinks nectar; female takes blood

from mammals. Larva eats small
aquatic animals.

Life Cycle: Egg masses are attached to plants
overhanging fresh water. Larvae drop
into water, feed, and then overwinter in
mud for 2 winters. They pupate in late
spring. Males are short-lived; females
survive until fall.

Although deer, moose, and domestic
livestock are this horse fly's usual
victims, it readily attacks people.

SNIPE FLIES
(Family Rhagionidae)

These long-legged black, gray, or
brown flies are usually found resting on
leaves in shady woods or other places
with dense vegetation. They are ¼–⅜″
(5–15 mm) long and have large
rounded heads and short tapering
abdomens. Some adults feed on nectar
and honeydew, while others are
predators. Some are suspected of
sucking blood from frogs, small
mammals, and people. Larvae develop
in rotting wood or moist soil, where
they prey on small insects.

489   Gold-backed Snipe Fly
(*Chrysopilus ornatus*)

Description: ⅜″ (8–10 mm). Head and thorax black
with *golden hair on top of thorax,*
sometimes appearing as a central spot.
Abdomen is pale yellow or gold with
narrow black midline above and black
crossbands on each segment. Legs long
and slender, brownish or yellow.
Wings clear.

Habitat: Damp meadows and swamps.

Range: Maine south to Florida, northwest to
Kansas, north to Manitoba.

Food: Adult feeds on aphids and other small

insects. Larva preys on small insects in moss and decaying wood.

Life Cycle: Eggs are dropped in wet sites, where larvae are likely to find food. Fully grown larvae creep out of water to pupate under leaf litter.

This fly usually rests head downward and, when approached, runs or sidles rather than flies. The similar Two-spotted Snipe Fly (*C. thoracicus*), ⅜–⅝″ (8–15 mm), has black legs, smoky wings, and a black abdomen with 2 greenish-yellow spots on each segment. It ranges from Ontario south to North Carolina, and west to Illinois.

STILETTO FLIES
(Family Therevidae)

Stiletto flies—named for their slender, tapering abdomens—are hairy, rather inactive flies usually seen resting on foliage or flowers in open, sunny areas. They are ⅛–½″ (4–12 mm) long. Adults are similar to some robber flies but have slender legs and lack the "beard" of bristles on the face and the depression between the eyes. Little is known of their habits, but at least some adults are predatory. The long, slender larvae prey on insects in the soil.

513 Stiletto Flies
(*Thereva* spp.)

Description: ⅜″ (8–10 mm). *Carrot-shaped,* widest where wings arise. Neck slender, short. Eyes large in male, almost meeting on midline; smaller in female. Thorax black, coated with short yellow hair. Antennae and collarlike mesonotum have dark gray, hairlike scales. Abdomen has distinct segments with yellow and black bands. *Legs long,*

usually pale. Wings whitish or clear.
Habitat: Meadows and forest edges.
Range: Throughout North America.
Food: Adult may drink nectar. Larva preys on other insects.
Life Cycle: Eggs are deposited in soil, in decomposing vegetables, or in decaying wood.

Adults are usually seen in open areas, visiting summer flowers.

FLOWER-LOVING FLIES
(Family Apioceridae)

Flower-loving flies are found in western desert regions and get their name from the fact that the adults are usually found hovering over flowers and sucking the nectar like bees. A small family, the flower-loving flies measure ⅛–⅜″ (4–8 mm) in length and have 3 simple eyes, 3-segmented antennae, and a tapering abdomen. The proboscis may be short or long and thin. These flies resemble robber flies but differ from them in lacking a "beard" of facial bristles and a depression between the eyes. Adults fly noisily. The larvae, which live in sand, are probably predacious.

397 Flower-loving Fly
(*Apiocera haruspex*)

Description: ⅛–¼″ (4–5 mm). Spindle-shaped, like a robber fly. Beige to tan or gray with dark brown to black patterns. Grayish hair under abdomen. *Head small.* Legs long. Wings clear with brown veins in distinctive pattern.
Habitat: Gardens, pastures, and open country, on flowers.
Range: Wyoming south to Arizona, west to California, north to British Columbia.

Food: Adult sips nectar. Larva feeds on decaying vegetation, other fly larvae, and perhaps other insects.

Life Cycle: Eggs are deposited on soil. Larvae feed, later pupate in soil.

This fly resembles other flies that frequent flowers. Some members of this family lay their eggs in loose sand close to the stems of desert plants.

MYDAS FLIES
(Family Mydidae)

Only a few species of mydas flies occur in North America. All are easily recognized by their large size, 1–1½" (25–38 mm) long, nearly hairless bodies, and distinctly clubbed, 4-segmented antennae. They have 1 or no simple eyes. The adults are predatory and closely resemble wasps and robber flies. The slender larvae, which reach 1½" (38 mm) when fully grown, live in soil or rotting wood and prey on beetle larvae.

459 Mydas Fly
(*Mydas clavatus*)

Description: 1–1⅛" (25–30 mm), wingspan to 2" (50 mm) or more. Stout, cylindrical, *hairless*. Velvety black with *orange or red band* across top of 2nd abdominal segment. Eyes black, prominent. *Antennae long with 4 segments and club at tip*. Legs long, black with red and yellow marks on hind pair. Wings bluish or brownish, long and broad with long 4th vein that curves forward and almost reaches tip.

Habitat: Meadows and shrubby margins of deciduous forests.

Range: Throughout the United States and Canada.

Food: Adult eats caterpillars, other flies, bees, and true bugs. Larva preys on other insects in soil, especially June beetle larvae.

Life Cycle: Eggs are laid singly on soil. Larvae feed until fully grown, then pupate in cells close to soil surface. Adults fly July–August.

This fly may appear to be sluggish, but it is a rapid flier and skillful in capturing prey. The Western Mydas Fly (*Nemomydas pantherinus*) has larger eyes, heavier antennal clubs, clear colorless wings, yellow hair on the head and thorax, and yellow bands across the back. It is found from Colorado west to California, north to Idaho and British Columbia.

ROBBER FLIES
(Family Asilidae)

Robber flies are common, swift-flying predators, ¼–1⅛″ (5–30 mm) long. They have stout, spiny legs, a dense "beard" of bristles on the face, and a depression in the forehead between the 3 simple eyes, which are usually on a prominent tubercle. Many species are slender, having tapering abdomens, but some are stocky and closely resemble bees. Adults pounce on resting insects from above and use the short, strong proboscis to drain their prey's body fluids. The larvae are found in loose soil, fallen leaves, or decaying wood, where they feed on other insects.

398 Bearded Robber Fly
(*Efferia pogonias*)

Description: ½–¾″ (13–19 mm). *Humpbacked.* Head and thorax sandy to dark brown, with *white or yellow beardlike bristles.*

Male's abdomen steel-gray with last segments silvery; female's lighter gray with 2 black spots in middle. Legs spiny silvery gray with orange and black tibiae. Wings clear to smoky.

Habitat: Pastures and open fields.

Range: Southern United States east of the Rocky Mountains.

Food: Adult feeds on other flies, especially deer flies and horse flies, as well as flying ants, small bees, true bugs, grasshoppers, butterflies, and moths. Larva preys on white beetle larvae and other insects in soil.

Life Cycle: Eggs are laid on soil, into which larvae burrow in search of prey. They overwinter underground and pupate in spring. Adults fly August–September.

The Bearded Robber Fly often overtakes horse flies or deer flies in flight, whirring at high speed. It captures them and sucks them dry while standing on a leaf or twig. The largest robber fly encountered in North America is usually the California Bearded Robber Fly (*E. californicus*), ¾–⅞" (19–22 mm), which is black with white pile. This species was formerly assigned to genus *Erax*.

501 Sacken's Bee Hunter
(*Laphria sackeni*)

Description: ⅝–⅞" (15–22 mm). *Male black with yellow beard, yellow tibiae, and yellow bands* of hair on thorax and abdominal segments 5–7. Female similar, but abdominal segments 6–7 mostly black.

Habitat: Open fields, meadows, and gardens near flowers visited by bees.

Range: California to Oregon.

Food: Adult feeds on Honey Bees. Larva eats insects in soil.

Life Cycle: Eggs are deposited on decaying wood, into which hatchling larvae burrow and feed. Fully grown larvae or pupae

overwinter, emerging in early summer.
1 generation a year.

Sacken's Bee Hunter flies rapidly and
suddenly descends on an unsuspecting
bee. It seizes its victim on the thorax so
the bee cannot use its stinger.

---

400 Giant Robber Fly
(*Proctacanthus rodecki*)

Description: 1–1⅛" (24–30 mm). *Thorax heavy;*
abdomen slender, tapering. Body and
legs beige with *many stiff, blackish-
brown or beige bristles.* Legs also covered
with white hair. Wings clear with
brown veins.

Habitat: Pastures and open fields.

Range: Kansas, Colorado, and New Mexico.

Food: Adult eats bees and other insects. Larva
feeds on many kinds of soil insects,
roots, and decaying plant matter.

Life Cycle: Eggs are thrust into soil crevices.
Larvae usually require more than 1 year
to attain full size before pupating in
soil. Adults emerge in summer.

Giant Robber Flies rival dragonflies in
their effectiveness as predators.

---

401 Bee Killer
"Giant Robber Fly"
(*Promachus fitchii*)

Description: ¾–1⅛" (20–30 mm). Elongate.
Thorax hairy. Brown with yellow
bands. *Last abdominal segment has silvery
hair.* Legs have brownish-yellow or
black bristles. Eyes reddish brown.
Wings smoky with dark brown
veins.

Habitat: Meadows and near Honey Bee hives.

Range: Massachusetts south to Florida, west to
Texas, north to Nebraska.

Food: Adult eats bees and wasps, particularly

Honey Bees. Larva feeds on subterranean larvae of June beetles.

Life Cycle: Eggs are deposited on soil near grass roots. Larvae burrow into soil to find prey, later pupate in unlined cells in soil. Adults are active June–August.

The Bee Killer often rests on leaves and branches with a clear view of flowers visited by Honey Bees. It seizes its victim from above, pierces its body and sucks out juices, then drops the emptied prey. A dozen or more bodies may pile up on the ground below a favorite perch. The dark brown False Bee Killer (*P. bastardii*), same size, ranges somewhat farther north into Canada and preys on wild bees. Other species in this genus attack different insects, including grasshoppers and katydids.

399 Robber Flies
(*Tolmerus* spp.)

Description: ⅝–¾" (15–20 mm). Abdomen slender, tapering to tip. *Gray with black markings* on thorax and abdomen. Legs mostly black, with black bristles. Wings clear.

Habitat: Pastures, open fields, and gardens.

Range: East of the Rocky Mountains in the United States and Canada.

Food: Adult sucks body juices from many flying insects. Larva feeds on insect larvae, especially of beetles.

Life Cycle: Female presses abdomen into holes in the soil and deposits eggs. Larvae tunnel downward in search of prey and pupate in the soil close to surface. Adults fly July–September.

Adults in mated pairs often rest on leaves or flowers, flying off quickly if disturbed. The more vigorous takes the other by the tail and tows the mate, which makes no attempt to fly away.

BEE FLIES
(Family Bombyliidae)

These stout-bodied, furry flies, ¼–⅝″ (7–15 mm) long, frequently resemble small, long-legged bumble bees. Many species have elongated mouthparts that form a conspicuous beak, and others have boldly patterned wings. They are usually seen hovering motionless in midair, or resting with outstretched wings on flowers or on the ground in open, sunny places. Adults feed on nectar, but the larvae are parasites of beetle larvae, solitary wasps and bees, and other burrowing or hole-nesting insects.

496 Bee Fly
(*Anthrax analis*)

Description: ¼–⅜″ (7–8 mm). Velvety black. Head large with eyes above and forward rather than on sides. *Wings clear in outer ⅓;* remainder opaque black or brownish black, conspicuous when wings outstretched at rest.

Habitat: Meadows, open fields, and gardens with sandy soil.

Range: Quebec and New England to Florida, west to Arizona, north to Montana.

Food: Adult drinks nectar. Larva is a parasite of tiger beetle larvae.

Life Cycle: Eggs are deposited singly on ground close to burrow of tiger beetle larva. Each fly larva enters beetle burrow, attaches itself to larva and overwinters there. When host pupates in spring, fly larva sheds its skin, cuts into beetle pupa, and completes growth as internal parasite. Fully grown fly larva pupates, working its way to soil surface. Adult emerges in early spring.

The related Tiger Bee Fly (*A. tigrinus*), ½–⅝″ (12–15 mm), is purplish black with mottled black patterns on its

wings. It is found from the Canadian
Maritime Provinces south to Florida
and the Gulf states, and from California
to southern Alberta.

515, 516 Large Bee Flies
(*Bombylius* spp.)

Description: ¼–½" (7–12 mm). Mostly black.
Mouthparts slender, beaklike, almost as
long as body. Thorax and abdomen
densely covered with pile of long
yellow, brown, gray, and black hairs,
making fly appear *fuzzy like a bee.*
Wings clear with black patterns from
base along front margin.

Habitat: Meadows, open fields, and gardens on
flowers frequented by solitary bees.

Range: Most of North America.

Food: Adult drinks nectar. Larva is a parasite
in nests of solitary bees.

Life Cycle: Female fly follows solitary bee female
from flower to nest, waits until bee
departs, then lays eggs in entrance
tunnel. Fly larvae feed on bee larvae,
pupate in nest, and emerge as adults in
early summer.

Bee flies are capable of hovering
motionless while waiting for a female
bee but can dart quickly in pursuit.
They often settle on foliage or bare
ground, but are difficult to capture
because they are so alert and quick.

495, 512, Progressive Bee Flies
514 (*Exoprosopa* spp.)

Description: ¼–½" (7–12 mm), wingspan about
twice body length. Dark brown to
black, densely covered with pale hair.
*Abdominal segments often have broad gray
stripes* across bases, but rear margin of
each segment usually bare and black.
Legs, antennae, and projecting

mouthparts black. Wings smoky gray with black veins and vague brown mottling, or distinctly patterned with brownish black. Face usually projects forward conically. Claws have long, sharp basal tooth.

Habitat: Meadows or on bare ground over low vegetation.

Range: Throughout North America.

Food: Adult may sip nectar. Larva preys on robber fly larvae and larvae of some bees, wasps, and ants.

Life Cycle: Incompletely known.

These flies hover over low flowers but quickly fly away if disturbed.

DANCE FLIES
(Family Empididae)

These small bristly flies, less than ⅜" (9 mm) long, have a round head, a stout humpbacked thorax, and a distinct neck. Males have a conspicuous genital capsule under the abdomen, but unlike male long-legged flies, the capsule is not folded forward. Adults prey on small insects, often lying in wait for them on flowers. The more spectacular members of this family form mating swarms that dance up and down in the air—hence their common name. The larvae live in water or in decaying vegetation and are predatory.

390 **Dance Flies**
(*Empis* spp.)

Description: ⅛–¼" (3–5 mm). Head small with *distinct neck.* Thorax and abdomen bulbous, making fly appear *humpbacked.* Blackish to grayish or yellowish. Legs brownish or yellowish, long and slender. Eyes brownish orange. Wings clear or smoky.

Habitat: Meadows, open fields, and gardens, on foliage and flowers.

Range: Throughout the United States.

Food: Male in some species preys on feebler insects and drinks nectar; other males may not eat. Female only eats food brought by prospective mates. Larva feeds on small insects.

Life Cycle: Mating swarms of males fly up and down in ellipses waiting to capture females that fly into the swarm. In some species, males bring females "nuptial balloons" enclosed in a mesh of secretions. After mating, females lay eggs on soil, decaying vegetation, or along moist edge of standing water, into which larvae crawl. Larvae pupate in soil or on rotting vegetation, emerging as adults in spring. 1 generation a year.

The male in some species captures prey, wraps it in a frothy "nuptial balloon," and carries it with him under the abdomen until mating. Whether the balloon contains a large or small gift, or nothing at all, varies among species.

LONG-LEGGED FLIES
(Family Dolichopodidae)

These small flies, mostly less than ⅜" (9 mm) long, have rather long legs. They are often bright metallic green or coppery. Males have a conspicuous genital capsule at the end of the abdomen—this capsule is folded forward and underneath, unlike the capsule in male dance flies. The adults are common in damp woods and fields, and along streams, where they may be found running over leaves and visiting flowers in pursuit of their small insect prey. The larvae, also predatory, live in damp soil, rotting wood, or water. Males of some species perform elaborate mating dances in slow motion in front of the females. Some males have fans or disks of black or white scales on their legs, which they wave like flags in front of the females. Males in other species have patterned wings that serve the same function in courtship. Unlike many other courting insects, the male fly is in no danger of being eaten by the female, because the female's mouthparts are too small.

---

414 Condylostylid Long-legged Flies
(*Condylostylus* spp.)

Description: ¹⁄₁₆–¹⁄₄″ (2–5 mm), wingspan to ³⁄₈″ (9 mm). *Brilliant metallic green or blue.* Eyes dark red. *Legs very long.* Wings smoky, mottled or spotted in outer ¹⁄₂.

Habitat: Meadows and marsh edges on waterside vegetation.

Range: Throughout the United States.

Food: Adult preys on smaller insects and mites. Most larvae feed on aquatic organisms; some larvae that live under bark or in soil eat small insects.

Life Cycle: Incompletely known.

In this wide-ranging genus there are more than 40 species, which only a specialist can identify.

---

413 Texan Long-legged Fly
(*Condylostylus sipho*)

Description: ¹⁄₈–¹⁄₄″ (4–5 mm), wingspan to ³⁄₈″ (9 mm). Bright metallic green or blue. *Legs long.* Wings dusky, mottled or spotted on outer ¹⁄₂.

Habitat: Meadows on foliage.

Range: Southern Canada to Florida, west to Texas.

Food: Adult eats small soft-bodied insects and spider mites. Aquatic larva preys on small aquatic organisms.

Life Cycle: Incompletely known.

This fly superficially resembles members of the long-legged flies in the genus *Dolichopus*. The flies in this family are all very similar and are so small that the genera and species can be distinguished only by such microscopic details as wing venation, bristles, or genitalia.

HOVER FLIES
(Family Syrphidae)

Hover flies are a large and abundant group of flies. The coloring and movements of most species mimic bees or wasps—they are either stocky and covered with hair or boldly patterned in black and yellow. They measure ¼–¾" (5–18 mm) long and have large eyes that appear to cover the entire head. Each wing has a very characteristic fold, or "false vein," which can be noticed without the aid of a magnifying lens. It is located behind the first vein that reaches the outer margin just before the wing tip. Adults are usually seen hovering over flowers—hence the name. When they alight, many species quiver their abdomens. The larvae live in varied habitats. Some are aquatic, others scavenge in soil or decaying plant matter, still others feed on live plant tissue or attack living insects, and a few are parasites in the nests of ants or bees.

504 Woolly Bear Hover Flies
(*Arctophila* spp.)

Description: ½–⅝" (13–16 mm). Thick. *Black covered by brownish-yellow pile* suggesting that of bumble bee. Legs black; hind femora thick; both femora and tibiae are curved. Wings blackish.

Habitat: Meadows and forest edges.

Range: Colorado and New Mexico to
California, north to Alaska.
Food: Unknown.
Life Cycle: Unknown.

The Conspicuous Hover Fly (*A.
flagrans*), ½" (13 mm), has a conical
face. It is found in Colorado and New
Mexico. The Black-faced Hover Fly
(*A. harveyi*), ⅝" (16 mm), has black
pile on its thorax and a black stripe on
its face. It ranges from California to
British Columbia.

---

454 Elongate Aphid Fly
(*Baccha elongata*)

Description: ⅜" (9–11 mm). Slender with broad
head. Abdomen so narrow in 1st several
segments that insect appears *wasp-
waisted*. Eyes red. Thorax black.
*Abdomen reddish brown, crossed by golden
lines resembling those of wasps.* Wings
smoky to clear. Larva is greenish with
red stripe down back.
Habitat: Pastures and fields, on foliage and
flowers.
Range: Quebec to Virginia, west to California,
north to Alaska.
Food: Adult drinks nectar. Larva preys on
young aphids, mealybugs, or scale
insects.
Life Cycle: Female deposits eggs singly on foliage,
usually near aphids or other prey. Fully
grown larvae drop to the ground and
pupate on soil. Adults fly July–
September in the North, earlier and
later in the South. 1 generation a year.

The related Four-spotted Aphid Fly
(*B. clavata*) is green and reddish brown
with smoky wings and has 2 pairs of
spots on the last 2 abdominal segments
It ranges from New Jersey to Florida,
west to California, northeast to
Wisconsin.

508 Drone Fly
(Eristalis tenax)

Description: ⅝" (16 mm). Stout, *resembling Honey Bee but no waist* between thorax and abdomen. Dark brown to blackish. Thorax and 1st abdominal segment have short brownish-yellow hair; 2nd abdominal segment has 2 yellow blotches on sides, which do not meet in the middle; 3rd segment has narrow yellow band. Wings clear or slightly blotched with brown, held apart, slightly raised at rest. *Movements mimic Honey Bee.*

Habitat: Meadows and fields, on flowers, particularly daisies and other composites.

Range: Labrador to Florida, west to California, north to Alaska.

Food: Adult feeds on nectar and pollen. Larva scavenges organic matter.

Life Cycle: Sticky, elongate white eggs are laid in groups of 20 or more in crevices near foul-smelling stagnant water. Aquatic larvae breathe through exceptionally long snorkel-like tubes at tip of abdomens. Fully grown larvae creep to soil and pupate for 8–10 days. Adults appear in late June but disappear in cool weather.

Larvae, called rat-tailed maggots, are usually found on wet carrion and in open latrines. They are responsible for numerous cases of intestinal myiasis in people. Adult flies sometimes emerge from carrion, a phenomenon that was probably the basis for the myth that Honey Bees develop in dead mammals, as told in the Biblical story of Samson and the lion, and in writings of Ovid, Vergil, and Solomon. The adults so closely resemble Honey Bees that people and insectivorous animals avoid them; nevertheless the robber fly known as the Bee Assassin catches Drone Flies as readily as it does Honey Bees.

507 Bulb Fly
"Narcissus Bulb Fly"
(*Merodon equestris*)

Description: ⅜–½" (10–13 mm). Stout. *Black with bands of long yellow, orange, or tan hair,* resembling a small bumble bee. Occasionally nearly all black or all yellowish with fine hair. Legs black. Wings clear.

Habitat: Gardens, wherever narcissus and other spring bulbs grow.

Range: Throughout North America.

Food: Adult drinks nectar. Larva eats inner tissues of bulbs of narcissuses, hyacinths, tulips, lilies, and other plants.

Life Cycle: Small white eggs are laid near surface of soil on crown of host plant. Larvae feed at center of bulbs, usually 1 larva to a bulb. Larvae grow to ⅝" (15 mm) and overwinter inside bulbs, pupating near soil surface in spring. Some larvae overwinter 2 years. Pupae are large, oval, brown, and have 2 short breathing tubes. 1 generation a year.

A brown splotch at the base of a flower bulb is often a sign of infestation by Bulb Fly larvae. An infested bulb is usually soft. This is one of the few syrphid flies with a detrimental rather than beneficial impact.

487 American Hover Fly
(*Metasyrphus americanus*)

Description: ⅜" (9–10 mm). Stout. *Black to metallic green. Abdomen has 3 broad yellow crossbands* not reaching margins of abdomen. Face yellowish with black stripe and black cheeks. Wings clear.

Habitat: Meadows and fields on flowers and foliage.

Range: Throughout North America.

Food: Adult drinks nectar. Larva preys on aphids and larvae of scale insects.

Life Cycle: Elongated white eggs are laid singly on plants infested by aphids. Pale grayish, sluglike larvae feed, then drop to soil to pupate under debris. Adults are active June–August.

This fly gets its name from the way it hovers in the air above flowers. It is considered to be highly beneficial because its larvae help eliminate insects that attack ornamental plants and commercial crops.

488 Toxomerus Hover Flies
"Flower Flies"
(*Toxomerus* spp.)

Description: ¼–½" (7–13 mm). Mostly black. Face yellow. *Thorax has yellow stripe on sides. Abdomen has 3–4 sets of broken yellow bands.* Legs brownish yellow. Wings clear.
Habitat: Meadows on foliage and flowers.
Range: Throughout North America.
Food: Adult drinks nectar. Larva feeds on aphids.
Life Cycle: Chalky white eggs are laid singly on plants, usually near aphids. Larvae feed and, when fully grown, pupate in soil cavities. Adults emerge in summer.

This fly cannot bite or sting. Its larvae are probably as important as ladybug beetles in controlling aphid populations.

THICK-HEADED FLIES
(Family Conopidae)

These slender flies, ⅜–½" (10–13 mm) long, resemble wasps—the long abdomen narrows at the base and usually has black and yellow bands. The bulbous head has a long, slender proboscis, which may be elbowed.

These flies often have darkened wings, increasing their resemblance to wasps. Adults are usually seen on flowers, feeding on nectar and awaiting the arrival of a bee or wasp. When one of these insects approaches, the female fly seizes it in midair and deposits an egg on the surface of its body. The larva then develops as an internal parasite.

451 Thick-headed Fly
(*Physocephala texana*)

Description: ⅜" (8–9 mm). *Wasplike waist.* Head yellowish to tan. Body reddish to yellow-brown. Antennae bulbous at tip. Wings have dark areas. Knoblike halteres golden yellow.

Habitat: Chiefly in open brushlands near flowers.

Range: Throughout the United States.

Food: Adult drinks nectar. Larva is an internal parasite of sand wasps.

Life Cycle: Female uses pincers to hold prey while in flight, then lays 1 egg on victim. Larva attacks internally, finally killing its food source. Fully grown larva pupates in holes in soil.

Victims of thick-headed flies offer little or no resistance to being caught. They are held captive during the short flight while the egg is being attached.

PYRGOTID FLIES
(Family Pyrgotidae)

These medium-sized to large flies, ¼–¾" (7–20 mm) long, have large, unusually shaped heads that protrude considerably in front of the eyes. Their long bristly antennae are held forward at the tip of the head. Wings are long and usually have brown mottling. Mostly nocturnal, adults fly to artificial lights in pursuit of adult May beetles.

Female flies lay eggs on the abdomens of May beetles while they are in flight, and the fly larvae develop as internal parasites of the beetles.

441 Pyrgotid Flies
(*Pyrgota* spp.)

Description: ⅜–⅝" (9–15 mm). Robust. Yellowish brown. *Head triangular, large, projecting forward.* Antennae held forward, 1st and 2nd segments bristly. Legs long, brown. Abdomen held with tip curled downward. Female has conical ovipositor with sharp point. Wings mottled brown.

Habitat: Meadows near deciduous forests.

Range: Throughout the United States.

Food: Adult may eat nothing. Larva is internal parasite of May beetle.

Life Cycle: Mated female seeks May beetle adult, then pounces on it. This excites beetle into spreading its wings and exposing its abdomen, where fly inserts ovipositor and introduces a slender egg. Larva feeds inside beetle, attaining length of ⅜" (10 mm) in 10–14 days, finally killing the host and scavenging the remains. Larva pupates and overwinters inside host. Adult emerges in spring.

Flies apparently follow beetles to artificial lights and mate there. They are only moderately wary. Several species exist but they can be distinguished only by specialists on the basis of inconspicuous details. The smaller Black Pyrgota Fly (*Sphecomyiella valida*), ¼–⅜" (7–9 mm), is black with black-specked wings. Formerly known as *P. valida,* it ranges from Massachusetts to Florida, northeast to Minnesota and Ontario.

FRUIT FLIES
(Family Tephritidae)

Fruit flies are small to medium-sized insects, ⅛–⅜″ (4–9 mm) long, with elaborate and often colorful patterning on the body and wings. They are sometimes called peacock flies because of their habit of rhythmically waving their wings up and down when walking about on leaves. The larvae feed on flowers or fruit, or form galls in plant stems. A few species are serious pests in orchards, attacking apples, cherries, and citrus crops.

436, 509 **Walnut Husk Fly**
(Rhagoletis completa)

Description: ¼″ (5–7 mm). Brown. Thorax has yellowish-white spot on top, yellowish-white stripe on sides. Abdomen has blackish spots on sides of each segment. Eyes blue to bluish green. *Wings have brownish crossbands.*

Habitat: Deciduous forests, walnut and peach orchards.

Range: Minnesota to Texas, west to California, north to Oregon.

Food: Adult may not eat. Larva feeds on walnut husks and occasionally on flesh of cultivated peaches.

Life Cycle: White eggs are deposited 12–15 at a time on stalk end of developing fruit. Larvae burrow inside. When fully grown, larvae tunnel out, drop to ground, overwinter and pupate in litter or top soil. Adults fly July–September. 1 generation a year.

This pest discolors walnut shells and makes the flesh of peaches "wormy." It was discovered in Arizona in 1920 and in California in 1926.

435 Apple Maggot Fly
(*Rhagoletis pomonella*)

Description: ¼" (5–7 mm). Shiny black to tan with yellowish-white lines on sides of thorax and on triangular scutellum. Abdomen has yellowish-white crossbands. Head orange. Eyes red. Legs yellowish and grayish. Antennae have bulbous 3rd segment and long antennal bristle (arista). *Wings have black F-shaped bands with curving tails.*

Habitat: Orchards and edges of deciduous woods with native hawthorns, blueberry bushes, or western snowberries.

Range: Nova Scotia to Florida, west to Texas, and north to Alberta.

Food: Adult scrapes particles from leaves and fruit. Larva feeds on pulp of fruit.

Life Cycle: Eggs are inserted singly into fruit skin. Larvae penetrate fruit, leaving a brown stain or small depression that is sometimes visible through the skin. Fully grown larvae tunnel out, drop to ground, pupate in soil, where they overwinter. Adults emerge in July; some overwinter for another year.

Males perform courtship dances, waving their wings, while the females watch. The White-banded Cherry Fruit Fly (*R. cingulata*), ⅛–¼" (4–5 mm), is shiny black with a white head; white lines on the sides of the thorax; white crossbands on the abdomen; yellow on the scutellum, tibiae and tarsi; and a black crossband and oblique line on each wing. Several species attack ripening cherries.

SEAWEED FLIES
(Family Coelopidae)

These dark brown or blackish flies, ⅛–⅜" (4–9 mm) long, are easily recognized by the somewhat flattened thorax and abdomen, and the stout, bristly legs. They occur only along the seashore, where adults are common on flowers and gather in swarms to lay their eggs on decaying seaweed just above the high tide line. Both the adults and the larvae, which feed on seaweed, provide an abundant food supply for shorebirds.

420 Californian Seaweed Fly
(*Coelopa vanduzeei*)

Description: ⅛–¼" (4–6 mm). Body black. Male's abdomen has yellowish-red border on last 3 segments, narrower border in female. Mouthparts and antennae reddish. Legs black with reddish-yellow tibiae and spines; *brush of yellow hair halfway to tarsal tip.* Eyes red. Wings smoky.

Habitat: Seacoasts.

Range: California north to southern Alaska.

Food: Adult feeds on flowers. Larva feeds on decaying seaweed.

Life Cycle: Eggs are laid on decaying seaweed washed ashore by storm waves in late fall. Larvae pupate under remaining seaweed. Adults appear in early spring and fly until late fall.

Adults and larvae are active in quite cold weather. The similar Arctic Seaweed Fly (*C. frigida*), same size, is found along arctic coasts from Hudson Bay to Labrador, south to Rhode Island.

MARSH FLIES
(Family Sciomyzidae)

These slender, yellowish or brownish flies, ⅛–⅜" (4–8 mm) long, have fairly prominent eyes, rather long antennae that project forward from the head, and bristles on the femora. Their wings are mottled. These flies are common along the edges of ponds and marshes, where the larvae prey upon or become parasites of snails and slugs.

434 Marsh Flies
(*Tetanocera* spp.)

Description: ⅛–¼" (4–6 mm). Body and legs brownish yellow. Eyes red. *Antennae project forward prominently. Femora bristly.* Wings yellow to amber, usually with brownish spot near middle of front margin and with other markings.

Habitat: Near ponds, streams, and wet meadows.

Range: Throughout North America, more common in the North than the South.

Food: Adult may drink dew and nectar. Larva believed to prey on freshwater snails.

Life Cycle: Incompletely known.

This small fly runs about on marsh plants along the banks of woodland streams and ponds.

SHORE FLIES
(Family Ephydridae)

These blackish flies, up to ¼" (6 mm) long, are abundant along the edges of ponds and lakes and at the seashore. They differ from seaweed flies in being smaller and not flattened, having fewer bristles, and running on the surface of the water rather than swarming at seaweed. The adults are predatory, seizing small insects that have fallen on the surface of the water. The larvae feed on decaying vegetation in the water or tunnel into the stems and leaves of aquatic plants.

417   **Yellowstone Brine Fly**
       (*Ephydra thermophila*)

Description: ⅛–¼″ (4–6 mm). Grayish black.
               Thorax dark gray with *iridescent purple
               above,* slightly brownish below. Legs
               black. Eyes have purple reflection. Long
               fine hair on thorax. Wings smoky.
Habitat: Thermal springs.
Range: Thermal area of Yellowstone National
               Park, Wyoming.
Food: Adult eats small insects. Larva feeds on
               heat-tolerant algae in hot springwater.
Life Cycle: Masses of elongate orange eggs encased
               in mucus are deposited on limy
               projections of hot-spring terraces close
               to the water. Larvae enter hot water to
               find food, tolerating high temperature.
               Site of pupation unknown. Adults are
               active in May.

Closely related are the Gray Brine Fly
(*E. cinerea*), common near the Great
Salt Lake in Utah, and in Texas,
California, and Mexico; and the
introduced European Brine Fly (*E.
macellaria*), now found from
Massachusetts to Virginia, west to
California. Other species in this genus
frequent the seashore, scavenging on sea
animals and organic debris.

POMACE FLIES
(Family Drosophilidae)

The pomace flies are a large group of
rather robust little flies, ¹⁄₁₆–⅛″ (2–
4 mm) long, and usually yellow or
brown. They feed on yeasts found in
flowers or on rotting fruit, where they
lay their eggs. The larvae burrow into
the rotting fruit. These flies have
evolved a very short life cycle—an
adaptation to the fact that decaying
fruit seldom lasts very long in nature.
A new generation occurs every 10 days
to 2 weeks. A few species are very

important in studies of genetics and heredity because they are easily reared in the laboratory and have large families and short life cycles.

433 Vinegar Fly
(*Drosophila melanogaster*)

Description: ¹⁄₁₆″ (2 mm). Body short in proportion to oval wings. Brownish yellow with dark crossbands on last 3 abdominal segments in male. *Eyes bright red.* Legs brownish yellow; front tarsi of male have black "sex combs," important in courtship. Male's abdomen rounded, female's pointed. *Feathery bristle (* arista) *at tip of antennae.*

Habitat: Ponds, marshes, and swamps on wet decaying plant matter; on rotting fruit; also in homes.

Range: Worldwide.

Food: Adult drinks nectar and other sugary solutions. Larva feeds on yeasts in fermenting juices.

Life Cycle: Female lays up to 200 slender grayish-white eggs, each with 2 short respiratory tubes projecting above surface of moist food. Eggs hatch in 2 days. Larvae creep to drier sites and transform to adults in 4–5 days. Adults are ready to mate in 2 days and live 2 weeks. Surviving male may mate with its own daughters, but female seldom produces fertile eggs with sons.

These flies are so named because of their attraction to the sour odor of fermentation and bacterial waste. They are alternately known as "Pomace Flies" because the sour odor comes from pomace—the liquid squeezed from crushed fruit or seeds. Thomas Hunt Morgan at Columbia University made these flies famous by discovering genetic principles in their reproduction, for which he received the Nobel Prize in 1933.

ANTHOMYIID FLIES
(Family Anthomyiidae)

A large and varied group of flies, the anthomyiids have a lobe, or calypter, at the base of each wing. Most anthomyiids, ⅛–½" (3–12 mm), closely resemble the House Fly but are often more slender. They differ from tachinid flies by lacking a well-developed postscutellum. Unlike blow flies and flesh flies, they never have bristles both above and below the wing base on the side of the thorax. Details of wing venation distinguish them from muscid flies. The larvae of many anthomyiids feed on plant tissues; some are agricultural pests, while other species breed in dung and other decaying matter. A few larvae are aquatic. Anthomyiids were formerly classified as Family Scatophagidae.

---

440 Dung Fly
(*Scatophaga stercoraria*)

Description: ⅜" (8–10 mm). Slender. *Male yellowish brown to shiny gold,* densely covered with short upright hair; female duller with shorter hair. Legs are long, hairy, and have reddish-brown tarsi. Eyes red. Wings yellowish brown in front, clear behind.

Habitat: Barnyards and pastures near mammal manure.

Range: Newfoundland south to Georgia, west to Mexico, north to southern Alaska.

Food: Adult preys on House Flies and other adult insects. Larva feeds on dung.

Life Cycle: Eggs are deposited on manure, into which larvae tunnel. They often pupate in soil nearby. Adults are active all summer.

The first European colonists inadvertently brought this fly to the New World on their livestock.

MUSCID FLIES
(Family Muscidae)

Flies in this large family closely resemble its most familiar member, the House Fly. These stocky, alert flies, ⅛–½" (3–12 mm) long, have large eyes and a well-developed lobe, or calypter, at the base of each wing. They differ from tachinid flies in lacking a postscutellum. Unlike blow flies, flesh flies, and tachinids, they rarely have bristles above and below the wing base on the side of the thorax. They are most similar to anthomyiid flies, but differ in wing venation. Adults of some species prey on other insects, and a few bite people, but most eat and breed in decaying material of many kinds. Muscids are important because they spread disease-causing bacteria.

---

421 House Fly
(*Musca domestica*)

Description: ⅛–¼" (3–6 mm). Gray with 4 black lengthwise stripes on thorax. Abdomen gray or yellowish with dark midline and irregular dark markings on sides. Eyes reddish. *Antennal bristle* (arista) *slightly feathery. Mouthparts expanded at end,* suitable for sponging up food. Legs hairy; each tarsus has adhesive pads and sharp claws. Wings clear, held level and straight back.

Habitat: Near horse manure, garbage, or exposed food.

Range: Worldwide, except in Antarctica and a few remote islands.

Food: Adult sucks liquids containing sweet or decaying substances. Larva feeds on moist food rich in organic matter.

Life Cycle: Female lays 5–6 batches of 75–120 oval, white eggs on moist manure or garbage. Eggs hatch in 10–24 hours. Larvae reach full size in 5 days emerging as adults about 5 days later.

Males live for 15 days, females up to 26 if they have access to milk, sugar, and water.

98 percent or more of the flies caught in houses are House Flies. Because they can transmit typhoid fever, cholera, dysentery, pinworms, hookworms, and some tapeworms, House Flies are regarded as a greater threat to human health than most other insects. The larger and darker Face Fly (*M. autumnalis*), 1/4–3/8" (6–8 mm), resembles the House Fly but settles on cow rather than horse manure. It creeps into the nostrils and eyes of cattle and into horse fly wounds.

---

426 Biting Stable Fly
(*Stomoxys calcitrans*)

Description: 1/4–3/8" (6–8 mm). Gray with 4 indistinct darker stripes down top of thorax. *Mouthparts adapted for piercing, held projecting forward.* Eyes reddish; male's almost meeting above, female's wider apart. Abdomen mottled with black spots on gray background. Legs black. Wings clear with brown veins.

Habitat: Along coasts, in pastures and barnyards, near garbage dumps.

Range: Worldwide.

Food: Adult feeds on nectar and sucks blood of large animals. Larva eats fermenting vegetation, but not manure or garbage.

Life Cycle: Female attaches ivory-colored eggs to moist surface; eggs hatch in 1–3 days. Larvae develop in 15–30 days and overwinter as larvae or as pupae. In summer, pupal stages may last only 6 days. Several generations a year.

Adults bite fiercely, often on ankles, or low on the legs of domestic livestock. They are not known to transmit diseases among people, but carry infectious anemia to horses.

418 LOUSE FLIES
(Family Hippoboscidae)

Louse flies are leathery, flattened insects, $\frac{1}{16}$–$\frac{3}{8}$" (2–8 mm) long, that are external parasites of birds and mammals and are seldom found away from their hosts. When exposed on the fur of a sheep or deer, or among the feathers of a bird, louse flies have a curious, almost crablike sideways gait and are adept at avoiding capture. Many species that occur on mammals are wingless. Females develop eggs within the abdomen and give birth to larvae, or occasionally pupae, which are glued to the hair or feathers of the host. Both the adults and larvae suck blood.

73 Sheep Ked
"Sheeptick"
(*Melophagus ovinus*)

Description: ¼" (5–7 mm). *Body leathery*. Abdomen round, *ticklike*. Light reddish brown. Legs strong; femora swollen; tibiae have spines at tip and large hooked claws. Antennae concealed in pits. Sucking mouthparts are held under body except when used. Wings rudimentary; knoblike halteres absent.
Habitat: On sheep.
Range: Worldwide.
Food: Blood of sheep.
Life Cycle: Young are produced one at a time and retained in female until ready to pupate. Puparium is attached to the wool of host. Adult appears 19–24 days later.

Both sexes pierce skin to draw blood and may inflict painful stabs on sheep handlers. The somewhat egg-shaped puparium on the wool is often mistaken for an egg.

BLOW FLIES
(Family Calliphoridae)

Blow flies, ¼–⅝" (5–15 mm) long,
have a large, lobe-shaped calypter at the
base of each wing. They resemble flies
in four other families, but can be
distinguished by minute anatomical
features. Unlike muscid and
anthomyiid flies, they have a row of
stout bristles on each side of the thorax
just above the base of the hind leg and
another row of bristles just under the
base of the wing. Muscid and
anthomyiid flies rarely have both of
these sets of bristles. Unlike tachinid
flies, which have these bristles, blow
flies lack a postscutellum. They are very
similar to flesh flies but have 2 or 3
rather than 4 bristles on top of the
thorax, and a feathery bristlelike tip of
the antenna, called the arista. They are
also frequently metallic green or blue,
while flesh flies are blackish. Most blow
flies breed in carrion or other decaying
matter, and often gather in large
numbers at a carcass. A few species are
bloodsuckers or have larvae that live as
internal parasites of mammals.

430 Blue Bottle Fly
"Common Blow Fly"
(*Calliphora vomitoria*)

Description: ½" (13 mm). Head gray with large red
eyes, black sides, and reddish beard.
Thorax dark gray. *Abdomen metallic blue.*
Legs dark gray to black, bristly. Wings
clear.

Habitat: Pastures and barnyards, near decaying
meat, exposed flesh wounds, and dung.

Range: Greenland south along the Atlantic
Coast to Virginia, southwest to Mexico,
north to Alaska.

Food: Female feeds on wounds and rotting
meat. Larva eats juice from decaying
flesh.

Life Cycle: Eggs require 90 percent relative humidity to hatch; if enough moisture is present, hatching occurs almost immediately. Larvae grow rapidly and often crawl many feet to pupate in drier places, either in soil or in crevices of buildings. Adults emerge in 2–3 weeks. Many generations a year.

Females are attracted to meat and often enter open houses. They buzz loudly when they can't find an exit. When feeding in wounds, larvae keep flesh essentially sterile but retard healing. On live animals the infestation is known as myiasis. The related Red-headed Blow Fly (*C. erythrocephala*), same size, is very similar in appearance and behavior but has orange sides to its head and a black beard.

432 Screw-worm Fly
(*Cochliomyia hominivorax*)

Description: ½–⅝" (13–15 mm). Short, somewhat flattened. Shiny bluish green with 3 faint, lengthwise black stripes on thorax. Head broad with *yellow to reddish-yellow face*. Eyes large, reddish brown. Lapping mouthparts like those of House Fly. Larva, to ¾" (18–20 mm), is yellow-white. Wings clear with light brown veins.

Habitat: Pastures and cattle-grazing lands with short grass.

Range: New Jersey south to Florida, west to California, north to Montana.

Food: Skin and nasal membranes of grazing animals and humans.

Life Cycle: Masses of 10–400 eggs are attached to dry skin of host, especially on the umbilical cord of newborn calves, colts, and fawns. Larvae hatch in 12–24 hours, feed at the edge of a wound, then invade the tissue, creating an open sore that attracts other blow flies. Full-grown larvae crawl out of wound and

drop to soil, pupating below surface.
There are 8–10 generations a year.

This fly has been successfully
eliminated in many areas as a result of
releasing radiation-sterilized males in
great numbers for several generations.
Because the female mates only once, the
eggs of increasing numbers of females
are not fertile and reproduction is
blocked.

431 Green Bottle Fly
(*Phaenicia sericata*)

Description: ⅜–½″ (10–14 mm). Stout, larger than
House Fly. *Brilliant metallic blue-green or
golden* with black markings. Thorax has
3 cross-grooves and black bristlelike
hair. Antennae and legs black. Wings
clear with light brown veins.

Habitat: In dead fish and other dead animals; in
fresh, unprotected wounds (or wool of
sheep); and in manure and garbage
containing animal matter.

Range: Southern California to Mexico.

Food: Decomposing animal matter.

Life Cycle: Mass of up to 180 eggs is deposited on
carrion or garbage. Larvae attain full
size in 2–10 days, drop to the soil, and
burrow shallowly before pupating.
There are 4–8 generations a year, the
last overwintering as larvae in the soil.

The Green Bottle Fly is not as attracted
to the smell of fresh meat as is the Blue
Bottle Fly, and is less likely to enter
homes. It is often found near
slaughterhouses and garbage cans. The
similar Caesar Green Bottle Fly (*Lucilia
illustris,* formerly known as *L. caesar*),
¼″ (6 mm), is common in the West.

FLESH FLIES
(Family Sarcophagidae)

Flesh flies are common and conspicuous flies, ¹⁄₁₆–½" (2–14 mm) long. They have a well-developed lobe, or calypter, at the base of each wing, but differ from tachinid flies in lacking a large swelling, the postscutellum, underneath the scutellum on the thorax. Unlike muscid and anthomyiid flies, they have a row of bristles on each side of the thorax just above the base of the hind leg, and another row of bristles just under the base of the wing. Muscid and anthomyiid flies rarely have both sets of bristles. They most resemble blow flies but are never metallic, often having sharply contrasting black and gray stripes on the thorax. Flesh flies also have 4 rather than only 2 or 3 bristles on top of the thorax, and the bristlelike arista of the antennae is feathery only at the base. Most flesh flies breed in carrion, dung, and other decaying material. A few lay their eggs in open wounds of mammals —inspiring their common name. Some larvae are internal parasites of other insects.

425 **Flesh Flies**
  (*Sarcophaga* spp.)

Description: ¼–½" (6–14 mm). Ash gray as though dusty. Compound eyes forward-facing, dark red. *Thorax often with 3 stripes. Abdomen* above gray *with* black patterns and *reddish-brown tip;* abdomen below black. Wings clear with brown veins.

Habitat: Lake shores where dead fish may be found, garbage dumps, and wherever carcasses or excrement are exposed to daylight.

Range: Almost worldwide; rare or absent in South America.

Food: Larva feeds on decaying flesh, dead

insects, excrement, crabs, snails, and spiders.

Life Cycle: Eggs are deposited on or near suitable food. Larvae complete growth in a few days, burrow into soil to pupate and overwinter, emerging as adults in summer.

Several large flesh flies can lay so many eggs on a carcass that when the eggs hatch, the carcass is transformed into a squirming mass of larvae. The Eastern Flesh Fly (*S. sarraceniae*), 3/8" (10 mm), is gray with a broad black stripe on the thorax and most of the abdomen and with other black markings on the sides. It sometimes drops eggs inside the leaves of pitcher plants, where the larvae feed on drowned insects. The larvae then cut holes through the leaves to pupate in soil, draining and thereby ruining the pitcher. The Western Flesh Fly (*Blaesoxipha kellyi*), 1/4–3/8" (6–10 mm), is gray with black stripes above and with reddish brown below its fifth abdominal segment. It was formerly known as *S. kellyi*.

422, 423   TACHINID FLIES
(Family Tachinidae)

The tachinids are a very large group of stocky, active flies, 1/8–1/2" (3–14 mm) long, with a well-developed lobe, or calypter, at the base of each wing. Unlike muscid flies, blow flies, and other flies with calypteres, tachinid flies have a large and conspicuous swelling, the postscutellum, underneath the scutellum on the thorax. They have a row of stout bristles on the side of the thorax just above the base of the hind leg, and another row just under the base of the wing. Blow flies and flesh flies also have these two sets of bristles, but muscid and anthomyiid flies rarely have both. In most tachinids, the

bristlelike tip of the antennae, called
the arista, is bare. Females lay their
eggs on the bodies of other insects, and
the larvae live as internal parasites,
eventually killing the host. Tachinids
are important in the control of many
insect pests.

439 Beelike Tachinid Fly
(*Bombyliopsis abrupta*)

Description: ⅜–½″ (10–13 mm). Stout. Head tan.
Eyes reddish brown to black. Antennal
bristle (arista) longer than rest of
antenna. Thorax is black to yellow. *Legs
and abdomen are orange with long, strong
black bristles,* in 3 tracts on abdomen.
Wings slightly smoky.

Habitat: Meadows and fields near flowers.

Range: Nova Scotia to Georgia, west to Mexico
and California, north to British
Columbia.

Food: Adult drinks nectar. Larva is internal
parasite of moth or butterfly caterpillar.

Life Cycle: Female deposits 1 or 2 eggs near or on a
suitable host. Larvae penetrate
caterpillar's body, often through a
breathing pore, and feed. Fully grown,
larvae emerge from host and drop to
ground to pupate. Adults fly June–
September and are most abundant in
midsummer.

This fly might be mistaken for a small
bumble bee but is much less active.
The 3 tracts of black bristles on its
abdomen leave bare spots that are quite
unlike the hair on bumble bees.

500 Early Tachinid Fly
(*Epalpus signifer*)

Description: ¼–⅜″ (6–10 mm). Head pale grayish
tan, hairy. Eyes brownish red.
Antennal bristle (arista) about as long

as rest of antenna. Thorax grayish with dark tan stripes above. Abdomen black except for *striking large yellow abdominal spot; abdomen has long black bristles,* especially prominent at rear. Wings amber.

Habitat: Fields and meadows near flowers.

Range: Nova Scotia to Georgia, west to California, north to British Columbia.

Food: Adult drinks nectar. Larva is internal parasite of moth or butterfly caterpillar.

Life Cycle: Female places 1 or 2 eggs on undersurface of caterpillar host, beyond reach of its jaws. Fly larvae burrow into host and feed. Fully grown larvae drop from host to pupate in soil. Adults fly late spring—early summer.

This fly is rather inactive but it is very alert and will fly away swiftly if approached.

437 Repetitive Tachinid Fly
(*Peleteria iterans*)

Description: 3/8–1/2" (8–13 mm). Head tan. Eyes reddish brown. Antennal bristle (arista) distinctly shorter than rest of antenna. Thorax grayish black. *Abdomen mostly dark reddish yellow with 2 dark middorsal spots, dark rear, and long sparse black bristles* that are not in obvious tracts. Legs are black. Wings are slightly smoky.

Habitat: Fields and meadows near flowers.

Range: Nova Scotia to Massachusetts, west to South Dakota, southwest to California and northern Mexico.

Food: Adult drinks nectar. Larva is internal parasite of moth or butterfly caterpillar.

Life Cycle: Female places eggs on a plant close to caterpillar and often lingers a few hours as though waiting for eggs to hatch. Larvae penetrate host and feed. Fully grown larvae emerge from host and drop to ground to pupate. Adults fly late June—August.

The color pattern formed by the contrasting black abdominal bristles aids greatly in recognizing this wary fly.

438 Tachina Fly
(*Trichopoda pennipes*)

Description: ¼–½" (7–13 mm). Thorax yellowish to brown with black patterns. Abdomen yellowish to amber. Eyes reddish. *Hind legs hairy.* Wings smoky black, yellowish near body and margin transparent.

Habitat: Meadows and crop fields.

Range: Maine to the Gulf states, west to California, north to Washington, east to Ontario.

Food: Adult may drink nectar. Larva is internal parasite of true bugs.

Life Cycle: Adults mate near flowers. Females attach eggs to medium-sized to large true bugs. Larvae enter host, feeding first on less essential tissues and later killing host. Those maturing in autumn pupate inside the host, emerging as adults in spring; others pupate in soil, emerging summer or fall.

This tachinid fly is often encouraged for biological control of squash bugs, stink bugs, leaf-footed bugs, cotton stainers, and other plant bugs.

485 Fringe-legged Tachinid Fly
(*Trichopoda subdivisa*)

Description: ¼" (5–7 mm). Thorax black with collar of yellow stripes near head. Abdomen honey-yellow to brown with black stripes. Eyes large, separated by velvety yellow hair. *Hind legs hairy.* Wings clouded amber.

Habitat: Meadows and forest glades.

Range: Throughout the United States.

Food: Adult sips nectar. Larva is internal
parasite of true bugs, especially stink
bugs.

Life Cycle: Eggs are laid on true bugs. Larvae bore
into host, eventually killing it. Larvae
that mature in autumn pupate inside
host, emerging as adults in spring.
Others pupate in soil and emerge in
summer or fall.

Fringe-legged Tachinid Flies are usually
found on flowers, where they search for
true bug hosts.

Caddisflies
(Order Trichoptera)

Caddisflies resemble moths in that their wings are often shaped and colored like those of certain moth species. But caddisflies lack the coiled proboscis of moths and usually have minute hair rather than scales on the wings and body. The Greek order name means "hairy wings." Like some moths, all caddisflies hold their wings rooflike over the body at rest. Most are poor fliers, flying to lights at night and hiding by day. These insects have slender legs with spurs and 2 strong claws, and long, multisegmented antennae. The chewing mouthparts have well-developed palps but reduced jaws.

Female caddisflies drop masses or strings of hundreds of eggs into fresh water, or they attach them to vegetation overhanging water, into which larvae drop. Eggs hatch in a few days but most larvae need a year to develop. Adults live only a month and rarely eat. All larvae, called caddisworms, are aquatic and undergo complete metamorphosis. The caterpillarlike body has a pair of hooklike appendages at the rear and filamentous gills on each abdominal segment. Some larvae construct portable cases around their bodies, which later become pupal shelters. Others creep freely over rocks in rapid streams in search of prey. Still others spin silken nets, which they attach to rocks in fast-moving streams, and eat plant food caught inside. All larvae pupate underwater. There are 1,000 species of caddisflies in North America divided into 18 families, and 4,500 species worldwide.

46 LARGE CADDISFLIES
(Family Phryganeidae)

This family contains some of the largest caddisflies, ½–1" (14–25 mm) in length and with a wingspan of ⅝–2" (15–50 mm). Most have mottled gray and brown wings. Males have 4-segmented maxillary palps, while those of females have 5 segments. Larvae, ¾–1⅛" (20–27 mm) long, construct cylindrical cases up to 2" (50 mm) long from leaf or twig fragments, which they cement together in a spiral pattern. They live mostly in ponds, marshes, and other quiet waters.

335 Ash-winged Large Caddisfly
(*Phryganea cinerea*)

Description: ¾" (18–20 mm); wingspan to 2" (50 mm). Body brown with gray hair. Fore wings have gray and brown hair in regular patterns forming *triangular patches at midline when wings are folded at rest*. Tibiae have 2 spurs at tip of fore legs, 4 at or near tip of middle and hind legs. Larva has yellow head, pronotum, and legs.

Habitat: Adult is found on foliage and tree bark close to ponds and lakes with shallow margins. Larva lives in ponds and quiet bays of lakes and rivers.

Range: East of the Rocky Mountains across northern states and Canada.

Food: Adult does not eat. Larva feeds on algae.

Life Cycle: Ring-shaped mass of eggs in mucus is attached to branch of aquatic plant just below water level. Larvae construct tubular cases of leaf fragments, cemented together in spiral pattern. They pupate in closed larval cases.

Other species, both larger and smaller, have different markings that can be distinguished only by fine details.

NORTHERN CADDISFLIES
(Family Limnephilidae)

Northern caddisflies, ¼–⅞" (7–23 mm) long, are mostly brown with patterned or mottled wings. Males have 3-segmented maxillary palps; those of females have 5 segments. These caddisflies are found near ponds and quiet waters, where the larvae, ⅝–¾" (15–20 mm) long, construct cases up to ¾" (20 mm) long. Cases made by older larvae do not resemble those of younger larvae.

337 Betten's Silverstreak Caddisfly
(*Grammotaulius bettenii*)

Description: ¼–⅞" (7–23 mm); wingspan 1⅜–1⅝" (35–41 mm). Body and wings dark yellowish brown. Head and pronotum raised into yellowish warts with white hair. Antennae yellowish except for 1st antennal segment, which is greenish white and as long as head. Legs pale yellowish; fore femora bear 2 black spines at tip; fore tibiae have about 20 yellow or black spines; fore tarsi spiny below. *Wings have short silvery streaks.*

Habitat: Near freshwater ponds and slow broad streams, on foliage, twigs, and tree bark.

Range: Oregon to British Columbia.

Food: Adult does not eat. Larva feeds on algae and organic debris.

Life Cycle: Eggs are deposited singly through surface film at edge of water. Larvae gather fragments of dead leaves and twigs, cement them with salivary silk until their bodies are encased in cylindrical tubes which they drag everywhere while feeding from the front opening. They molt several times inside tubes and, when fully grown, close tube ends and pupate inside. Adults break through tubes, rise to the water surface, and fly off.

This species was named in honor of Dr. Cornelius Betten of Cornell University, whose book on the trichoptera of New York State appeared in 1934.

47 LONG-HORNED CADDISFLIES (Family Leptoceridae)

These slender caddisflies have very long, threadlike antennae, which are often twice the length of the fore wings. Most are pale and ¼–¾" (5–20 mm) long. They have 2 warts separated by a deep groove on the pronotum. The larvae, ¼–½" (5–12 mm) long, have a pair of bar-shaped, hard plates above the thorax. They live in quiet or running water and construct cases of sand or twigs up to ¾" (20 mm) long.

338 Tawny Brown-marked Longhorn (*Triaenodes* spp.)

Description: ⅜–½" (9–12 mm). Slender. Pale fore wing has pattern of darker brown on outer third; at rest, dark brown meets at midline. Pale antennae with conspicuous dark bands at each segment on first fourth.

Habitat: Spaces at upper levels among submerged, dense aquatic plants.

Range: Most American species occur in East; a few range to West Coast.

Food: Fragments of vascular plants.

Life Cycle: Female flies over water, washing off extruded eggs that separate, sink, and hatch quickly. Larva builds tapering case of plant fragments in lengthwise spiral, to 1¾" (33 mm) long; propels self and case freely with hindmost legs.

A related species that occurs in Europe is a pest of cultivated rice in Italy, where it tolerates water temperatures as high as 87°F (33°C).

Butterflies and Moths
(Order Lepidoptera)

Admired for their often magnificently
colored wings, butterflies and moths
constitute a large order with at least
125,000 known species worldwide—
including 12,000 species in North
America—and many more awaiting
classification. These insects have 4
membranous wings covered with
delicate pigmented or prismatic scales
that rub off easily. The mouthparts of
most adults form a long, coiled tube,
or proboscis, used for drinking liquids
such as nectar or fermenting tree sap.
In some species the adults do not feed,
and the proboscis is reduced or absent.
A very few primitive moths have jaws
and feed on pollen. Although they are
all quite uniform in structure,
butterflies and moths vary greatly in
wingspan, ranging from $\frac{1}{8}''$ to $10\frac{5}{8}''$
(3–270 mm).
Butterflies fly only during the day and
tend to be brightly colored with
knobbed antennae. Most hold the
wings together vertically over the
thorax at rest. The majority of moths
are night fliers and are colored in
generally somber hues, but some moths
fly by day and have brighter colors.
Moths have various types of antennae.
Unlike butterflies, moths hold the
wings rooflike over the body, curled
around the body, or flat against a
support when at rest.
Metamorphosis is complete. The larva,
or caterpillar, has a well-developed head
with powerful jaws, functional legs on
the thorax, and up to 5 pairs of
abdominal prolegs—leglike structures
that disappear during pupation. Most
caterpillars feed on plant tissue but a
few eat dried animal matter, and a still
smaller number are predators. The pupa
of a moth may be enclosed in a silken
cocoon, or it may lack a cocoon and lie
hidden in leaf litter, in an earthen cell,

or in a cavity within a plant stem. The pupa of a butterfly, called a chrysalis, lacks a cocoon and is attached to the larval food plant or some nearby protective support. While some species are agricultural pests, most moths and butterflies are harmless. Some are valuable as pollinators of flowers and others as the source of commercial silk.

Incurvariidae

FW and HW: venation
fairly complete.
HW: subcosta with
strong basal fork.

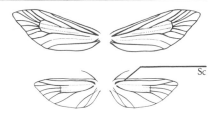

Sc

Tineidae

FW and HW: wing
membranes without
spines.
FW: R5 runs to
costa or apex.

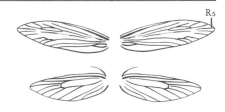

R_5

Psychidae

FW: 1A and 2A either
fused or joined by a
cross vein.
HW: nearly as wide
as long.

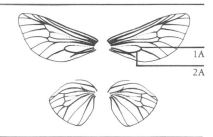

1A

2A

Gelechiidae

FW and HW: 1A
absent.
FW: R4 and R5 stalked
at base.
HW: outer margin
concave with apex
pointed.

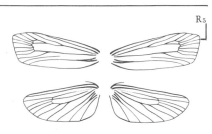

R_4

R_5

Yponomeutidae

FW and HW: all veins
well separated.
FW: R5 runs to
outer margin.

R_5

Sesiidae

FW: anal veins much reduced or absent.
HW: anal veins well developed.

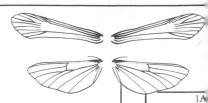

1A
2A
3A

Tortricidae

FW: widened at base with 1A absent.
CU₂ arises in inner ¾ of discal cell.

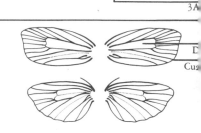

D
Cu₂

Hesperiidae

HW: frenulum absent; expanded humeral angle.

Papilionidae

HW: only 1 anal vein (2A).

2A

Pieridae

FW: M₁ stalked with radius.

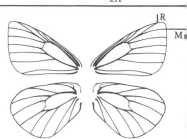

R
M₁

ycaenidae

W: M$_1$ unstalked;
adius never more
han 4-branched.
HW: costa not
hickened; no
umeral vein.

R$_1$ R$_2$ R$_3$ R$_4$ + R$_5$

M$_1$

C

Riodinidae

HW: humeral vein
resent; costa
hickened to humeral
ngle.

C
h

Libytheidae

W: distinctive wing
hape with outer
margin deeply
otched.

Nymphalidae

W: nearly triangular
with radius 5-branched;
ubitus appears
-branched; 3A absent.
HW: 2 anal veins.

R$_1$ R$_2$

R$_3$
R$_4$
R$_5$
M$_3$
Cu$_1$
Cu$_2$

3A   2A

Apaturidae

HW: humeral vein
resent; discal
ell open.

h
D

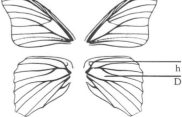

Satyridae

FW: vein bases
swollen.

Danaidae

FW: 3A short.
HW: 2 anal veins.

3A

2A
3A

Zygaenidae

HW: subcosta and
radial sector fused
toward outer end of
discal cell; 2 or 3 anal
veins.

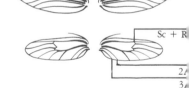

Sc + R

2A
3A

Limacodidae

FW and HW: all anal
veins retained.
FW: cubitus appears
4-branched.
HW: subcosta and
radius fused near
base.

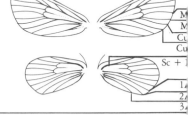

M
M
Cu
Cu
Sc + R

1A
2A
3A

Pyralidae

FW: 1A absent.
HW: subcosta and
radius close or
fused; 3 anal veins.

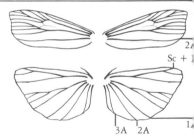

2A
Sc + R

1A

3A   2A

Pterophoridae

FW and HW: deeply cleft wings and peculiar venation.

Geometridae

HW: reduced anal veins; sharp angle and brace vein at base of subcosta.

Sc

2A

Lasiocampidae

FW: Cu2 arises in basal ½ of discal cell.
HW: frenulum absent, replaced by expanded humeral lobe.

D
Cu2

Saturniidae

HW: subcosta and radial sector not connected and humeral angle expanded; frenulum absent.

Sc
Rs

Sphingidae

FW: 3A strong, running into 2A.
HW: subcosta and radial sector closely parallel, connected by oblique cross vein.

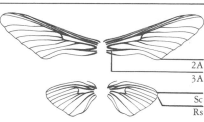

2A
3A

Sc
Rs

Notodontidae

FW: M₂ arises from
mid-end of discal cell;
cubitus appears
3-branched.
HW: Sc and R₁ and Rs
close together and
parallel; Rs and
M₁ stalked beyond
the discal cell.

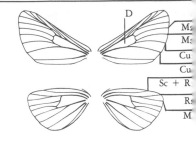

Arctiidae

HW: subcosta and R₁
fused along a portion
of discal cell (at least
⅛), never beyond
end of cell; cubitus
appears 4-branched.

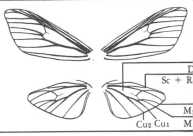

Ctenuchidae

HW: subcosta absent; 2
anal veins.

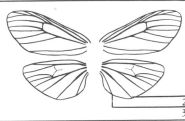

Lymantriidae

HW: subcosta and
radius connected by
cross vein, or
touching at a point
more than ⅓ out on
discal cell; cubitus
appears 4-branched.

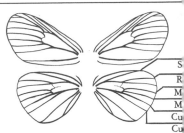

Noctuidae

FW and HW: cubitus
appears 4-branched.
HW: subcosta and
radius touching or
shortly fused less than
⅓ way out on discal
cell.

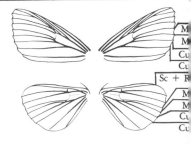

YUCCA MOTHS AND KIN
(Family Incurvariidae)

These small moths have wingspans of 1⅜" (35 mm) or considerably less. Many are black or metallic blue, sometimes marked with gold or silver; others are pure white. Some species have threadlike antennae that may be 4 times as long as the wings. The wing membranes are covered with tiny hairlike spines under the scales. This family includes both day-flying and night-flying species. The caterpillars feed on foliage or seeds. Many are leaf miners, while others build cases of leaf fragments.

522 Fairy Moths
(*Adela* spp.)

Description: Wingspan ½–⅝" (12–16 mm). Antennae of male much longer than wings. Fore wings variously colored: shiny black or greenish to purplish bronze with white bands and spots, or yellowish brown with rows of black dots and silver bands.

Habitat: Edges of shady woods.

Range: Throughout North America.

Food: Caterpillar feeds in flowers or seeds of milkweed and other plants when young, later on foliage.

Flight: April–July in the North.

Fairy moths are easily recognized by their incredibly long, delicate antennae. The caterpillars live in cases, which they make from oval pieces of leaves.

530 Yucca Moth
(*Tegeticula yuccasella*)

Description: Wingspan ¾–1" (19–25 mm). Wings unspotted white; undersides and

bottom margins of fore wings bronzy.
Maxillary palps very long, held curled
backward below head.

Habitat: Arid areas, wherever yucca grows.
Range: New York to the Midwest, south to the
Southwest and Mexico.
Food: Caterpillar eats seeds of yucca plants.
Flight: May–July.

Without attendant moths, yucca plants
cannot set seeds and can reproduce only
by root offshoots. The female moth uses
its maxillary palps to form the sticky
pollen of yucca flowers into a ball-
shaped mass, which it carries to another
yucca flower. Female inserts a few eggs
through the side of the ovary, pressing
the pollen into the cup-shaped stigma.
This ensures that the plant will be
cross-pollinated and produce a
maximum crop of future seeds for the
caterpillars to eat. There is 1 generation
a year. The Yucca Moth and its
caterpillars should not be confused with
the False Yucca Moths (*Prodoxus* spp.),
wingspan $\frac{5}{8}-\frac{3}{4}''$ (15–19 mm), which
lack the long maxillary palps. Their
caterpillars feed in the flower stems and
fleshy parts of yucca fruits.

534 CLOTHES MOTHS AND KIN
(Family Tineidae)

The tineids are small, pale moths, often
with a dull golden sheen on their
wings. The threadlike antennae have a
ring of scales on each segment, and the
head bears a tuft of dense bristly hair.
These moths have wingspans of $\frac{3}{8}-1\frac{3}{8}''$
(8–35 mm). The adults are short-lived
and do not feed; the palps are short,
and the proboscis is small or absent.
The caterpillars of many species live in
movable cases made of silk and bits of
food material, or in silk-lined tubes in
or near a food source. They feed on
leaves, fungus, vegetable fibers, or hair.

This family includes over 130 North American species, but is named for only 3 species that feed on woolen materials.

10 BAGWORM MOTHS
(Family Psychidae)

These moths are best known for the cocoons the caterpillars spin and carry around with them while feeding. The cocoons are covered with bits of leaves, grass, or twigs. At the end of the larval stage the caterpillars fasten the cocoons to branches and pupate inside. Inconspicuous adults emerge and fly briefly in summer and fall. Males are small or medium-sized and have blackish or transparent wings with wingspans of ⅜–1" (8–25 mm). In most species the females are wingless and legless and never leave their cocoons. Adults do not feed. After fertilization, the female lays eggs in the cocoon and dies. The cocoon, which has served to protect the caterpillar and house the pupa, now becomes an overwintering egg case. At least 1 species is a serious pest, frequently causing defoliation of shrubs and trees in urban areas. Caterpillars of most species measure ⅝–1⅝" (16–40 mm).

GELECHIID MOTHS
(Family Gelechiidae)

These small moths are often brilliant metallic colors. They usually have wingspans of less than ⅝" (16 mm). The hind wings, which are broader than the fore pair, are pointed at the tips and have a wide fringe along the outer margins. The caterpillars include leaf miners, leaf rollers, gall makers, and seed eaters. Many are pests capable of causing serious damage to crops.

533 Potato Tuber Moths
(*Phthorimaea* spp.)

Description: Wingspan ⅜–⅝″ (9–16 mm). Fore wings silver-gray or brown with fine dark speckling or streaks. Hind wings dirty white, about ½ as wide as fore pair, but broadly fringed along outer and rear margins. Body silver-gray or brown. Caterpillar, to ½″ (13 mm), is pinkish yellow to pale green with brown head and prothoracic shield.

Habitat: Potato fields and storage areas.

Range: Southern and southwestern United States.

Food: Caterpillar feeds on foliage, stems, and tubers of potatoes in fields and in storage.

Flight: Throughout the year.

The Potato Tuberworm (*P. operculella*) apparently reached the United States from Australia in the 1890s. The female moth lays 150–300 pearl-white eggs on leaves, stems, or exposed tubers which hatch in a few days. The caterpillars spin silken cocoons in the soil, potato sacks, or storage bins. There are usually 4–6 overlapping generations a year. The caterpillars may also feed on other solanaceous plants in the nightshade family and weeds. Several very similar and closely related species occur in the United States.

ERMINE MOTHS
(Family Yponomeutidae)

These small moths, with wingspans of 1⅛″ (29 mm) or less, have mostly spotted or boldly patterned wings. When at rest, adults wrap their wings around the body, with the antennae projecting straight out in front of the head. Caterpillars of most species spin webs over the foliage of the food plant; a few bore into fruit or are leaf miners.

552 Ailanthus Webworm Moth
(*Atteva punctella*)

Description: Wingspan 1⅛″ (27–29 mm). Fore
wings and thorax shiny orange; fore
wings have greenish-yellow spots,
bordered with black. Hind wings pale
gray, broadly bordered with darker gray
around outer margin; veins dark gray.
Abdomen and appendages brownish
gray.

Habitat: Meadows.

Range: Southern New England to Michigan,
south to the Gulf.

Food: Caterpillar eats foliage of deciduous
trees and shrubs, including introduced
ailanthus.

Flight: Late spring–fall.

When disturbed from their resting
position on grasses and low shrubs,
these moths fly to new sites displaying
their fore wing color. The caterpillars
spin loose cocoons low on the host plant
or among litter on the ground. There is
1 generation a year.

CLEAR-WINGED MOTHS
(Family Sesiidae)

These slender moths bear a strong
resemblance to wasps, having narrow,
partly transparent wings, dark bodies
banded with yellow or red, and fringes
of hair on the legs. They have a
wingspan of ½–2⅜″ (13–60 mm). The
antennae taper at both ends and bear a
tuft of bristles at the tip. Adults are
frequent visitors at flowers, where their
similarity to wasps protects them from
predators. The caterpillars bore into the
stems, roots, and bark of trees and
other plants and sometimes do
considerable damage in orchards. This
family was previously called Aegeriidae.

570 **Manroot Borer**
(*Melittia gloriosa*)

Description: Wingspan 1⅝–2⅜″ (40–60 mm). Fore wings brownish to gray. Hind wings feathery, reddish orange with bluish purple; sides of abdomen and parts of hind legs orangish. Wing membrane opaque in female but transparent in male. Caterpillar, to 1½″ (38 mm), is white with a brown head.

Habitat: Dry open areas.

Range: Kansas and western Texas to southern California, north to Oregon.

Food: Caterpillar bores into roots of several of the large-rooted wild members of the cucumber family, including manroot.

Flight: May–October.

One of the largest and most spectacular clear-winged moths, the Manroot Borer is greatly prized by collectors. These moths become more active as the temperature increases. Males are easily frightened, flying off quickly. The caterpillar, which gives its name to this species, usually overwinters in a silken cocoon beneath the soil.

447 **Squash Vine Borer**
(*Melittia satyriniformis*)

Description: Wingspan 1–1½″ (25–38 mm). Body to ⅝″ (17 mm). Wasplike. Fore wings metallic olive-green. Hind wings transparent with narrow brown margins. Abdomen red with a row of black dots on back. Legs black with thick reddish fringes. Caterpillar, to 1″ (25 mm), is white with brown head.

Habitat: Open areas under cultivation.

Range: Southern Canada to Mexico, except Pacific Coast.

Food: Caterpillar eats pith in stems of plants of gourd family, especially squash, pumpkin, cucumber, and musk-melon.

Flight: Late June–October.

These caterpillars bore into vines and eat out the pith, causing the plant to wilt or die.

443 Doll's Clearwing Moth
(*Paranthrene dollii*)

Description: Wingspan 1⅛–1⅝″ (30–40 mm). Body to ⅞″ (23 mm). Wings reddish brown, banded and streaked with yellow; fore wings partly clear, hind pair reddish brown, partly clear with dark veins and discal spot.

Habitat: Open woods, tree-lined urban streets, and along roads.

Range: Atlantic Coast from Massachusetts to the Appalachian Mountains in Virginia.

Food: Caterpillar bores into branches and trunks of willow and poplar.

Flight: June–July.

These moths are often discovered near wood-boring beetles. Both the beetle larvae and moth caterpillars burrow into wood, but the caterpillars fill their burrows with small wood fragments and excrement, while the beetle larvae make shavings. Caterpillars usually feed for 2 years before reaching full size.

473 California Sycamore Borer
(*Synanthedon resplendens*)

Description: Wingspan ¾–1″ (20–25 mm). Body to ¾″ (20 mm). Fore wings clear with prominent black spot; veins black with yellow interspaces near tip. Hind wings transparent with black veins and black and yellow fringe. Body black, banded with yellow. Middle and hind legs yellow, black at joints.

Habitat: Forests with California sycamore and oaks.

Range: Idaho to California, north to Washington.

Food: Caterpillar eats living trees, usually California sycamore and oaks, and usually only older trees.

Flight: April—August.

This moth lays its eggs in bark crevices, where caterpillars hatch in a few days. The caterpillars bore into the bark, expelling quantities of waste. They overwinter in side galleries and later pupate close to the bark surface. Often the tips of the pupae can be seen protruding from the bark.

TORTRICID MOTHS
(Family Tortricidae)

These small, mottled moths are brown, tan, yellow-gray, or black and white. They have short palps and wingspans of ⅜–1⅜″ (8–35 mm). Many have square-tipped fore wings, so that when the wings are folded back at rest the moth is shaped rather like a shield. The caterpillars live in rolled leaves held together with silk, or they bore into fruits, nuts, and the roots of herbaceous plants, or even feed openly on foliage. Many of these caterpillars are destructive pests. The adults do not feed. They are sometimes called "millers."

536 Fruit-tree Leaf Roller Moth
(*Archips argyrospila*)

Description: Wingspan ¾″ (19–20 mm). Fore wings pale yellowish brown, mottled with brown; cream-colored spots along front margin. Caterpillar, ¾–1″ (20–25 mm), is green with dark reddish-brown head; body covered with fine hair.

Habitat: Forests and orchards, among deciduous and fruit trees.

Range: Throughout North America.

Food: Caterpillar eats foliage of a great variety of trees, including apple, pear, quince, walnut, and other valuable crops.
Flight: June.

This moth flies readily to lights at night. In fall the female affixes a mass of eggs to a tree trunk or branch, then covers them with secreted cement. The eggs hatch in spring. When the caterpillars are disturbed, they often wriggle to the rear or side, then, somewhat like spiders, let themselves down to the ground on long silk threads that they can also use to return to their nests. They pupate in rolled-up leaves. There is 1 generation a year. The related Cherry Leaf Roller (*A. cerasivoranus*), slightly larger, builds nests of silk and dead leaves on terminal branches of cherry trees. Their nests should not be confused with nests of tent caterpillars, which are located in crotches of larger branches of cherry and related trees, and which do not incorporate dead leaves.

535 Orange Tortrix
(*Argyrotaenia citrana*)

Description: Wingspan ½–⅝" (14–16 mm). Wings tan to rusty brown; when folded at rest, a black diagonal band across fore pair forms a V-shaped mark on surface above. Dark brown triangle near front margin of each fore wing about halfway between the diagonal mark and wing tip; another faint, diagonal, brownish mark parallel to black one, but nearer wing base. Caterpillar is pale green to straw-yellow with pale brown head and thoracic shield.
Habitat: Orchards and shrubby fields.
Range: Wyoming to California, north to British Columbia in greenhouses.
Food: Caterpillar feeds on buds, leaves, and blossoms of orange and other citrus

trees, also apple, pear, apricot, plum, and other shrubby plants.

Flight: Year-round.

This moth lays masses of pale yellowish eggs that resemble overlapping shingles on twigs and branches. Caterpillars make nests among twig tips, where they usually overwinter and pupate the following spring. When caterpillars cut through the skin of ripe oranges to feed on pulp, they seriously injure plants. In some areas there are 2–4 generations a year.

542 Codling Moth
(*Laspeyresia pomonella*)

Description: Wingspan ⅝–¾" (15–20 mm). Body and fore wings brownish gray; fore wings have fine netlike lines; outer and basal ⅓ darker, many metallic scales in outer ⅓. Hind wings brown, bordered by paler fringes.

Habitat: Fruit orchards, especially apple and pear.

Range: Throughout North America.

Food: Caterpillars of 1st generation feed in developing fruit; 2nd generation feeds in mature fruit.

Flight: Spring–fall.

The small, grublike caterpillars destroy the commercial value of fruit. The caterpillars overwinter in cocoons in tree crotches, bark crevices, among grass and weed clumps, in litter on the ground, or in other objects near trees. They pupate in spring, with adults emerging in about 20 days, often when apple trees begin to expand leaves and open flowers. The presence of caterpillars can be recognized by holes in the fruit skin and quantities of fecal wastes deposited in the core. Many biological control agents are known, but are not as effective as using recommended pesticides and rigorously sanitizing the orchard to eliminate overwintering caterpillars.

SKIPPERS
(Family Hesperiidae)

Skippers have some characteristics of both butterflies and moths. They are stocky with large heads, thickset bodies, and short wings, spanning ½–2″ (14–50 mm). The antennae are set rather far apart on the head and end in a small hook. In contrast to true butterflies, which hold both wings either vertically or horizontally at rest, skippers often rest with fore wings open at about a 45° angle and hind wings horizontal. Their rapid, direct, and bouncing flight is aptly suggested by the name "skipper." There are two groups of skippers commonly represented in North America. The larger group consists mainly of small, tawny-brown species, whose caterpillars feed almost exclusively on grasses and sedges; these species are often very similar and difficult to distinguish. The other group is made up of more boldly marked species that are much easier to tell apart; the caterpillars of this group feed on a variety of plants and very often on legumes. Adults of both groups either sip nectar or extract organic nurtrients from decaying matter and wet mud. Unlike most butterflies, skippers pupate in a cocoon of leaves and silken strands.

606 **Least Skipper**
(*Ancyloxypha numitor*)

Description: Wingspan ¾–⅞″ (18–22 mm). Fore
wings blackish brown above, usually
with streak or patch of dull orange from
base outward. Hind wings brownish
orange with black borders. Fore and
hind wings below yellow-orange; fore
wings often blackish toward body.
Caterpillar is grass-green with dark
brown head.

Habitat: Wet meadows.

Range: Eastern United States and southern
Canada.

Food: Caterpillar eats grasses.

Flight: May–October.

This skipper, in marked contrast to
others, flies low with a weak, dancing
flight. It is easy to see as it moves
among the grasses. There are 3 or more
generations a year.

9, 574 **Silver-spotted Skipper**
(*Epargyreus clarus*)

Description: Wingspan 1¾–2″ (43–50 mm). Wings
chocolate-brown. Fore wings have an
irregular golden band below and
smaller yellow areas above. Hind wings
plain above except for white fringe;
large, silvery-white irregular spot
below. Caterpillar, to 2″ (50 mm), has
brownish-red head with 2 orange-red
spots and yellowish to green spindle-
shaped body with narrow dark green
bands.

Habitat: Gardens, roadsides, and open areas.

Range: Throughout most of North America.

Food: Caterpillar eats foliage of leguminous
plants, including locust trees, wisteria,
alfalfa, and stick-tights.

Flight: Throughout the summer.

This is one of the most conspicuous
skippers, partly because of its size and

partly because of its distinct silvery markings, which show while the insect rests. The caterpillars hide all day in silken nests among foliage, emerging to feed at night. There is 1 generation a year in the North; 2 or more in the South.

---

575 Long-tailed Skipper
(*Urbanus proteus*)

Description: ¾" (18–20 mm). Wingspan 1⅝–2" (40–50 mm). Fore wings dark brown; iridescent green-blue at base, with a distinctive pattern of white spots from the middle to tip. Hind wings have extensive iridescent green-blue overlaid on dark brown background. Prominent long tails. Caterpillar has yellowish-green, spindle-shaped body with black dorsal line and 2 reddish or yellow side stripes; head is black and brown with 2 yellow spots.

Habitat: Open areas and fields, roadsides, and gardens.

Range: Connecticut to Florida, west to California, south to Mexico; rare in the Northwest and extreme West.

Food: Caterpillar feeds on wild beans, mesquite, wisteria, cultivated beans, and a variety of other leguminous plants.

Flight: Throughout the year.

In the South, this caterpillar attacks cultivated beans and is known as the "Bean Leaf Roller." There are usually 3 or more generations a year. The related Dorantes Skipper (*U. dorantes*), wingspan 1⅝–1⅞" (40–48 mm), has similar tails but no greenish iridescence above. It is common in southern Texas.

SWALLOWTAILS AND KIN
(Family Papilionidae)

Among our most striking butterflies, swallowtails are named for the long tails that project from the hind wings of many species. They are boldly patterned in black, yellow, or white, and often have red or blue spots. Their wingspan is 2⅜–5⅞" (60–150 mm). The front legs are fully developed in both sexes. Most species visit flowers. The caterpillars of true swallowtails have a Y-shaped retractile organ, the osmeterium, on the top of the thorax, which gives off a foul-smelling odor when the caterpillar is disturbed. Most caterpillars feed on the foliage of trees. Some caterpillars overwinter partially grown and pupate in loose cocoons hidden among fallen leaves. Others overwinter as pupae; the chrysalises are attached to supports by silken girdles. Swallowtails are considered such typical butterflies that the word *papillon* means butterfly in French.

578 Green Swallowtail
"Pipevine Swallowtail"
(*Battus philenor*)

Description: Wingspan 3–4½" (75–115 mm). Wings black often with blue-green iridescence; hind wings have a row of white crescentlike marks (lunules) along the outer edge. Caterpillar is purplish black with paired slender tentacles and paired rows of dark red warts on its back.

Habitat: Thickets near deciduous woods and gardens where Dutchman's pipe grows.

Range: Throughout the United States, except for the extreme North; common in the South, rarer in the North.

Food: Caterpillar eats Virginia snakeroot and Dutchman's pipe.

Flight: Late April–September.

These butterflies are distasteful to birds, which learn to recognize their coloring and avoid them. The female butterfly is unusual because it lays clusters of eggs instead of single eggs. The young caterpillars feed side by side at the edge of the same leaf. They overwinter as chrysalises. There are usually 2 generations a year. The Tailless Swallowtail (*B. polydamus*), wingspan 3–3¾" (75–95 mm), has black fore wings with yellow spots near the outer margin and shiny greenish-black hind wings with yellow or greenish-yellow spots. It occurs in Florida, the Gulf states, and Mexico.

593 Zebra Swallowtail
(*Graphium marcellus*)

Description: Wingspan 2½–4½" (65–115 mm). Variable size and color. Wings dark brown or black with white stripes, tails very long and narrow with bright red at base. Caterpillar is pale green with black and yellow transverse bands.

Habitat: Moist, shaded lowland woods.

Range: Eastern half of the United States, more common in the South.

Food: Caterpillar eats pawpaw foliage.

Flight: April–August.

There are several generations a year of Zebra Swallowtails. Early adults are generally smaller and whiter, while later adults in the same generation are larger and have broader white bands and longer tails, often with a white tip.

584 Giant Swallowtail
(*Papilio cresphontes*)

Description: Wingspan 4–5½" (100–140 mm). Body and wings rich brown with paired yellow spots and stripes; each tail has a yellow to orange area surrounded by brown. Abdomen yellow with brown midline. Caterpillar is brown with dirty white markings.

Habitat: Deciduous forests and citrus orchards.

Range: Atlantic Coast west to Arizona.

Food: Caterpillar feeds on leaves of prickly ash, hop trees, and citrus trees.

Flight: Early May–August.

The caterpillar, known as "Orange Dog" or "Orange Puppy," is sometimes a pest in citrus orchards. There are 2 generations in the North, 3 in the South. A slightly larger Western Citrus Swallowtail (*P. thoas*) lacks the yellow spots near the outer margin of the fore wings and ranges from Colorado to southern Texas and Mexico. It and the Giant Swallowtail are two of the largest butterflies in North America.

594 Pale Swallowtail
(*Papilio eurymedon*)

Description: Wingspan 4–5⅜" (100–135 mm). Wings yellowish white with black tigerlike stripes and slender black tails sometimes marked with orange and blue at the base. Fore wings below have continuous band of whitish marginal spots. Caterpillar is green and has orange eyespots with black dot and orange and black band.

Habitat: Dry mountains.

Range: Rocky Mountains to Pacific Coast, from Mexico to British Columbia.

Food: Caterpillar eats foliage of California coffeeberry, California lilac, and others.

Flight: June–July.

Like many other swallowtails, this butterfly is often found sipping water at puddles. Its caterpillars spin silken mats on leaves and then curl the leaves into shelters. There is 1 generation a year.

22, 580 Eastern Tiger Swallowtail
(*Papilio glaucus*)

Description: Wingspan 4–5⅞″ (100–150 mm). All males and many females have yellow wings with black tigerlike stripes in pale phase. Wings of dark-phase females, common in southern parts of range, are brownish black with yellow and blue spots near outer margins. Caterpillar, to 2½″ (65 mm), is green and has yellow eyespots with black dot, orange and black bands, and blue dots.

Habitat: Mixed and deciduous forests; open areas, even in urban areas.

Range: East of the Rocky Mountains.

Food: Caterpillar eats foliage of wild cherry, birch, poplar, ash, and tulip trees.

Flight: Spring–September.

This butterfly's caterpillars are solitary feeders. Each spins a silken mat on a leaf, curling the leaf into a shelter. There are 2 generations in the North; several in the South. The Western Tiger Swallowtail (*P. rutulus*), wingspan 3½–4⅜″ (90–110 mm), has only a pale phase and differs in having a yellow band instead of a row of spots at the outer margin of the under portion of the fore wing. Its caterpillars feed on alder, poplar, willow, and sometimes hop vines in isolated moist areas west of the Rocky Mountains.

583 Eastern Black Swallowtail
"Parsley Swallowtail"
(*Papilio polyxenes asterius*)

Description: Wingspan 2¾–3½″ (70–90 mm). Wings black; fore wings have a double row of yellow spots, the inner row larger and more triangular in male, much smaller in female; hind wings have a row of yellow submarginal crescentlike marks (lunules) and inner row of much larger yellow spots (male) or small spots (female); area between yellow spots suffused with blue in varying degrees. Caterpillar is green with black head and a black band on each segment incorporating a series of golden spots.

Habitat: Open meadows.

Range: East of the Rocky Mountains from southern Canada to the Gulf, southwest to Arizona.

Food: Caterpillar eats foliage of plants in the carrot family, including wild and cultivated carrot, celery, and parsley.

Flight: Spring–fall.

This common eastern swallowtail flies low, alighting on clover and cultivated flowers. Its caterpillars are sometimes garden pests. There are 2 generations a year in the North, at least 3 in the South. The Anise Swallowtail (*P. zelicaon*), wingspan 2½–3½″ (65–90 mm), has wings with broad yellow spots that extend almost to the base. It is generally the most common swallowtail west of the Rocky Mountains, where it ranges from Baja California and northern Mexico to British Columbia, east to Idaho and Nevada.

576 Spicebush Swallowtail
"Green Clouded Swallowtail"
(*Papilio troilus*)

Description: Wingspan 4–4⅞″ (100–125 mm). Wings dark with yellow to pale crescentlike marks (lunules) along margin. Male has greenish iridescent wash over outer ½ of hind wings; female bluish in corresponding area. Caterpillar is green with orange-red head, a pair of large eyespots on thorax, a pair of yellow ovals on 1st abdominal segment, and yellow stripe along sides, orange below.

Habitat: Moist deciduous and mixed forests.

Range: Eastern half of United States and southernmost Canada; scarce in the Canadian Maritime Provinces and northern states.

Food: Caterpillar eats foliage of spicebush, sassafras, sweet bay, prickly ash, and other plants.

Flight: Mid-spring to August.

These familiar swallowtails often congregate in large groups around mud puddles, especially in the South. The large eyespots may serve to frighten predators. The caterpillar curls the leaf in which it hides, brushes out its fecal pellets, and often eats each cast skin except the hard head capsule. There are 2 generations in the North, 3 in the South.

591 Phoebus
(*Parnassius phoebus*)

Description: Wingspan 2⅜–3″ (60–75 mm). Antennae black, ringed with white. Hind wings almost as long and broad as fore wings, rounded without angles or tails. Fore and hind wings ivory-white with red spot near front margin of fore wing and 2 red spots on hind wings; each spot encircled with black. Female has larger red spots; wings more heavily marked with gray. Caterpillar is black or dark brown with many pale dots.

Habitat: Tundra, on mountains at low latitudes, or lowlands in high Arctic latitudes.

Range: West of the Great Plains, from mid-California and New Mexico north to Alaska, Canadian Northwest Territories, and Alberta.

Food: Caterpillar eats alpine arctic stonecrops and saxifrages.

Flight: Mid-June to August.

These tailless swallowtails are much admired but difficult to find in their lofty mountain habitats. Males appear 8–10 days before the females. They mate soon after the female emerges, often even before the female is able to fly. The caterpillars spin a few silk threads among litter on the ground for pupal shelters. There is 1 generation a year. The related Clodius Parnassian (*P. clodius*), same size, has black antennae and lacks red spots on the fore wings. It flies in midsummer from Alaska to central California and in the northern Rockies.

WHITES, SULPHURS, AND ORANGE TIPS
(Family Pieridae)

Many of our most familiar and easily identified butterflies belong to this family. They are white, yellow, or orange and have rounded wings and simple wing patterns, often consisting merely of a narrow, black border. Their wingspan is ⅞–2¾" (22–70 mm). The antennae end in a distinct club. Unlike members of other families, these butterflies have full-sized fore legs that are used for walking. Adults of most species frequent open, sunny places and are easy to see while they visit flowers or gather at mud puddles to drink. The caterpillars are smooth, slender, and green, and often have a few narrow, lengthwise stripes. The chrysalis is attached to a support both by a silken girdle and by hooks at the end of the abdomen caught in a pad of silk.

592 Sara Orange Tip
(*Anthocharis sara*)

Description: Wingspan 1–1¼" (25–33 mm). Wing white above with black-bordered orang tip on fore wing of male; female simila

but with black-bordered yellow stripe
on outer part of orange tip. Hind wings
marbled with greenish below.
Caterpillar is green with small black
tubercles.

Habitat: Meadows and edges of woods.

Range: Western United States from the Rocky
Mountains to the Pacific Coast.

Food: Caterpillar eats mustards, other low-
growing herbs of same family.

Flight: Spring–summer.

There are 2 generations a year; adults of
the second have paler marbling under
the hind wings. The Falcate Orange
Tip (*A. midea*), wingspan 1⅛–1⅜″
(30–35 mm), has an orange tip on the
fore wings of the male; in both sexes
the outer margin of the fore wing is
concave, producing a hooked wing tip.
It is found locally in rich woods
throughout the eastern United States in
early spring.

589 Great Southern White
(*Ascia monuste*)

Description: Wingspan 1¾–2⅜″ (45–60 mm).
Male's wings white with black margin
around tip of fore pair; female's wings
either brownish gray or white like
male's. Fore wing sometimes has black
spot. Caterpillar is yellow with 4
lengthwise dark green or purple stripes.

Habitat: Meadows and truck gardens.

Range: Virginia to Gulf states along the
Atlantic Coast.

Food: Caterpillar eats foliage of plants of
mustard family, including cabbage.

Flight: Throughout the year.

Dark-form females fly in steady streams
along the Florida Coast and Keys, then
venture out over open water, where
they have little chance of reaching land.
White females show no readiness to
travel far. The related Giant White

(*A. josephina*), wingspan 2½–2¾" (65–70 mm), has white wings with a black dot on each fore wing in the male and several in the female. It occurs in Texas.

579 California Dog Face
(*Colias eurydice*)

Description: Wingspan 1¾–2" (45–50 mm). Male's fore wing purplish to pinkish orange with dark purplish brown along front and outer margin, giving orange area pattern of dog's face in profile, the "eye" usually linked to the outer dark band. Female's fore wings entirely yellow above except for a black dot on fore wing. Hind wings of both yellow. Caterpillar is green with an orange lateral stripe bordered above by a broken dark line, numerous minute black dots, and short hair.

Habitat: Open woods.

Range: Coastal California south to Baja California.

Food: Caterpillar eats false indigo.

Flight: Summer–late fall.

This is the state butterfly of California. There are 2 generations a year. The larger Southern Dog Face (*C. caesonia*), wingspan 2¼–2½" (58–65 mm), has yellow wings; the fore pair in both sexes has a black outer border defining a yellow "dog's face." It occurs in the southern half of the United States, becoming rare and sporadic in the Northwest.

581 Cloudless Sulphur
(*Phoebis sennae*)

Description: Wingspan 2⅛–2¾" (55–70 mm). Wings pale yellowish green; male's unmarked, female's fore wings have

small red dot encircled by black,
usually a blackish narrow border, and a
few vague dark dots near tip.
Caterpillar is pale green and marked by
a yellow stripe on each side and black
spots in rows across each abdominal
segment.

Habitat: Tropical forests and areas of regrowth
with woody members of the pea family.

Range: Southern United States; in eastern
United States migrating north to New
York or beyond.

Food: Caterpillar feeds on cassia and other
woody legumes.

Flight: Throughout the year; northern
migrants fly late summer–fall.

This butterfly breeds continuously in
the tropics, and there are usually 2
generations a year in the southern
United States. The Giant Orange
Sulphur (*P. agarithe*), wingspan 2–2¾"
(50–70 mm), has deep orange on its
wings above and below; the male's are
unmarked, but female's may have a
brown spot and marginal markings or
at least cloudy markings. It is found
in the Gulf states, straying north to
Illinois. The Orange-barred Sulphur
(*P. philea*), wingspan 2–2¾" (50–
70 mm), has yellow wings with a vague
orange bar on the fore wings and orange
shading around the outer margin of the
hind wings. It has the same general
distribution and has recently become
more abundant in southern Florida.

590 **Checkered White**
"Southern Cabbage Worm Butterfly"
(*Pieris protodice*)

Description: Wingspan 1⅜–1⅝" (35–40 mm).
Wings white with diffuse gray or black
markings; fewer and darker markings in
male, extensive markings particularly
along veins in the female. Caterpillar is
pale green to bluish, speckled with

black dots, and marked with 4 yellow and green lengthwise stripes.

Habitat: Open areas, wherever plants of the mustard family are common.

Range: Coast to coast in most of North America; uncommon in the Northeast.

Food: Caterpillar feeds on wild crucifers, including mustards, rock cress, cultivated cabbage, and radishes.

Flight: Early spring–August.

This insect used to be more common, but it has been unable to compete well with the introduced European Cabbage Butterfly. There are several generations a year. Early spring adults are small, and the males are pale to almost unmarked; in later months adults are larger and females are darker. The last generation overwinters as a chrysalis. The California White (*P. sisymbrii*), wingspan 1⅜–1⅝″ (35–40 mm), has a black oblique bar in the cell halfway to the tip of the fore wings; hind wings below have most veins diffusely striped with dark greenish brown. It occurs west of the Great Plains, where the caterpillars eat wild crucifers on foothills and mountains.

21, 582   European Cabbage Butterfly
(*Pieris rapae*)

Description: Wingspan 1⅛–2″ (30–50 mm). Wing yellowish white with black tip on fore wing; female's fore wings have 2 small black spots, male's have 1. Both sexes have 1 black spot well out along front margin of hind wings. Caterpillar, to 1⅜″ (35 mm), is green, marked with 5 lengthwise yellow lines, and covered with fine velvety hair.

Habitat: Open areas in a wide variety of habitats.

Range: Throughout North America.

Food: Caterpillar eats cabbage and many other crucifers and related plants.
Flight: Early spring—September.

Introduced accidentally near Montreal in the 1860s, this species has become an important pest. Bacterial and viral diseases now provide some biological control. At least 3 generations are known in southern Canada, more in southern states. The related Mustard White (*P. napi*), same size, has veins underneath that are outlined with dark scales. It is found throughout the northern United States and Canada, southwest at higher elevations to Arizona.

GOSSAMER-WINGED BUTTERFLIES
(Family Lycaenidae)

This family includes the hairstreaks, coppers, blues, and harvesters—all sprightly, delicate little butterflies with wingspans of ⅜–1⅞″ (11–47 mm). Most species have brilliant blue, violet, or coppery wings, sometimes with tiny tail-like projections from the margins of the hind wings. Females have normal fore legs; those of males are somewhat smaller and lack claws. Adults commonly visit flowers. They are easy to identify because of their small size and their habit of holding the wings tightly together over the back, exposing the speckled undersurface of the wings. The caterpillars are stocky and sluglike. They feed on foliage or live as guests in anthills, feeding on debris or ant larvae. The compact chrysalis is suspended by a silken girdle, usually in leaf litter on the ground.

627 Great Purple Hairstreak
(*Atlides halesus*)

Description: Wingspan 1⅜" (34–36 mm). Wings
above brilliant iridescent, metallic
blue-green with dark borders and long
twinned tails on hind wings; male's fore
wings have prominent black discal
spot. Wings below purplish black with
several red spots near bases; brilliant
multicolored area along hind margin
near tail. Caterpillar is green with dark
green bands and yellowish stripes and
narrow green midline stripe.

Habitat: Woods and open areas.

Range: New Jersey to Florida, west to
California; rare in the North.

Food: Caterpillar eats the foliage of mistletoe,
which grows on live oaks.

Flight: September–November; also April in
Florida.

The hairstreak butterflies are named for
the thin hairlike tails on the hind
wings. The Great Purple Hairstreak
flies rather slowly and is easy to observe
while it rests on flowers.

577 Spring Azure
"Common Blue"
(*Celastrina argiolus*)

Description: Wingspan ¾–1¼" (20–33 mm).
Male's wings pale sky-blue above with
checkered fringe; below brownish gray
with a pattern of small dark dots on the
spring generation; later generations
much paler with even smaller spots.
Female's fore wings above have dark
blackish-gray front and outer margins
and blue area often fading to nearly
white; hind wings above pale bluish
gray with marginal row of dark spots.
Caterpillar is sluglike with variable
colors.

Habitat: Rich, moist woods.

Range: Coast to coast from the subarctic to

Mexico; rare or absent in Florida.

Food: Caterpillar eats flower buds and flowers of many kinds, including dogwood, viburnum, spiraea, milkweed, and willow.

Flight: May–midsummer.

The Spring Azure is one of the earliest butterflies to emerge, sometimes appearing even when snow is still on the ground. Its caterpillars exude a honeydew that attracts ants. There are 2 or more generations a year; the spring generation is of smaller size. Formerly known as *Lycaenopsis argiolus*.

588 Early Hairstreak
(*Erora laeta*)

Description: Wingspan ¾–⅞" (20–23 mm). Wings above black with bluish iridescence; gray-green to grayish white below with a line of brick-red spots across middle of both wings and a few red spots along edge of hind wings.

Habitat: Open deciduous woodlands, where beech trees grow.

Range: Southeastern Canada and New England, south in the Appalachian Mountains to Tennessee and Kentucky.

Food: Caterpillar probably eats beech foliage.

Flight: May–July.

Because of its scarcity, this butterfly is a special prize for collectors. Adults often alight on bare ground.

586 Eastern Tailed Blue
(*Everes comyntas*)

Description: Wingspan ⅞–1⅛" (23–28 mm). Male's wings above iridescent pale blue with brownish gray along outer margin; fore pair with a short, oblique black bar near middle; hind pair with a row of

submarginal black spots and a small
orange spot at the base of each
projecting tail. Female's wings larger
with longer tails; gray above on body
and wings, 2 or 3 small orange spots
with black dots near margin of hind
wings. Wings of both sexes below
silvery gray with small dark spots and a
few orange spots near margin of hind
pair. Caterpillar has hairy and pebbly
dark green body with dark brown
stripes and a small black head.

Habitat: Meadows, roadsides, and forest paths.

Range: Throughout most of the United States
and southern Canada; abundant in the
East.

Food: Caterpillar eats leguminous plants, such
as clover.

Flight: Spring–fall.

Unlike most butterflies, this species has
thrived where its habitat has been
encroached upon by human activities.
It is common along freshly mowed
roadsides, flying to puddles. Usually
there are 2 or more generations a year.
Many lose their tails. The Western
Tailed Blue (*E. amyntula*) is larger, has
fewer wing markings, and is paler
underneath. It occurs throughout most
of the western half of North America.

609 Harvester
"The Wanderer"
(*Feniseca tarquinius*)

Description: Wingspan 1⅛–1¼″ (29–33 mm).
Wings above brownish orange. Fore
wings irregularly dark brown around
margins; hind wings dark at base and
part way around outer margin, where
lightly dotted with brown. Wings
below paler with brown spots.
Caterpillar is greenish brown with faint
olive stripes and covered with fine hair.

Habitat: Wooded areas near streams, particularly
among alders.

Range: East of the Great Plains in southern
Canada and the United States.

Food: Caterpillar eats aphids, preferably
aphids on alder, and other unarmored
scale insects.

Flight: May–September.

Ironically this butterfly was named
for the Roman tyrant Tarquin before
the carnivorous habits of its caterpillars
were known. The adult flits and
abruptly alights anywhere there are
plants that attract aphids. The
caterpillar creeps to the nearest aphid
and eats it, transferring the waxy
filaments taken from its prey to sticky
hair on its back. Concealed, the
caterpillar continues feeding, adding to
its covering, and often attains full
growth in less than 2 weeks. Seen from
above, the chrysalis resembles a
monkey's face. There are usually 2
generations a year in the North, 4–5
generations a year in the South.

601 Silvery Blue
(*Glaucopsyche lygdamus*)

Description: Wingspan ⅞–1⅛″ (23–29 mm).
Male's wings above silvery blue edged
with black. Female's wings above duller
blue, smoky at base and margins; dark
spot in cell of fore wing. Wings of both
below brownish gray with row of black
spots edged in white toward outer
margin. Caterpillar is pale green with
white spots, dark green stripes edged
with yellowish green, and black head.

Habitat: Open woods.

Range: Coast to coast in Canada and the
northern United States, south along
mountain ranges in both the East and
West.

Food: Caterpillar feeds on everlasting peas,
vetch, lupine, greasewood, and various
other leguminous plants.

Flight: April–June.

Because they secrete a sugary liquid called honeydew, caterpillars of this and related species are often seen on leaves surrounded by many ants. There is 1 generation a year.

605 Brown Elfin
(*Incisalia augustinus*)

Description: Wingspan ⅞–1″ (23–25 mm). Wings grayish brown (male) to orange-brown (female), basal ½ to ⅔ below darker, remaining part has a row of somewhat blurred dark brown spots. Hind wings slightly scalloped along outer margin, without tails. Caterpillar is bright green with yellowish-green narrow midline, a pattern of spots and dashes along sides, yellowish-brown spiracles, and grayish to brownish-yellow head.

Habitat: Forest edges and clearings.

Range: Throughout most of the United States and Canada, except the extreme Southeast and Plains states.

Food: Caterpillar feeds on members of the heath family, especially wild blueberries and sheep laurel.

Flight: April–May.

Like the hairstreaks, the Brown Elfin holds its fore wings together stationary above the back at rest, while alternately raising and lowering its hind wings. But it flies slower than most hairstreaks and generally closer to the soil. Details of markings below the hind wings and of genitalia distinguish this species from other elfins. The Brown Elfin is often locally abundant.

595 Marine Blue
(*Leptotes marina*)

Description: Wingspan ⅝–1⅛″ (17–28 mm). Male's wings above purplish blue, sometimes whitish along hind wing margins; white below with many undulating gray lines and iridescent metallic submarginal spots on hind wing. Female's wings bluish near base, blackish at margins; underside markings visible through thin scaly hind wings. Caterpillar is variously colored from brown with dark marking to reddish or light green without marks.

Habitat: Open woods, meadows, and fields.

Range: Illinois to California, south to Texas and Mexico and the tropics.

Food: Caterpillar eats flowers and buds of sweet pea, wisteria, leadwort, and vetch.

Flight: Spring—fall in the North, year-round in the extreme South.

This delicate blue flutters around flowers, shrubs, and flowering trees, sometimes straying along flower-bordered roadways. There are 2 or more generations a year.

616 American Copper
(*Lycaena phlaeas americana*)

Description: Wingspan 1" (25 mm). Fore wings above shiny metallic orange with dark brown to black or gray outer border and about 8 well-defined dark spots in orange area. Hind wings black, brown, or gray with row of orange spots, each with a black dot at center, along outer margin. Caterpillar is sluglike, dull, pinkish red, with minute head.

Habitat: Meadows and roadsides.

Range: Southeastern Canada and the eastern United States, except the Coastal Plain and Gulf states.

Food: Caterpillar feeds on sorrel.

Flight: May—midsummer.

The first adults to appear in spring are more brightly colored than the second or third generations and have smaller or fewer spots. The related Bronze Copper (*L. thoë*) is one of the largest coppers in North America, with a wingspan of 1¼–1⅜″ (33–35 mm). Males have brown fore wings; females are yellower with more prominent black dots. This species does not occur as far south as the American Copper.

587   Gray Hairstreak
(*Strymon melinus*)

Description: Wingspan 1⅛″ (28–30 mm). Wings above dark grayish brown, hind wing with prominent orange spot at outer margin close to shorter of 2 blackish tails. Hind wing below whitish gray with marginal black-eyed orange spot, a small patch of blue before tail, and 2 broken crossbands of black and white spots. Caterpillar is sluglike, reddish brown, and has small head.

Habitat: Meadows, neglected roadsides, and crop fields.

Range: Southern Canada through Mexico.

Food: Caterpillar feeds on fruits and seeds of knotweed, St. Johnswort, hawthorn, mallow, cultivated beans, and hops.

Flight: May–September.

This butterfly is a pest of beans and hops. Adults fly quickly; they are best seen at rest when their wings are folded together over the back, one hind wing sometimes raised while the other is lowered. There are 2 generations a year in the North, 3 or more in the South.

METALMARKS
(Family Riodinidae)

These little butterflies get their name from the small metallic spots and lines on the wings of many species. In North America, metalmarks, with wingspans of ⅝–1½″ (16–37 mm), are usually dull-colored but in the tropics, where they are more abundant, most have brilliant hues. All species have long, slender antennae that arise from a small notch in the margin of each eye. Fore legs are reduced in males, and full-sized in females. Metalmarks are weak fliers, and when they alight they hold their wings out flat as do moths. The caterpillars are short, sluglike, and have small tubercles bearing fine bristles. The stout pupae often are covered with short hair and are supported by silken girdles.

613, 626 **Northern Metalmark**
(*Calephelis borealis*)

Description: Wingspan 1–1⅛″ (25–30 mm). Wings above dark brown marked with 2 narrow metallic lines and a row of small blackish dots. Wings below pale orange with metallic lines and a speckling of small blackish dots. Caterpillar is green with long white hair that conceals black spots along its back.

Habitat: Open woods and dry hilly pastures.

Range: Southern New England to Virginia, west to Ohio.

Food: Caterpillar eats ragwort or groundsel.

Flight: Early July.

Usually scarce and local, these butterflies alight on the undersurfaces of leaves or, occasionally, on the ground, where they spread their wings flat. There is 1 generation a year. The Little Metalmark (*C. virginiensis*), wingspan ⅝–¾″ (16–20 mm), has orange-brown wings above and is red-brown underneath with blue metallic spots. It is found in the southeastern United States from Virginia to Texas along the Gulf Coast.

SNOUT BUTTERFLIES
(Family Libytheidae)

This family gets its name from the long
labial palps which project in front of
the head like a snout. The wings,
spanning 1⅜–2″ (35–50 mm), are dark
brown with a few white and orange
markings, and there is a prominent
"tooth" on the outer margin of each
fore wing. The front legs are normal in
females and reduced in males. Only a
single species occurs regularly in North
America.

597   **Common Snout Butterfly**
      (*Libytheana bachmanii*)

Description: Wingspan 1¾–2″ (45–50 mm). Fore
             wings red-brown with 3 conspicuous
             white spots near dark brown curving
             tip. Hind wings brown and tawny
             above, brownish gray below. Caterpilla
             is downy green with yellow stripes on
             each side and on back; enlarged thorax
             makes it appear humpbacked.
Habitat: Roadsides, gardens, open country, lake
          margins, and muddy stream banks.
Range: Eastern United States west to the
        Rocky Mountains; less frequent in the
        North.
Food: Caterpillar eats hackberry leaves.
Flight: April–fall.

Depending on the climate, there may
be 3 or 4 generations a year. The Pale
Snout Butterfly (*L. carinenta*), same
size, is less pigmented and has straight
outer margins on the hind wings. It
occurs as a stray from Mexico in the
Southwest.

BRUSH-FOOTED BUTTERFLIES
(Family Nymphalidae)

The many members of this family vary greatly in color and size, having wingspans of 1–4" (24–100 mm). They are covered with long, hairy scales and the fore legs are greatly reduced in both sexes. Most species are strong fliers. Several are migratory, and many overwinter in the adult stage. Open country species usually visit flowers, while woodland species feed on the juices of rotting fruit, organic wastes, or on the fermenting sap of trees. The caterpillars, many of which feed only at night, are generally dark and have branching spines. The chrysalises are suspended by the tip of the abdomen and do not have a silken girdle. They overwinter as eggs, caterpillars, or adults, depending on the species. The heliconian butterflies, previously grouped in the separate Family Heliconiidae, are now included as nymphalids.

3, 614 Gulf Fritillary
(*Agraulis vanillae*)

Description: Wingspan 2½–3⅛" (65–80 mm). Fore wings almost twice as long as wide. Wings above brownish orange (male) or brownish (female) with 3 white spots on fore pair, numerous black markings on both pairs. Hind wings below have brilliant metallic silver spots. Caterpillar, to 2" (50 mm), is yellowish to greenish brown with brown to orange stripes, 6 rows of branching spines on its body, and 2 longer, backward-curving horns on its head.

Habitat: Tropical forests and adjacent open areas.

Range: Southern United States along the Gulf Coast, from the Atlantic to the Pacific coasts, rarely straying northward.

Food: Caterpillar eats passion flower foliage.

Flight: Spring–fall; throughout the year in Florida.

Because caterpillars eat passion flower foliage, both the caterpillars and butterflies contain a poisonous chemical, and their predators soon learn to leave them alone. There are continuous generations. The Julia (*Dryas julia*), wingspan 3⅛–3¾" (80–95 mm), has orange wings above (male) or pale yellowish brown (female), and pale dull brown below. Occasionally it swarms from Latin America into south Texas and Florida. Both species and their relatives were previously placed in the Family Heliconiidae.

608 White Peacock
(*Anartia jatrophae*)

Description: Wingspan 1¾–2¼" (45–58 mm). Wings mostly white, delicately and variably marked with 1 dark eyespot on fore wing and 2 smaller eyespots on hind wing; orange-yellow outer border on both wings also varies in intensity. Caterpillar is black with large silvery spots; head glossy; body has 4 rows of spines.

Habitat: Open woods and gardens.

Range: Florida and southern Texas, rarely straying northward.

Food: Caterpillar eats plants of spurge and vervain families.

Flight: Throughout the year.

This handsome, tropical insect is not especially wary or fast flying. Its nearest kin in the United States is the Fatima (*A. fatima*), wingspan 2" (50 mm), which has rows of pale spots across both pairs of wings and a line of 4 reddish spots on the hind wings. This butterfly is well established in southern Texas and Mexico.

610 Harris' Checkerspot
(*Chlosyne harrisii*)

Description: Wingspan 1⅜–1¾" (35–45 mm).
Wing color and pattern variable; above
mostly light orange-brown marked
with brownish black; below paler with
yellowish-white bands and white
submarginal crescentlike marks
(lunules). Caterpillar is orange with
black bands, black dorsal stripe, and
6 rows of short black spines.

Habitat: Moist, brushy meadows, where white
aster grows abundantly.

Range: Southern Canada to mid-Atlantic
states, west to the Great Plains.

Food: Caterpillar eats white aster leaves.

Flight: June.

These caterpillars feed together until
half grown. Their barbed spines
effectively repel birds. There is 1
generation a year. The related Silvery
Checkerspot (*C. nycteis*), wingspan 1½–
1⅝" (38–40 mm), is similarly marked
above but has paler, less contrastingly
marked hind wings below and only a
single white crescentlike mark on the
margin of each wing. Its caterpillar
feeds on various members of the aster
family in southern Canada and most of
the United States, west to the Rocky
Mountains. Harris' Checkerspot and the
Silvery Checkerspot were formerly
included in the genus *Melitaea*.

11, 596 Painted Lady
"Thistle Butterfly"
(*Cynthia cardui*)

Description: Wingspan 2–2¼" (50–58 mm). Fore
wings above orange with rosy tinge,
numerous black markings and several
white spots near tips. Fore wings below
grayish, marbled with white and black
in front ⅓, pink at base. *Hind wings
above orange, below with 4 eyespots*

(2 small eyespots between 2 moderate-sized ones). Caterpillar, to 2″ (50 mm), is greenish yellow to pink with 7 rows of spiny yellow tubercles, yellow side bands, and scattered black spots; hairy black head.

Habitat: Meadows, gardens, and open spaces.

Range: Worldwide.

Food: Caterpillar eats thistle, burdock, sunflower, sometimes hollyhock, and other herbaceous plants.

Flight: Spring–fall in the North, year-round in the South.

The caterpillar transforms to a greenish or bluish-white chrysalis with gold-tipped tubercles and black and brown marks. There are 2 generations a year in the North, probably more in the South. It was formerly assigned to the genus *Vanessa*. The West Coast Lady (*C. annabella*), wingspan 1⅝–1¾″ (40–45 mm), resembles the Painted Lady but has an orange instead of a white bar at the end of the fore wing cell. Its caterpillars feed on mallows west of the Rocky Mountains, from Mexico to British Columbia.

619 American Painted Lady
"Painted Beauty"
(*Cynthia virginiensis*)

Description: Wingspan 1¾–2″ (45–50 mm). Almost identical to the Painted Lady. Fore wings above orange with rosy tinge, numerous black markings and white spots near tips. Fore wings below grayish, marbled with white and black in front ⅓, pink at base. *Hind wings* above orange, *below with 2 large eyespots*. Caterpillar is velvety black with black spines, narrow yellow-green crossbands, and white spots on sides.

Habitat: Meadows and open areas.

Range: Throughout southern Canada and the United States.

Food: Caterpillar feeds on foliage of composite flowers, especially everlastings and forget-me-nots.

Flight: Spring–fall.

This species, like its close relatives, fluctuates greatly in abundance from year to year—occasionally it is extremely abundant, other times fairly scarce. Its caterpillars fashion compact, silken nests around leaves and blossoms, and often pupate inside. Formerly assigned to genus *Vanessa*.

623 Baltimore
(*Euphydryas phaeton*)

Description: Wingspan 1¾ 2½" (45 65 mm). Wings blue-black above with an outer row of orange spots, rows of smaller white spots nearer base, dull orange areas near base. Wings below with 2 rows of orange-yellow spots. Caterpillar is black with reddish-orange bands and short black spines.

Habitat: Wet meadows in the North, dry hillsides in the South.

Range: Southern Canada to Georgia, west to Kansas.

Food: Caterpillar eats foliage of turtlehead, honeysuckle, and occasionally other plants, but usually a particular plant in a given area.

Flight: May–July.

Baltimore caterpillars spin communal webs around the food plant, in which they overwinter. In spring they scatter in search of food, often feeding upon substitute host plants after exhausting the available supplies of preferred food. The pearl-gray chrysalis is marked with yellowish-brown and black spots. There is 1 generation a year. The related Chalcedon Checkerspot (*E. chalcedona*), wingspan 1¾–3" (45–75 mm), is variably marked, from predominantly brick-red to mainly black with rows of yellow spots. It occurs from southern California to Oregon. Its caterpillars feed on monkey flower, Indian paintbrush, and related plants.

585 **Zebra**
"Yellow-barred Heliconian"
(*Heliconius charitonius*)

Description: Wingspan 3–3⅜″ (75–85 mm). Wings long, narrow, and black with 3 yellow lines crossing fore wings; hind wings with broad yellow band and row of yellow spots near hind margin. Both wings below grayish brown with crimson at base. Abdomen extends back as far as hind wings. Caterpillar is gray to white with brown or black spots and 6 rows of branching black spines.

Habitat: Moist tropical forests.

Range: Gulf states, rarely wandering to Kansas or southern South Carolina.

Food: Caterpillar eats passion flower foliage.

Flight: Throughout the year.

This butterfly makes a creaking sound by wriggling when disturbed. Large gatherings congregate at nightfall and roost in groups on bare twigs. There are continuous generations throughout the year. The black to brown pendant chrysalis is oddly shaped because the abdomen of the developing butterfly is not in line with the thorax and head. Other related species occur in tropical America. They have the same narrow long wings and feebly developed fore legs and fly at slow speed, moving their wings only slightly. The Zebra was formerly placed in the Family Heliconiidae.

621 Viceroy
(*Limenitis archippus*)

Description: Wingspan 2½–2¾" (65–70 mm). Wings above brownish orange with blue-black or dark brown margins and veins. Fore wings have 2 white spots near tip along front margin and a row of white dots in a black band along outer margin. Hind wings have a narrow, curved black line crossing middle of wing, parallel to outer margin. In the South, orange color becomes quite dark. Caterpillar is white and olive-brown and has 2 prominent spiny tubercles and several smaller tubercles, giving it an irregular shape.

Habitat: Wet habitats, wherever willow or poplar grow.

Range: Throughout the United States and southern Canada.

Food: Caterpillar eats foliage of willow and poplar.

Flight: Throughout the summer.

The irregular shape and color of the caterpillar produce a striking resemblance to bird droppings, giving the insect considerable protection from predators. The adult butterfly is also somewhat protected because its coloration mimics the Monarch. Birds avoid Viceroys if they have previously sampled a Monarch or a Queen, which are poisonous. However, those birds that have not had this experience readily eat Viceroy butterflies. The Viceroy glides with its wings held horizontally, not at an angle as does a Monarch or Queen. There are usually 2 generations in the North, more in the South.

624   White Admiral
"Banded Purple"
(*Limenitis arthemis*)

Description: Wingspan 3⅛–3⅜" (80–85 mm).
Wings above black, overlaid with
metallic blue or iridescent green and
with a prominent band of white
interrupted by narrow black veins. A
series of red spots may or may not be
present near margins, especially on
hind wings. Wings below have more
prominent red spots. Caterpillar is
white and olive-brown with prominent
spiny tubercles.

Habitat: Deciduous and mixed forests,
particularly close to water.

Range: Coast to coast in Canada and the
northern United States, south to
Pennsylvania.

Food: Caterpillar feeds on foliage of willow,
poplar, hawthorn, birch, shadbush, and
basswood.

Flight: Mid-June to July.

The White Admiral caterpillar is nearly
indistinguishable from the Viceroy
caterpillar and has a similar life cycle,
except that there is only 1 generation a
year. The related Weidemeyer's
Admiral (*L. weidemeyerii*), wingspan
2½–3" (65–75 mm), is banded
similarly in white but without blue or
green iridescence. It occurs from the
western Great Plains to eastern
California.

2, 625   Mourning Cloak
(*Nymphalis antiopa*)

Description: Wingspan 2¾–3⅜" (70–85 mm).
Wings above velvety purplish brown,
broadly edged with dirty yellow beyond
a black band enclosing separate blue
spots. Wings below dark brownish gray
with pale gray freckling resembling
weathered wood. Caterpillar, to 2⅜"

(60 mm), is blue-black and has a row of red spots on back, many minute white dots on back and sides, and branching spines.

Habitat: Deciduous and mixed woods, meadows, and suburban and urban parks.

Range: Throughout North America.

Food: Caterpillar feeds on willow, elm, and poplar.

Flight: Warm winter days; early spring–summer.

This butterfly presumably got its name from its somber color, resembling the dark funeral shawl worn by widows. Its caterpillars feed in large groups, often defoliating trees. There are 2 or more generations in the South.

617 Milbert's Tortoise Shell
(*Nymphalis milberti*)

Description: Wingspan 1¾" (45 mm). Wings above dark brown on basal ½, except for 2 reddish-orange spots separated by a black bar near front margin of fore wing; broad yellow-orange band across both wings; metallic blue crescentlike marks (lunules) on margins of hind wing. Wings below brownish black; basal portion darker, marbled with gray and white. Caterpillar is black above with white dots, yellowish sides, and branched spines.

Habitat: Wet meadows and swampy edges of deciduous forest.

Range: Coast to coast in Canada and the northern United States, in mountains south to West Virginia.

Food: Caterpillar feeds on nettles.

Flight: Warm winter days; spring and summer.

Like many other nymphalid caterpillars, these feed in groups, sometimes severely defoliating trees. There are often 3 generations a year.

612 **Pearl Crescent**
(*Phyciodes tharos*)

Description: Wingspan 1¼–1¾″ (33–45 mm).
Wings above mostly brownish orange,
marked with dark brown. Wings below
brighter orange with fewer marks on
fore wing (summer generation) or
darker marks (spring generation). Hind
wings are more silvery and dull brown
in 1st generation, more yellow in 2nd
generation. Caterpillar is black, banded
with yellow on each side, dotted with
yellow on top, and has 8 rows of
brownish-yellow spines.

Habitat: Open areas.

Range: Most of North America, except the
Pacific states.

Food: Caterpillar feeds on asters and other
members of this family.

Flight: Spring–fall.

One of our most common small
butterflies, the Pearl Crescent is often
seen along roadsides and at mud
puddles. It is especially fond of
butterfly weed or orange milkweed
flowers. Spring butterflies are usually
darker than summer generations.

7, 615 **Question Mark**
(*Polygonia interrogationis*)

Description: Wingspan 2½–2¾″ (65–70 mm).
Wings above tawny brownish orange
with numerous black spots and pale
lilac around margins. Hind wings
darker, especially in 1st generation.
Wings below are brownish gray with a
silvery marking resembling a question
mark or semicolon near center of hind
pair. Caterpillar, to 2½″ (65 mm), is
spiny, cylindrical, and reddish brown
with lighter dots and streaks, and with
a roughly cubical head.

Habitat: Deciduous forests and adjacent shrubby
areas.

Range: Throughout most of southern Canada and the United States, except the Southwest and Pacific Coast.

Food: Caterpillar eats foliage of elm, nettle, hackberry, hops, basswood, and many other members of nettle family.

Flight: June–fall.

This butterfly lays flat-topped, cone-shaped eggs in clusters or short rows. There are 2 generations in the North, and up to 4 in the South. The Comma (*P. comma*), wingspan 1¾–2″ (45–50 mm), lacks the silver dot on the underside of the hind wings and occurs in the eastern United States and southern Canada, south to Georgia, west to the Great Plains. Its caterpillar, the "Hop Merchant," is a pest on hops. At rest both butterflies hold the wings at an angle.

1, 607 Buckeye
(*Precis coenia*)

Description: Wingspan 2⅛–2½″ (55–65 mm). Wings above mostly light olive-brown; fore wings have yellowish-white mark at tip enclosing 1 eyespot toward the hind margin and 2 red chevrons in discal cell; hind wings have 1 large and 1 small eyespot. Wings below similarly marked but paler. Caterpillar, to 2″ (50 mm), is black-gray, striped yellow-orange or whitish along sides, and has many paired branching spines.

Habitat: Open areas.

Range: Throughout most of the United States, south to the tropics; rare in the North, except along the Atlantic Coast.

Food: Caterpillar feeds on snapdragon, monkey flower, plantain, stonecrop, and other low herbs.

Flight: Midsummer–fall.

One of our most strikingly colored butterflies, the Buckeye flies swiftly if disturbed. Males dash after others of their own and different species and even chase Carolina Locusts. There are 2 or more generations a year.

611 Great Spangled Fritillary
(*Speyeria cybele*)

Description: Wingspan 3–3¾" (75–95 mm). Wings above orange-brown, marked with many black spots. Fore wings below paler with brownish-black markings and a few silvery-yellow spots. Hind wings darker below basally with silvery spots partly edged in black and a tawny band separating 2 outermost rows of silver spots. Females have bolder dark markings on wings above. Caterpillar is velvety black with 6 rows of branching orange-red and black spines.

Habitat: Moist open woods and meadows.

Range: Throughout southwestern Canada and most of the United States, except the Deep South.

Food: Caterpillar eats violet foliage.

Flight: Mid-May to October.

The Great Spangled Fritillary is the most common of the larger fritillaries. Its caterpillars eat only at night and hide during the day away from their food source. In the western portion of their range, females become much paler; at the extreme they are straw-yellow, resulting in a striking difference between males and females that is not especially noticeable in the East. Several other related species occur in various parts of the East and west of the Great Plains; all have similar life cycles. They can be difficult to identify correctly, presenting a challenge to even an entomologist.

618 Red Admiral
(*Vanessa atalanta*)

Description: Wingspan 2–2⅜" (50–60 mm). Fore wings dark brown with darker tips; several white spots and a broad orange crossband on darker part. Hind wings dark brown with orange outer margin. Caterpillars vary in color—black, mottled white, pale green, or brown— and have yellow spots and raised white tubercles.

Habitat: Deciduous forest edges and adjacent meadows.

Range: Throughout North America, except in arctic regions.

Food: Caterpillar eats leaves of hops, nettles, false nettle, and pellitory.

Flight: Spring and midsummer.

Males are territorial and will often chase away other males straying into areas that they claim. The caterpillars feed alone in folded leaves held together by silk and may pupate in this nest. There are 2 generations a year.

HACKBERRY AND GOATWEED BUTTERFLIES
(Family Apaturidae)

Apaturids are medium-sized, orange-brown to tawny butterflies with wingspans of 1–3" (25–75 mm). In some ways they are intermediate between the satyrids and the brush-footed butterflies and have often been included with the latter. The caterpillars resemble those of the satyrids, having branching horns on the head and a forked abdominal tip. The adults, however, lack the swollen, bulbous bases of the primary veins in the fore wings, which are characteristic of the satyrids, and are strong fliers, like the brush-footed butterflies. The hackberry butterflies have numerous eyespots on the wings, like the satyrids; the goatweed butterflies do not. Pupae of the goatweeds are compact, smooth, and barrel-shaped, resembling those of the milkweed butterflies. The pupae of the hackberry butterflies are unlike those of satyrids, brush-footed butterflies, or milkweed butterflies. Adults are commonly attracted to decaying fruit, dead animals, and excrement. Unlike other butterflies, they do not feed at flowers. The caterpillars feed on hackberry and goatweed. They overwinter as partly grown caterpillars or as adults. Most North American species are found in the South.

---

604 Hackberry Butterfly
(*Asterocampa celtis*)

Description: Wingspan 1¾–2⅛" (45–55 mm). Male's fore wings narrower and more pointed than female's. Wing coloration varies in each sex. General color tawny brown with larger white spots and black spots on fore wings than on hind wings, which have a broken row of small black eyespots encircled by pale brown; 1 large black eyespot on fore wings. Caterpillar is spindle-shaped, striped lengthwise with green, yellow, and white, and has 2 large spiny horns on rear of head; forked abdominal tip.

Habitat: Open woodlands and forest edges, wherever hackberry trees occur.

Range: Throughout most of the eastern half of the United States, more common in the South.

Food: Caterpillar feeds on hackberry foliage.

Flight: June–September.

Adults fly rapidly and erratically, alighting on tree trunks, twigs, leaves, and sometimes on people. Caterpillars are gregarious when young. They overwinter when partly grown. There are 2 or more generations in the South. The Tawny Emperor (*A. clyton*), wingspan 1⅞–2⅝″ (47–66 mm), lacks the eyespot on the fore wings, has larger eyespots on the hind wings, and an unbroken black streak beyond the middle of the fore wings. It ranges from the Atlantic Coast to slightly west of the Mississippi River, almost to the Canadian border.

SATYRS, NYMPHS, AND ARCTICS
(Family Satyridae)

Most members of this family are dull brown or tawny, often with 1 eyespot or more on each wing. Their wingspan ranges from 1 to 2¾″ (25–70 mm). In both sexes the fore legs are small and useless for walking. The dull colors and erratic, bobbing flight make the adults difficult to follow as they fly, usually close to the ground, in wooded areas or in open, brushy places. Adults seldom visit flowers, feeding instead on sap oozing from wounds in tree trunks or on the juice of decaying fruit. The caterpillars are spindle-shaped and have a fork at the end of the abdomen; they feed almost entirely on grasses and sedges. The chrysalis is usually concealed on or near the ground, either suspended in a cavity under a rock or lying in fallen leaves.

600 **Wood Nymph**
"American Grayling"
(*Cercyonis pegala*)

Description: Wingspan 2–2¾″ (50–70 mm). Wings chocolate-brown; fore wings often have yellow patch above and below near outer margin, usually containing 2 blue eyespots edged with black. (Northern and western butterflies lack the yellow patch.) Both wings below have fine netlike brown lines. Caterpillar is yellowish green with 4 pale lengthwise stripes and a red forked abdominal tip.

Habitat: Deciduous and mixed woodlands, adjacent meadows.

Range: Throughout most of the United States and southern Canada.

Food: Caterpillar feeds on grasses.

Flight: Late June—summer.

This insect flies erratically, often alighting on tree trunks. The species is highly variable geographically and has been divided into a number of named populations. All females lay ivory-white cone-shaped eggs singly. Caterpillars overwinter and feed again in early spring.

599 Little Wood Satyr
(Euptychia cymela)

Description: Wingspan 1¾–2″ (45–50 mm). Wings above light brown, paler areas near outer margins enclosing 2 prominent eyespots on each wing. Wings below paler, similarly marked, several additional smaller eyespots along outer margin of both wings. Caterpillar is finely hairy, pale greenish brown with dark lengthwise stripes.

Habitat: Open deciduous forests and adjacent shrubby meadows.

Range: Eastern half of the United States and southern Canada.

Food: Caterpillar feeds on grasses.

Flight: May—September.

One of the most common satyrs, this butterfly erratically swoops close to the ground or darts through gaps in the shrubbery. Several other related species are much less common and occur throughout the Southeast.

598 Eyed Brown
"Grass Nymph"
(*Lethe eurydice*)

Description: Wingspan 1⅜–2" (40–50 mm). Wings olive-brown to yellowish brown; with a paler band beyond middle surrounding 4 small dots, each encircled by a still paler circle—these correspond to 4 eyespots on undersurface. Hind wings have 6 encircled dark spots above (1 tiny eyespot near inner margin, 1 large eyespot near front edge, 4 along outer margin)—all corresponding to conspicuous eyespots on undersurface. Caterpillar is pale green with narrow dark green stripes and red-tipped horns on its head and last segment.

Habitat: Marshes and open wet meadows.

Range: Central Quebec to Northwest Territories, south to mid-Atlantic states; local and spotty in distribution.

Food: Caterpillar eats sedges.

Flight: June–August.

This butterfly flies so weakly and close to the ground that it is easy to observe. Its caterpillar overwinters partly grown and pupates in late spring. The chrysalis is olive-green. There is 1 generation a year.

602 Pearly Eye
(*Lethe portlandia*)

Description: Wingspan 1⅜–2" (40–50 mm). Wings yellowish brown; fore wings above have a row of 4 dark brown spots outlined in yellow; hind wings above have 6 similar spots. Wings below have dark brown spots in eyelike patterns. Caterpillar is yellowish green with 2 red-tipped horns on head and longer, forked red-tipped horns on last abdominal segment.

Habitat: Deciduous forests and woodland margins.

Range: Eastern half of North America, from

southern Canada to the Gulf of Mexico.

Food: Caterpillar eats grasses.

Flight: June and July.

Pearly Eyes fly rapidly. Males alight on tree trunks, dash at other butterflies, and battle furiously with males of their own species for territory. The female lays flattened, spheroid eggs singly on grasses. The caterpillar overwinters and then transforms to a pale green chrysalis, which is attached to grasses close to the soil. There is 1 generation a year in the North, 2 in the South. The Creole (*L. creola*), same size, has fore wing veins outlined in pale scales, and the female has 5 instead of 4 eyespots below its fore wing. It occurs in the Southeast, except for Florida.

603 Chryxus Arctic
(*Oeneis chryxus*)

Description: Wingspan 1⅜–2½″ (35–65 mm). Wings brownish yellow with dark brown margins; 2 or more small eyespots above and below on fore wings, and often 1 eyespot on hind wings. Body and appendages brown. Caterpillar is pale green or pale brown with lengthwise stripes.

Habitat: Subarctic grassy meadows.

Range: Quebec to British Columbia, south in the West to California and New Mexico.

Food: Caterpillar feeds on grasses.

Flight: Every other year in midsummer.

The caterpillars of this butterfly often need 2 years to complete growth in their cold environment. The stout chrysalis is unattached and usually found in a slight depression in soil, between grass roots, or under stones. The Jutta Arctic (*O. jutta*), wingspan 1¾–2⅛″ (45–55 mm), has dull orange near the hind wing margins and some orange around the dark eyespots on the fore wings. It is found near Hudson and James bays in northern Ontario, south into Newfoundland, northern Nova Scotia, northern Maine, and New Hampshire, and occasionally in northern Michigan. The Great Arctic (*O. nevadensis*), wingspan 2–2⅜" (50–60 mm), has rusty orange wings with black margins, several eyespots with white centers, and a dark smudge on the male's fore wings above. It ranges from the Arctic to California.

MILKWEED BUTTERFLIES (Family Danaidae)

The few North American members of this family are large, strong-flying, orange or brown butterflies with wingspans of 3–4" (75–100 mm). The margins of the wings are black with white speckles, and the front legs are reduced in both sexes. Males have a scent pocket on each hind wing, which releases an odor attractive to females. Adults are usually seen feeding at flowers or flying in a leisurely manner in open places. Both adults and caterpillars are toxic to predators, since the caterpillars feed on milkweeds and other poisonous plants. Like the adults, the caterpillars are boldly patterned.

17, 622 **Queen**
(*Danaus gilippus berenice*)

Description: Wingspan 3⅛–3⅜" (80–85 mm). Wings dark reddish brown; fore wings have white spots near tip; fore and hind wings have smaller white or light brown spots in dark brown marginal bands. Each hind wing of male has black scent pocket. Caterpillar, to 2½" (65 mm), is whitish with brown and yellow bands, a yellow stripe along sides, and 3 pairs of black filaments.

Habitat: Meadows and roadsides.

Range: Georgia and Florida to Kansas.

Food: Caterpillar feeds on milkweed foliage.

Flight: Throughout the year.

Unlike the Monarch butterfly, the Queen does not migrate but remains all year in the extreme South, where it is especially common in the Florida Everglades. There are usually 3 generations a year. The Mexican Queen (*D. g. strigosus*), same size, has veins on its hind wings that are narrowly edged with grayish white. It is found from Mexico to Texas, New Mexico, Arizona, and California.

16, 620 **Monarch**
(*Danaus plexippus*)

Description: Wingspan 3½–4″ (88–100 mm). Wings brownish orange; black to dark brown on margins with 2 rows of orange and/or white spots and veins outlined in black. Each hind wing of male has a black scent pocket. Head and body black with white spots. Caterpillar, to 2¾″ (70 mm), is black with white and yellow bands and has a pair of flexible black filaments on its thorax and next-to-last abdominal segment.

Habitat: Meadows, roadsides, and sandy areas wherever milkweeds grow.

Range: Throughout North America, except far North.

Food: Caterpillar feeds on milkweed foliage, flower buds, and milky juice.

Flight: Late May–fall.

Monarchs are capable of flying 2,000 miles from Canada to Mexico and back again to the southern United States. Millions migrate every autumn, often stopping in the same rest spots each year. Some even fly as far as Hawaii and eastern Australia. In early spring and summer, returning females travel north in relays, new generations replacing old, laying their eggs along the way. The fully grown caterpillar changes to a barrel-shaped, leaf-green chrysalis studded with gold dots, then shows the colors of the developing butterfly inside. The change from egg to butterfly takes about 4 weeks. There are many generations a year. The Canadians call this butterfly "King Billy" because its orange and black colors are those of King William of Orange.

SMOKY MOTHS
(Family Zygaenidae)

Smoky moths are small, black or brightly colored, and have wingspans of ⅝–1⅛" (16–28 mm). They have rounded wings with a thin covering of scales and a well-developed proboscis. Some species are nocturnal. Members of diurnal species visit flowers and strongly resemble ctenuchids, but they can be distinguished by wing venation. Most caterpillars feed on the foliage of Virginia creeper or grape. Several often eat side by side, devouring an entire leaf before moving on to another one. This family was formerly known as Pyromorphidae.

520 Grape Leaf Skeletonizer
(*Harrisina americana*)

Description: Wingspan ¾–1" (18–25 mm). Wings long and narrow, rounded at tip, dull metallic blue to black. Body, antennae, and legs same color. Collar orange. Antennae narrowly feathered. Caterpillar, to ½" (13 mm), is yellow with black spots and spines.

Habitat: Moist, deciduous woods and marginal areas.
Range: Eastern half of the United States.
Food: Caterpillar eats grape foliage.
Flight: May–August.

Tiny caterpillars often line up side by side to feed on leaves. While young they do not eat veins, but leave them as a skeleton—hence the common name. As they grow, the caterpillars eat small veins, leaving only the coarse ones. Fully grown caterpillars disperse over the vine, then spin tough, flat white cocoons, emerging as adults about 2 weeks later. There are probably 2 generations a year.

SLUG CATERPILLAR MOTHS (Family Limacodidae)

Adults of this family are stout, rather hairy moths with stumpy, rounded wings. Most are brownish with green, white, or silver markings and wingspans of ⅜–1⅛″ (10–30 mm). The caterpillars are short and stocky and do not have prolegs. They creep about on leaves in a sluglike manner. In some species they have tufts of short, stinging bristles that protect them from predators. These bristles are incorporated into the firm-walled cocoon, so that the pupa is similarly protected. Adults do not eat; caterpillars feed on many plants.

30 Saddleback Caterpillar Moth
(Sibine stimulea)

Description: Wingspan 1⅛″ (30 mm). Fore wings glossy dark brown with black streaks and several white dots. Hind wings paler. Body relatively heavy, densely covered with dark brown scales. Caterpillar, to 1″ (25 mm), is green at middle with brown at each end and a brown or purple saddle-shaped mark edged with white on midback; many tufts of bristles.

Habitat: Forest edges, orchards, and gardens.

Range: Eastern United States from Massachusetts to the Gulf states, north to Illinois.

Food: Caterpillar eats foliage of a wide variety of orchard trees, horticultural shrubs, corn, and other plants.

Flight: Midsummer.

Saddleback caterpillars are easier to recognize than adult moths. The spines on their sides are mildly poisonous and, if touched, will sting.

PYRALID MOTHS
(Family Pyralidae)

Members of this large and varied family of small to medium-sized moths have long palps that form a snoutlike projection in front of the head and wingspans of ⅜–2⅛″ (10–55 mm). They differ from other similar-looking families in having a flat, round hearing organ, or tympanum, on each side of the abdomen. The caterpillars feed mainly on foliage or dried plant debris, including stored grain products. A few are aquatic, and in one species the caterpillar eats wax in beehives.

525 Mediterranean Flour Moth
(Anagasta kuehniella)

Description: Wingspan ⅞″ (22 mm). Body and fore wings dark gray. Fore wings crossed by zigzag black markings and wavy dark bands. Hind wings grayish white. Caterpillar is cream-colored to reddish with a honey-colored head.

Habitat: Mostly indoors, in flour mills, grain elevators, and storage bins.

Range: Worldwide.

Food: Caterpillar feeds on whole or cracked grain, flour, and bran.

Flight: Throughout the year.

This moth and its caterpillars live in total darkness. The full-grown caterpillars spin silken tubes and webbing among cereal products, taking 8–10 weeks to pupate. The Tobacco Moth (*A. elutella*), wingspan ⅝″ (15 mm), has several brown areas and 1 partial whitish crossbar on its fore wings. The caterpillars attack tobacco and various stored food products.

532 Sod Webworm Moths
(*Crambus* spp.)

Description: Wingspan ½–1½″ (12–39 mm). Wings grayish white to brown, often with silvery streaks. At rest, wings curl around the body, giving insect an almost cylindrical form; pattern of darker and paler areas on fore wings inconspicuous until wings are spread. Caterpillar is dirty white to pale brown with pairs of dark spots on back and side and a few coarse hairs.

Habitat: Grasslands, open fields, and pastures.

Range: Throughout North America.

Food: Caterpillar feeds in stems, crowns, and roots of grasses.

Flight: Throughout the summer.

Adults rest on grass blades, head down or up, and fly away if approached closely. A single caterpillar may produce an irregular brown spot in a new lawn; many may cause severe damage by destroying grass roots—a situation often aggravated by extended periods of dry weather. There are 2–3 generations a year.

524 Indian Meal Moth
(*Plodia interpunctella*)

Description: Wingspan ⅝" (15 mm). Fore wings silvery-gray, outer ⅔ coppery brown. Hind wings dirty white. Caterpillar is yellow-green or pinkish with brown head and cervical shield.

Habitat: Many places, including grain storage centers and homes.

Range: Worldwide.

Food: Human food in varying degrees of preparation, especially types high in carbohydrates and proteins, such as packaged cereals, candies, and dried fruits.

Flight: Throughout the year.

This widespread moth is a major pest, because one female lays 300–400 eggs that hatch in a few weeks. The caterpillars spin silken webs among food and pupate in them, emerging as adults in a few months. Several generations a year are possible. These moths are controlled by fumigation with gases that do not harm packaged food or leave harmful residues on food products.

537 Meal Moth
(*Pyralis farinalis*)

Description: Wingspan ⅝–1" (15–25 mm). Fore wings pale brown at middle crossed by narrow white lines, dark brown at base and tip. Hind wings gray with 2 curving white lines and a broad pale fringe. Caterpillar, to 1" (25 mm), is whitish with black head and orange-bordered cervical shield.

Habitat: Mostly indoors, in storage bins and granaries.

Range: Worldwide.

Food: Caterpillar eats grains, meal, bran, husks, straw, and moist stored hay.

Flight: Throughout the year.

Cleanliness and frequent emptying of storage bins are the best means of controlling this insect, which is a common household pest. Its caterpillars spin tubular webs among food, eating from an open end. Then they abandon the feeding tubes, spin cocoons, and pupate 6–8 weeks after eggs hatch. There are several generations a year.

PLUME MOTHS
(Family Pterophoridae)

These slender moths have narrow wings that are deeply divided into feathery lobes, and long legs bearing spines at the joints. They have wingspans of ½–1⅝″ (12–40 mm) and are often seen resting quietly on leaves with their wings held out stiffly at right angles to the body. The caterpillars are hairy and bore into stems or live in rolled leaves, feeding on foliage. A few species are agricultural pests.

538 Artichoke Plume Moth
(*Platyptilia carduidactyla*)

Description: Wingspan ¾–1⅛″ (20–28 mm). Fore wings narrow, shallowly notched at outer margin; mostly pale brown with 1 pale spot just beyond 1 dark spot at the front margin toward the tip. Hind wings somewhat broader, twice notched on outer margin; mostly grayish. Body and appendages tan to gray. Caterpillar is yellowish with shiny black head, black cervical shield, and black plate on last abdominal segment.

Habitat: Meadows and cultivated fields.

Range: Coast to coast.

Food: Caterpillar feeds on tender foliage and buds of artichoke, carduus, and thistles.

Flight: Midsummer.

This graceful little plume moth rests
easily on long slender legs, ready to
flutter away from any disturbance. The
caterpillars spin webs on plants while
feeding. The pupating caterpillars
suspend themselves by the cremaster,
and the pupae retain this connection,
resembling many butterfly pupae.

MEASURINGWORM MOTHS
(Family Geometridae)

Moths in this large and varied family
are rather delicate with slender bodies
and broad, flimsy wings spanning ⅜–
2½" (8–65 mm). They are easily
recognized by their habit of spreading
their wings out flat when at rest,
exposing the similarly patterned fore
and hind wings. Geometrids have a
hearing organ, or tympanum, on each
side of the abdomen. In a few species
the females are wingless. Some species
feed as adults, others do not. The larvae
are the familiar measuringworms or
inchworms—slender caterpillars with
1 or 2 pairs of prolegs at the end of the
abdomen and a characteristic looping
method of locomotion. They feed on
many different plants and are often seen
hanging by a strand of silk from the
foliage of trees.

560 Tulip Tree Beauty
(*Epimecis hortaria*)

Description: Wingspan 1½–2" (38–50 mm). Body
and wings light gray mottled with
brown. Fore wings above have pattern
of scalloped brown crosslines, a brown
dot or bar in discal cell, streaks on
outer margin. Hind wings above
similar with light brown fringe. Both
wings below light brown with gray-
brown dots; outer margin of hind wing

dusky. Caterpillar, ¾–1⅛″ (19–30 mm), is pale brown above with dark lines and spots and black lateral band; whitish below.

Habitat: Moist woods and forests.

Range: Southern Ontario to Florida, more common in the South and Appalachian region.

Food: Caterpillar feeds on the foliage of sassafras, yellow poplar, and red bay.

Flight: May–July.

The Tulip Tree Beauty is one of the largest geometrids in North America. It rests with its wings fully spread and pressed flat against the bark of a tree, where it is so well camouflaged that it is almost invisible. In 1936 its caterpillars severely defoliated sassafras trees in Connecticut.

543 California Cankerworm Moth
(*Paleacrita longiciliata*)

Description: Wingspan 1–1⅜″ (25–35 mm). Male's fore wings grayish brown or gray with brownish or blackish scales, small grayish black spot in discal cell. Hind wings above light grayish brown, somewhat translucent. Female's wings vestigial, less than ⅟₁₆″ (1 mm) long; body grayish brown with some gray scales.

Habitat: Open scrub areas and deserts.

Range: California.

Food: Caterpillar eats foliage of shrubs in the rose family.

Flight: November–May.

In this species and its eastern relations, females are wingless. The caterpillars, called cankerworms, occasionally cause severe defoliation of forest and shade trees.

14, 539 Large California Spanworm Moth
(*Procherodes truxaliata*)

Description: Wingspan 1⅜–1¾" (35–45 mm). Fore wings yellowish brown, slightly incurved from tip to middle of outer margin producing 2 projecting points. Hind wings paler with 1 projecting point midway along outer margin. Wings are variably speckled with gray-brown; fore wings crossed by 2 transverse bands, which vary from strongly contrasting dark gray to an almost completely indistinguishable yellow-brown. Caterpillar, to 2⅛" (55 mm), is white with black lengthwise stripes, a yellow stripe along each side, and an orangish head.

Habitat: Sandy, arid regions to mountain forests; not specialized.

Range: Colorado and New Mexico to California Sierras, south to Baja California.

Food: Caterpillar feeds on a great variety of trees and shrubs.

Flight: May–August.

This species is occasionally abundant enough in southern California to become a minor pest of cultivated trees and shrubs.

15 TENT CATERPILLAR MOTHS
(Family Lasiocampidae)

These heavy-bodied, rather hairy, dull brown moths have wingspans of ⅞–4⅛" (22–105 mm). Unlike owlet moths, the adults do not feed and have only a very small proboscis or none at all. They have shorter wings and more feathery antennae. The males' antennae are bipectinate, having 2 feathery branches on each segment. The caterpillars, 1½–3" (37–75 mm), are slender and hairy. In many species they are social, living together in silken tents and feeding on the foliage of trees. The cocoon is frequently spun in some protected place—often under eaves of houses, outdoor furniture, or loose bark.

546 Western Tent Caterpillar Moth
(*Malacosoma constrictum*)

Description: Wingspan 1¼−1⅝″ (32−41 mm). Fore wings beige to faint brown crossed by 2 dark bands parallel to outer margins. Body brown, hairy. Antennae narrowly feathery in both sexes. Mouthparts vestigial. Caterpillar is orange-brown with blue dots on the back and sides.

Habitat: Woods and on isolated trees.

Range: California to Washington.

Food: Caterpillar feeds almost exclusively on the foliage of oaks, and only occasionally on other deciduous broad-leaved trees and shrubs, after available oak has been defoliated.

Flight: May−July.

These tent caterpillars appear in early spring. They group together to spin a tent of gray silk between branches usually at a fork. This tent is used as a molting mat, unlike the much larger tents of some related species, whose caterpillars hide by day, emerging to feed at night. Western Tent caterpillars pupate in pale cocoons spun in bark crevices or surrounded by curled leaves. There is 1 generation a year. The American Tent Caterpillar Moth (*M. americanum*), wingspan 1⅛−1⅝″ (30−41 mm), is dark brown with 2 oblique whitish lines nearly parallel to the outer margins of the fore wings. It is widespread in eastern North America, and lays eggs in cufflike clusters around twigs of apple, pear, wild cherry, and hawthorn. Its caterpillars defoliate and sometimes kill the tree, emerging only at intervals from large communal silken tents.

GIANT SILKWORM MOTHS
(Family Saturniidae)

Giant silkworms, with wingspans of
1⅛–5⅞″ (30–150 mm), are our largest
moths. Most are brightly colored, and
some species have large, transparent
eyespots on the wings. The antennae
are large and often feathery. These
moths do not have hearing organs, or
tympana. The short-lived adults have
vestigial mouthparts and do not feed.
They are usually seen at night, clinging
to window screens or fluttering like
bats around streetlights. Some species
lay eggs singly, others in small groups,
and still others in large masses. The
caterpillars are smooth or spiny and
generally feed on the foliage of trees. In
many species the caterpillars spin a
tough cocoon, which may be attached
to a twig or hidden in fallen leaves, but
in a few cases the pupa occupies a small
chamber in the soil instead of a cocoon.
Most species overwinter as pupae.
These large moths are not closely
related to the true Asiatic silkworm,
although largely unsuccessful attempts
have been made to utilize commercially
the silk from their cocoons.

24, 573 **Luna Moth**
(*Actias luna*)

Description: Wingspan 3⅛–4½″ (80–115 mm).
Wings pale green; fore wings have purple
front margins, *hind wings have long tails.*
Caterpillar, to 3⅛″ (80 mm), is green
with a yellow stripe on each side, spiny
tubercles, and hair.

Habitat: Deciduous forests.

Range: Eastern half of the United States and
southern Canada.

Food: Caterpillar eats foliage of hickory,
walnut, sweet gum, persimmon, birch,
and sometimes other trees.

Flight: April and June.

The Luna caterpillar pupates in a thin
cocoon, which may include a flexible
leaf, usually loose on the ground. There
are 2 well-defined generations a year in
most of the range. This beautiful moth
is only found in North America. It is
now considered an endangered
species because many have been
killed by pollutants and pesticides.

23, 567   **Polyphemus Moth**
          (*Antheraea polyphemus*)

Description: Wingspan 3½–5½″ (90–140 mm).
               Wings brownish yellow or ocher with a
               black and white irregular line parallel
               to the paler outer border. Fore wings
               have eyespot edged in yellow. Hind
               wings have larger eyespot edged in
               yellow and surrounded by black and
               blue. Caterpillar, to 3½″ (90 mm), is
               plump, bright green, with yellow
               bands, and has red and silver tubercles.

Habitat: Deciduous forests.

Range: East of the Rocky Mountains from
               Canada to Mexico.

Food: Caterpillar eats leaves of many different
               trees, including alder, basswood, birch,
               chestnut, elm, hickory, maple, poplar,
               and sycamore.

Flight: July.

Because of the conspicuous eyespot on
each hind wing, this moth is named
after Polyphemus, the one-eyed giant of
Greek myths. At night adults often fly
to artificial lights. The fully grown
caterpillars spin tough egg-shaped
cocoons, which may remain attached to
branches, but usually fall with the
leaves in late autumn. There are 2
generations a year in the South; 1 in
the North.

27, 566 Io Moth
(*Automeris io*)

Description: Wingspan 2⅜–2¾″ (60–70 mm).
Male's fore wings yellow; female's
reddish brown. Hind wings of both
sexes yellow with reddish-orange
submarginal band, reddish-orange
shading near inner margin, and central
black eyespot ringed with black.
Thorax yellow. Caterpillar, to 3″ (75
mm), is green with reddish-pink and
white stripes on each side and clusters
of branching spines.

Habitat: Open woods and meadows, occasionally
cornfields.

Range: East of the Rocky Mountains from
Canada to Mexico.

Food: Caterpillar eats foliage of a great variety
of plants, especially wild cherry.

Flight: July.

The caterpillar's spines cause a painful
stinging if they penetrate human skin.
These caterpillars spin thin, rather
flimsy cocoons among debris on the
ground.

563 Promethea Moth
"Spicebush Silkmoth"
(*Callosamia promethea*)

Description: Wingspan 2¾–4″ (70–100 mm). Sexes
dissimilar. Male's body, legs, and
wings largely brownish black with faint
spots (or no spots), rather obscure pale
crossband, reddish tip enclosing
prominent eyespot, and tan outer
margin with wavy dark brown lines and
spots. Female's body and wings reddish
brown with markings similar to those
of male but much more prominent and
contrasting. Caterpillar, to 3″ (75 mm),
is bluish green with 2 pairs of red
tubercles on thorax, 6 rows of small
black tubercles on thorax and abdomen,
1 yellow tubercle on the 11th

abdominal segment, yellow rear and legs, and a yellowish-green head; sometimes entire body appears frosted.

Habitat: Deciduous woods and open areas.

Range: Southeastern Canada and the eastern half of the United States, west to the Great Plains.

Food: Caterpillar usually feeds on spicebush, wild cherry, and sassafras, but also on a wide variety of other trees and shrubs.

Flight: Summer.

Male moths may fly in the late afternoon like butterflies, but females fly only at night. The caterpillars spin silken cocoons on plant stems, incorporating leaves. They were once considered to be a possible source of raw silk, but finding cheap labor to unreel the cocoons proved impossible in North America and the idea was abandoned. The slightly larger Tulip Tree Silkmoth (*C. angulifera*) resembles Promethea in pattern and has prominent spots on the basal half of the wings of both sexes. Its caterpillars feed on tulip tree foliage from Massachusetts to Florida, west to the Mississippi River.

28, 568   **Regal Moth**
"Royal Walnut Moth"
(*Citheronia regalis*)

Description: Wingspan 4¾–5⅞" (120–150 mm). Wings bright orange with gray stripes and yellowish spots. Body orange with yellowish bands. Caterpillar is variably colored, often blue-green, orange, or brown with several pairs of long, curved black-tipped orange horns on its back.

Habitat: Deciduous forests.

Range: Eastern United States and adjacent Canada, most common in the southern part of range.

Food: Caterpillar feeds on foliage of hickory,

walnut, butternut, ash, sumac; and in the South, on sweet gum and persimmon.

Flight: Midsummer.

This moth's caterpillar is called the "Hickory Horned Devil" because of its black-tipped orange horns. It pupates without a cocoon in an earthen cell. There is 1 generation a year. The Pine Devil Moth (*C. sepulchralis*), wingspan 3⅛–4″ (80–100 mm), has dull gray-brown wings and its caterpillars feed on pines from Massachusetts to Georgia.

572 Rosy Maple Moth
(*Dryocampa rubicunda*)

Description: Wingspan 1⅛–2″ (30–50 mm). Wings and body pale sulfur-yellow; fore wings pastel pink at base and along outer margin. Caterpillar is green with lengthwise white stripes, reddish-brown head, pink below rear abdominal segments, and has short black spines.

Habitat: Deciduous forests, or open brushy areas wherever maples grow.

Range: Eastern United States and southern Canada, west to the Great Plains.

Food: Caterpillar eats foliage of red and silver maples.

Flight: May–September.

This moth's caterpillars, called "Green-striped Maple Worms," are sometimes so abundant that they strip trees of all foliage.

19, 550 Imperial Moth
(*Eacles imperialis*)

Description: Wingspan 4–5⅞″ (100–150 mm). Wings yellow with purple-gray marks and dark speckles. Fore wings have pointed tips. Caterpillar, to 5⅛″ (130),

is green, orange, or brown with yellow
head, whitish-yellow spiracles, 6 short
yellow horns on thorax and smaller
horns down back, and is sparsely
covered with coarse white hair.

Habitat: Deciduous and mixed forests.

Range: East of the Rocky Mountains from
Canada to Mexico.

Food: Caterpillar feeds on many different
kinds of broad-leaved foliage or
coniferous needles.

Flight: June–August.

Imperial Moth caterpillars pupate in
earthen chambers. Adults often fly to
artificial lights, basking in the
illumination until dawn. Many also
remain there throughout the day and
are eaten by birds, so that the species is
unfortunately becoming rare in areas
where artificial lights are abundant.
Close relatives inhabit Latin America.

571 Sheep Moth
(*Hemileuca eglanterina*)

Description: Wingspan 2⅜–2¾″ (60–70 mm).
Color and pattern variable: fore wings
usually pinkish gray, hind wings
yellow, both with black stripes and
spots. Caterpillar, to 5⅞″ (150 mm), is
dark brown with red spots along
midline above, red line on each side,
and covered with tan and black short
spines.

Habitat: Mountain meadows and sheep pastures.

Range: Rocky Mountains west to Pacific Coast,
from southern California to southern
Canada.

Food: Caterpillar eats foliage of a variety of
plants, often those of rose family.

Flight: Summer.

Adults fly swiftly by day and close to
the ground. The female lays eggs in
masses. Full-grown caterpillars pupate
in late summer, and moths emerge the

next or following summer. The moth's common name comes from its association with areas where domestic sheep graze.

20, 564 Cecropia Moth
(*Hyalophora cecropia*)

Description: Wingspan 4¾–5⅞" (120–150 mm). Wings speckled gray-brown with rusty shading, especially near body, and have white crescents (lunules), white and red crossbands, tawny outer margin, and pale lilac tip with eyespots. Body dull red-orange with white collar and white rings on abdomen. Caterpillar, to 4⅜" (110 mm), is green with bluish shading on sides and back and covered with numerous red, yellow, and bluish tubercles.

Habitat: Open areas, often in and around cities.

Range: East of the Rocky Mountains in the United States and southern Canada.

Food: Caterpillar eats the foliage of many trees and shrubs, including ash, birch, alder, elm, maple, wild cherry, willow, apple, and lilac.

Flight: May or June.

This is the largest North American moth. Its caterpillars spin large brown cocoons that weather to gray. The cocoon is attached along one entire side to a branch, sometimes incorporating the branch and even twigs into its structure. There is 1 generation a year.

12, 562 Cynthia Moth
(*Samia cynthia*)

Description: Wingspan 3–5⅜" (75–135 mm). Body olive-brown with 3 tufts of white hair on each segment. Wings olive-brown; basal ½ darker with transverse middle bands of pink, white, and black; bold white crescents (lunules) bordered in front with black, behind with yellow. Basal part of both wings has curved white line bordered with black. Fore wing has small eyespot near tip. Caterpillar, to 4″ (100 mm), is green and has black dots, blue tubercles, black spiracles with white dots, yellow head, legs, prolegs, and rear; it is usually covered with whitish, powdery, waxlike substance.

Habitat: Only near major metropolitan areas, such as parks, railroad yards, and vacant lots, where ailanthus grows.

Range: Atlantic Coast.

Food: Caterpillar eats foliage of ailanthus, also known as tree-of-heaven.

Flight: Midsummer.

A native of the Orient, this moth was introduced into Philadelphia in the 1860s, from where it spread with the ailanthus tree to other cities on the East Coast. Its caterpillars devour shed skins and pupate in cocoons wrapped in leaves fastened with silk to branches. There is 1 generation a year. No closely related moths occur in the New World.

SPHINX MOTHS
(Family Sphingidae)

These stout moths have stiff, powerful wings that are often boldly or colorfully patterned. Their wings, spanning 1¼–6⅛″ (32–155 mm), beat so rapidly that some day-flying species resemble hummingbirds or large bees. These moths lack hearing organs, or tympana. Adults of most species visit flowers for nectar and have a long proboscis that is coiled under the head when not in use. In a few species, adults feed on decaying fruit and fermenting tree sap. The caterpillars are stout, usually hairless, and often bright green or reddish. In most species there is a large hornlike projection at the end of the abdomen that gives the caterpillar a formidable appearance although it is harmless. The pupa usually occupies a chamber in the soil, without a silken cocoon, but in some species a flimsy cocoon is constructed under organic litter at the soil surface. In some species the developing moth's proboscis is so long that the pupa has a sheath curving out from the head away from the body and back, resembling a jug handle. In Canada and Europe members of this family are known as "Hawk Moths."

549 Virginia-creeper Sphinx
"Hog Sphinx"
(*Darapsa myron*)

Description: Wingspan 2–2¾" (50–70 mm). Fore wings tan to greenish gray, banded broadly with dark brown or olive. Hind wings brownish orange. Caterpillar leaf-green, bluish green below, streaked obliquely on each side with greenish white.

Habitat: Forest edges, river margins, and vineyards.

Range: Quebec and southern Ontario, and eastern half of the United States.

Food: Caterpillar eats foliage of Virginia creeper and grape.

Flight: Late spring–fall.

Unlike most adults, these moths do not visit flowers but feed on decaying fruit and fermenting tree sap. Their caterpillars spin loose silken cocoons on the ground among soil litter where they overwinter as pupae. There are 2 generations a year. The fully grown caterpillar is often parasitized by internal wasp larvae, which bore out through its skin and spin white silken cocoons on its back.

547   **Pandora Sphinx**
(*Eumorpha pandorus*)

Description: Wingspan 3⅜–4″ (85–100 mm). Fore
and hind wings both patterned with
olive-green and paler green, suffused in
areas with pink and gray. Caterpillar is
green or reddish brown with 5 large
white spots encircling spiracles and has
a shiny eyelike button at the rear.

Habitat: Forest edges, river margins, and
vineyards.

Range: Eastern United States and southeastern
Canada.

Food: Caterpillar feeds on foliage of Virginia
creeper and wild and cultivated grapes.

Flight: June–August in the North, longer in
the South.

This moth visits flowers at dusk and
before dawn, but rarely feeds in total
darkness. Like most sphinx moths, it is
strongly attracted to artificial lights.
Formerly assigned to genus *Pholus*.

569   **Hummingbird Moth**
"Common Clearwing"
(*Hemaris thysbe*)

Description: Wingspan 1½–2″ (38–50 mm). Wings
initially plum-red to brownish black,
but scales drop off after 1st flight,
leaving clear areas devoid of scales,
except along veins. Body spindle-
shaped, mostly olive-green with plum-
red bands across abdomen and rear
tufts. Caterpillar is yellowish green
with darker green lines and reddish-
brown spots on abdomen and with
yellow tail horn.

Habitat: Forest edges, meadows, and cultivated
flower gardens.

Range: Coast to coast in the North; also east of
the Great Plains south to the Gulf.

Food: Caterpillar feeds on foliage of plants of
the honeysuckle family.

Flight: May–September.

This moth hovers over flowers in full
sunlight, producing a buzz with its
wings similar to but softer than that of
a hummingbird similarly engaged.
There are 2 generations a year.

554 White-lined Sphinx
(*Hyles lineata*)

Description: Wingspan 2½–3½" (65–90 mm). Fore
wings brown with a buff-colored band
from near base to tip; veins outlined in
white. Hind wing mostly pink, dark
brown to black near body and along
outer margin. Head prominent, brown
between eyes above. Thorax brown with
6 white stripes. Abdomen brown with
paired dark spots on each segment,
separated by 3 pale dorsal lines broken
lengthwise. Caterpillar is bright green
with yellow head, 2 lengthwise side
rows of pale spots bordered by black
lines, 3 yellow lines, and orange or
yellow rear horn.

Habitat: Meadows and gardens, especially where
portulaca grows.

Range: Southern Canada to Latin America,
from Atlantic to Pacific coasts.

Food: Caterpillar feeds on foliage of portulaca
or that of common weeds, such as
chickweed, purslane, evening primrose;
forage plants, especially buckwheat;
apple, pear, plum, and grape; and
truck crops, such as turnip tops,
tomato leaves, and melon foliage.

Flight: Midsummer.

These moths whir like hummingbirds
as they visit garden flowers at dusk or
in darkness. Often they fly in numbers
to artificial lights. Sometimes they seek
nectar in daylight. There are 2 or more
generations a year, one overwintering as
pupae underground. The Galium
Sphinx (*H. gallii*), wingspan not more
than 3" (75 mm), is similar except it
lacks the white stripes on thorax and its

veins are not outlined in white. Its
caterpillars feed on bedstraw, spurge,
fifeweed, and other plants from
Labrador to British Columbia, south in
the East from New England to Georgia,
and in the West to Mexico. These
moths were formerly included in the
genus *Celerio*.

26 Tomato Hornworm Moth
(*Manduca quinquemaculata*)

Description: Wingspan 3½–4⅜" (90–110 mm).
Wings mouse-gray, streaked with black
and brown. Hind wings strong,
pointed. Fore wings paler with 2 zigzag
dark lines crossing middle of wing.
Body heavy, torpedo-shaped. Head
prominent, eyes and palps large,
extended tongue at least as long as
body. Abdomen above has 5 pairs of
yellow spots. Caterpillar, 3–4" (75–
102 mm), is pale or dark green with a
black horn at rear, V-shaped white
mark at side of each segment.

Habitat: Open areas, particularly those under
cultivation.

Range: Eastern half of the United States and
southeastern Canada.

Food: Caterpillars feed on foliage of plants in
the nightshade family, such as tomato
and tobacco.

Flight: Summer–fall.

The caterpillars are seen much more
frequently than adult moths. They feed
mostly at night and later pupate in
unlined cells in the soil. Although they
are called tomato worms or hornworms,
they also attack the foliage of potatoes,
eggplants, green peppers, and various
weeds. Persistent rumors that
caterpillars can "sting" with their
horns are totally false. These moths are
known in southern tobacco-growing
states as "tobacco flies."

25, 558 **Tobacco Hornworm Moth**
(*Manduca sexta*)

Description: Wingspan 3½–4½" (90–115 mm).
Wings blackish gray, mottled and
striped with brown and white.
Abdomen has a row of 6 orange-yellow
spots on each side. Caterpillar, to 3¾"
(95 mm), is green with diagonal white
stripes and red horn at rear.

Habitat: Wild or cultivated fields where
herbaceous plants of the nightshade
family are numerous.

Range: Southern New England to Minnesota,
south to the Gulf, southwest to
Arizona; most common in the South.

Food: Caterpillar eats foliage of plants in the
nightshade family, such as potato,
tomato, and tobacco.

Flight: July–October.

This moth is also known as the
"Carolina Sphinx" and "Six-spotted
Sphinx." In southern tobacco-growing
states the adult is called a "tobacco fly."
Its caterpillars hatch from large green
eggs and grow rapidly in 4–5 weeks.
The pupae have a distinctive juglike
handle. The related Rustic Sphinx
(*M. rustica*), wingspan 4–5" (100–125
mm), is chocolate-brown, mottled with
white, black, and yellow on its fore
wings; it lacks distinct lines on its hind
wings, and has 6 pairs of yellow spots
on the abdomen. It is found from
Virginia to Central America but often
strays into northern states.

548 **Big Poplar Sphinx**
(*Pachysphinx modesta*)

Description: Wingspan 3½–5½" (90–140 mm).
Fore wings with alternating bands
colored mouse-gray to velvet-brown.
Hind wings suffused with red tinge;
blue-gray in wing base. Caterpillar, to
4" (100 mm), is bluish green with

small yellowish-white dots, oblique white lines on sides, and short tail horn.

Habitat: Deciduous and mixed forests.

Range: Coast to coast in southern Canada, south to Gulf in the East; in the West, south to Colorado in the Rockies.

Food: Caterpillar feeds on foliage of poplar and willows.

Flight: May–August.

This moth has one of the greatest wingspans of all North American sphinx moths.

---

565 Cerisy's Sphinx
(*Smerinthus cerisyi*)

Description: Wingspan 2⅜–3⅜" (60–85 mm). Fore wings variable, marked with contrasting light and dark gray or tawny brown. Hind wings rosy pink with tawny outer margin and black and blue eyespot. Caterpillar is bluish green and has a few diagonal yellow streaks and a green and yellow horn at rear.

Habitat: River margins and low ground, where willows grow.

Range: Coast to coast in the northern United States and Canada, south along the Rocky Mountains to Arizona, and along the Sierra Mountains to California.

Food: Caterpillar eats willow foliage.

Flight: May and June.

The more widespread Twin-spotted Sphinx (*S. jamaicensis*), wingspan 2–2¾" (50–70 mm), has red rather than pink on its hind wings, and the blue area in the eyespot is divided by a black line—hence "twin-spotted." The Blind Sphinx (*Paonias excaecatus*), wingspan 2⅜–3⅛" (60–80 mm), has no black center in the blue eyespot—hence "blind"—and has 6–8 toothlike lobes on the outer margin of its fore wings. Its caterpillars feed on apple, birch, and a variety of other trees.

523 Wild-cherry Sphinx
(*Sphinx drupiferarum*)

Description: Wingspan 3–4½″ (75–115 mm). Fore wings dark bluish gray, whitish along front and outer margins. Hind wings black, crossed by 2 pale gray bands. Caterpillar is dark green with 7 oblique violet lines on sides and a purplish horn at rear.

Habitat: Deciduous forests and open areas.

Range: Coast to coast across southern Canada and the United States.

Food: Caterpillar eats foliage of wild or cultivated cherry, plum, and apple.

Flight: May and midsummer.

The Great Ash Sphinx (*S. chersis*), wingspan 4–5″ (100–128 mm), has several fine, black, lengthwise lines on each ash-gray fore wing. Its caterpillars feed on wild cherry, ash, lilac, and privet, ranging throughout most of the United States and southern Canada. The Elegant Sphinx (*S. perelegans*), wingspan 3⅜–4⅜″ (85–110 mm), is similar but less contrastingly marked. It is found from Colorado and Arizona west to the Pacific Coast, north to British Columbia. The Apple Sphinx (*S. gordius*), wingspan 2½–3½″ (65–90 mm), lacks the band along its dark fore wing margin. Its caterpillars feed on foliage of plants of the rose family, ash, myrtle, and sweet fern in southern Canada and the eastern United States.

PROMINENTS
(Family Notodontidae)

These common, medium-sized moths, 1–2⅜″ (25–60 mm) long, are varying shades of brown, gray, olive-green, or yellowish tan, and often spotted or streaked with black. In some species the fore wing has a toothlike projection at the middle of the inner margin,

which shows prominently when the
wings are folded rooflike over the body
at rest. Many members of this family
somewhat resemble noctuids but can be
distinguished by the venation of the
fore wings. Males have comblike
antennae, unlike many noctuids.
Caterpillars are mottled or striped and
many have lumpy tubercles on their
backs. They feed on the foliage of many
kinds of trees and shrubs—most
feeding singly, a few in large groups.
Some caterpillars are serious orchard
and forest pests. If disturbed, the
caterpillars often "freeze," raising the
front and rear of the body and holding
on to their support by 4 pairs of
prolegs. They overwinter as mature
caterpillars or pupae within earthen
cells or in flimsy cocoons spun among
litter at the surface of the soil.

---

29 **Tentacled Prominents**
"Puss Moths"
(*Cerura* spp.)

Description: Wingspan 1–1¾" (25–45 mm). Body
stout, densely covered with white or
grayish hairlike scales. Fore wings
white or gray, variously marked with
contrasting dark gray to black median
bands and with multiple zigzag
crosslines and series of dark dots. Hind
wings generally unmarked white with 1
terminal row of dark dots. Caterpillar,
to 1¼" (33 mm), is green with reddish-
brown saddlelike mark in middle of
abdomen.

Habitat: Deciduous and mixed forests and river
drainages.

Range: Throughout most of North America.

Food: Caterpillar feeds on willows, poplars,
and wild cherry.

Flight: April–August.

The plump caterpillar can retract its
head so far into its body that it seems

to disappear. When disturbed, the caterpillar extends whiplike filaments from each of the 2 fleshy hornlike projections at the tip of the abdomen and waves these filaments. It can eject an irritating fluid from glands on the thorax. The caterpillar pupates in a tough, brown silken cocoon mixed with wood chips in a cavity in rotten wood or bark.

8 Red-humped Appleworm Moth
(*Schizura concinna*)

Description: Wingspan 1⅛–1⅜″ (30–35 mm). Male's fore wings tawny with gray and brown streaks, reddish brown along inner margin, black discal spot; hind wing dirty white with dark spot close to rear margin. Female's wings reddish brown with few or no marks. Caterpillar, to 1½″ (38 mm), is yellow with brown and white stripes on each side, red hump on 4th segment, many black spines, and red head.

Habitat: Woodlands and marginal areas.

Range: Throughout North America.

Food: Caterpillar feeds on foliage of apple, cherry, pear, rose, blackberry, and other members of rose family, as well as many other trees.

Flight: Summer.

Caterpillars spin loose silken cocoons on the ground among litter. They overwinter and pupate in late spring. There is 1 generation a year.

TIGER MOTHS
(Family Arctiidae)

The tiger moths are small or medium-sized, with stout, furry bodies and broad wings, spanning ½–3⅛″ (12–80 mm). Some are largely white; others are boldly patterned in black and white or yellow; and still others have different colors. Tiger moths are similar in size and shape to owlet moths, but are usually lighter and more brightly colored. They resemble some boldly patterned ctenuchids, but can be distinguished from these and other similar-looking moths mostly through differences in wing venation. They also have a well-developed hearing organ, or tympanum, on each side of the thorax. Many tiger moths contain toxic substances, and their conspicuous patterning serves as a warning to predators. Adults generally do not feed. Eggs are laid in flat masses or loosely scattered over low vegetation. The caterpillars are hairy or bristly and like the adults are boldly marked and toxic. They pupate in loose cocoons made of silk and their own hair.

555 Ornate Tiger Moth
(*Apantesis ornata*)

Description: Wingspan 1⅛–1⅝″ (30–40 mm). Fore wings black, crosshatched with ivory-white. Hind wings pink to red, spotted with black. Body black, thorax streaked with ivory; dull brownish yellow often along sides of abdomen. Caterpillar is mostly black with yellow midline stripe and often has pale spots above each side of stripe.

Habitat: Fields and roadsides.

Range: Great Basin to Pacific Coast.

Food: Caterpillar feeds on various herbaceous plants.

Flight: Spring–summer.

Male moths are attracted to artificial lights at night, but females generally remain close to breeding areas and food plants. This large genus of attractive moths has numerous species throughout North America with different geographic ranges and different markings on the wings; many have yellow hind wings.

531 Yellow Woolly Bear Moth
(*Diacrisia virginica*)

Description: Wingspan 1½–2″ (38–50 mm). Wings snow-white with small gray dots; 1 on each fore wing, 3–4 on each hind wing. Abdomen yellowish with black dots on midline and along each side. Front femora and coxae are orange or yellowish brown. Caterpillar has short and long hair on pale yellow to white body, and yellow legs.
Habitat: Meadows, roadsides, and crop fields.
Range: Throughout most of North America.
Food: Caterpillar feeds on herbaceous plants.
Flight: May–summer.

This moth's caterpillars are sometimes pests on crops and garden flowers. There are 2 generations; the second overwinters as pupae.

4, 527 Acraea Moth
(*Estigmene acraea*)

Description: Wingspan 1⅞–2″ (48–50 mm). Thorax and fore wings above white with 4–6 black dots on front margin, with further speckling elsewhere. Hind wings brownish yellow (male) or white (female). In both sexes, hind wings have 3 black spots near outer margin, middle spot largest. Caterpillar, to 2½″ (65 mm), is black covered with dense, long rusty-red hair.
Habitat: Fields, pastures, and marshes.
Range: Throughout North America, except northern Canada.
Food: Caterpillar feeds on herbaceous plants, including cord grasses in salt marshes.
Flight: June–July.

Because these caterpillars often feed on grasses in salt marshes, they are sometimes called "Salt Marsh Caterpillars."

6, 529 Milkweed Tiger Moth
(*Euchaetias egle*)

Description: Wingspan 1–1¾" (25–45 mm). Wing silvery to grayish brown. Abdomen yellow with 3 lengthwise rows of black spots. Wings disproportionately long for body, but of characteristic tiger moth shape—fore wings slant backward to form a nearly smooth contour with shorter hind wings. Caterpillar, to 1⅝" (40 mm), is densely hairy with long tufts of black and white hair at each end, shorter tufts along sides, and orange hair down midline, which has short black hair on each side.

Habitat: Meadows and roadsides, where milkweed is common.

Range: Ontario and northeastern United States to North Carolina mountains, west to Great Plains.

Food: Caterpillar eats milkweed foliage.

Flight: June–early August.

The colorful black, white, and orange caterpillar is often called the "Harlequin Caterpillar." The related Oregon Euchaetiad (*E. oregonensis*), same size, has unmarked white wings and a yellow abdomen with only a single midline of black dots. It ranges coast to coast across the northern United States and southern Canada, south to Mississippi and Texas. Its hairy caterpillars feed on various weeds.

540 Spotted Tiger Moth
(*Halisidota maculata*)

Description: Wingspan 1⅜–2″ (35–50 mm). Fore wings beige with brown mottling and spots. Hind wings translucent yellowish. Caterpillar is yellowish white in the North and mostly black in the South with long pencils of white hair on thorax and abdomen; head black.

Habitat: Deciduous forests and woods.

Range: Quebec to North Carolina, west to California.

Food: Caterpillar eats foliage of poplar, maple, and other trees.

Flight: Midsummer.

The Hickory Tiger (*H. caryae*), same size, has silver-spotted fore wings and occurs east of the Rocky Mountains in the northern United States and southern Canada, to Texas and Arizona. The Silvery Tiger (*H. argentata*), wingspan 1¾–2″ (45–50 mm), has reddish-brown fore wings spotted with silver. Its caterpillars feed on conifers west of the Rocky Mountains.

553 Colona
(*Haploa colona*)

Description: Wingspan 2–2¼″ (50–58 mm). Fore wings white with variable number of brown markings. Hind wings and abdomen sulfur yellow. Thorax pure white or with dark brown stripes. Caterpillar is dark brown to black and has large bluish tubercles; it is marked by broken lengthwise stripes and covered with very short black hair.

Habitat: Rich, moist southern woodlands.

Range: Gulf states.

Food: Not known.

Flight: Midsummer.

This moth is the largest North American member of the genus.

5, 541 **Woolly Bear Caterpillar Moth**
"Banded Woolly Bear"
(*Isia isabella*)

Description: Wingspan 1⅝–2″ (40–50 mm). Fore
wings yellow-brown with a series or
row of small black dots. Hind wings
slightly paler, slightly pinkish with
several indistinct gray dots. Abdomen
has 3 black spots above on rear edge of
each segment. Caterpillar, to 2⅛″ (55
mm), is black, covered with stiff
bristles, and has a broad band of red-
brown bristles around the middle.

Habitat: Meadows, pastures, uncultivated fields,
and road edges.

Range: Throughout North America, except
northern Canada.

Food: Caterpillar feeds on low herbaceous
plants of many kinds, mostly wild; it
seldom attacks crops or ornamentals.

Flight: June–August in the North, February–
November in the South.

Familiar since Colonial times as the
"Woolly Bear," the caterpillar is often
seen crossing roads and paths on warm
days in late fall. According to
superstition, the amount of black in the
caterpillar's bristle coating forecasts the
severity of the coming winter.
Actually, the coloration indicates how
near the caterpillar is to full growth
before autumn weather stimulates it to
seek a winter shelter.

165 **Lichen Moth**
(*Lycomorpha pholus*)

Description: Wingspan 1–1⅛″ (25–30 mm). Body
to ⅜″ (10 mm). Fore wings long,
narrow, rounded at tip; yellow, orange,
or red on basal ½; black beyond. Hind
wings similarly colored along front ½.
Body, antennae, and legs black.
Caterpillar is lichen-colored.

Habitat: Mixed and deciduous forests and marginal open areas.

Range: Throughout much of North America into the far North.

Food: Caterpillar eats lichens.

Flight: July–September.

This diurnal moth is easily mistaken for a net-wing beetle. In northern parts of the range, caterpillars sometimes feed for several years before attaining full size. The hairy cocoons are attached to rocks or tree trunks, close to the former food supply. Moth formerly assigned to family Ctenuchidae or Amatidae.

551 Rattlebox Moth
(*Utetheisa bella*)

Description: Wingspan 1¼–1¾" (33–46 mm). Fore wings orange to yellow, speckled with black on small areas of white. Hind wings bright pink with an irregular black outer border. Antennae black. Caterpillar is yellow with red head, white side stripes, and segmental crossbands behind large black areas; it has a few long black hairs in tufts and more bare skin showing than in most other tiger moth caterpillars.

Habitat: Sandy pine lands where rattlebox and related plants grow wild.

Range: New England to Florida along the coast; scattered localities inland to Kansas and Texas.

Food: Caterpillar feeds on rattlebox, sweet clover, and occasionally sweet fern.

Flight: July–September in the North, year-round in the South.

This day-flying moth often seems to disappear suddenly. It settles on grass and wraps its wings around the blade, thereby fully concealing the pink color, which is so evident when the moth flies. All stages of this insect are repugnant to insectivorous birds.

CTENUCHID MOTHS
(Family Ctenuchidae)

The ctenuchids are small or medium-sized, day-flying moths that have relatively long, narrow fore wings, rather small hind wings, and wingspans of 1⅛–2″ (28–50 mm). Many have brilliant metallic-colored bodies and boldly patterned wings. Some resemble tiger moths. A few southern species have transparent wings, superficially resembling clear-winged moths; others somewhat resemble smoky moths. Ctenuchids can be distinguished from these and other moths by their wing venation. All ctenuchids visit flowers, where their resemblance to wasps or toxic beetles protects them from predators. The caterpillars are hairy and feed on grasses and various other vegetation. They pupate in cocoons, incorporating their hair in the silk. This family was previously called Amatidae.

544 Virginia Ctenuchid Moth
(*Ctenucha virginica*)

Description: Wingspan 1⅜–2″ (35–50 mm). Wings dark olive-brown, partly fringed with white. Body metallic blue-green. Head orange-yellow. Antennae feathery. Caterpillar, to 1″ (25 mm), is yellowish tan and has many tufts of short white hair.

Habitat: Wet meadows.

Range: Northeastern United States and adjacent Canada, south to North Carolina coast.

Food: Caterpillar feeds primarily on grasses.

Flight: May–July.

The southeastern states harbor the similar Brown Ctenuchid (*C. brunnea*), same size, which has orange marks resembling epaulets on its thorax, dark

veins on somewhat paler olive-brown
fore wings, and almost black hind
wings. Additional species, mostly with
yellow veins on the fore wings, occur in
the Southwest.

TUSSOCK MOTHS AND KIN
(Family Lymantriidae)

These stout, rather hairy, brownish or
whitish moths have wingspans of ¾–
2¾" (20–70 mm). They lack simple
eyes, and males have feathery antennae
—these characteristics along with the
wing venation distinguish tussock
moths from related families. The adults
are short-lived and do not feed; they
have a reduced proboscis or none at all.
The females of some species are
wingless; in others the females have
wings but are poor fliers. Eggs are often
deposited in masses, usually covered
with hair from the female's abdomen.
The slender caterpillars have long tufts
of hair at each end of the body and
shorter, often brightly colored tufts on
the back. The hair of some species can
cause an irritating rash if touched. The
pupae are enclosed in a loose cocoon of
silk. This family was previously named
Liparidae.

528 Gypsy Moth
(*Lymantria dispar*)

Description: Male wingspan ¾" (20 mm); female
wingspan 1⅛–2¾" (28–70 mm).
Male's fore wings brownish gray with
brown irregular lines more or less
parallel to outer margin; hind wings
dark around outer margin. Female's
wings yellowish white with narrow
wavy lines paralleling outer margin and
with series of dark brown dots around
outer margin of hind wing. Male has

feathery antennae and slender, conical abdomen. Female has threadlike antennae and bulbous abdomen. Caterpillar is gray with long dark hair along sides and 5 pairs of blue tubercle and 6 pairs of red tubercles on its back.

Habitat: Deciduous, coniferous, or mixed forests.

Range: Northeastern United States and extrem southern Canada; recently introduced ir many other parts of the United States.

Food: Caterpillar feeds on many kinds of deciduous and evergreen trees, including oak, pine, and hemlock.

Flight: July–August.

In 1869 this species was accidentally carried to Massachusetts from Europe. Their caterpillars have become major pests of forest and shade trees. They denuded millions of acres of trees in the early 1970's. The male flies strongly, while the flightless female only flutters along the ground. One female produces masses of about 400 eggs, depositing them on tree trunks and buildings or in other protected areas. The caterpillars feed at night. There is 1 generation a year.

13, 545   White-marked Tussock Moth
(*Orgyia leucostigma*)

Description: Male wingspan 1⅛–1¼" (28–33 mm); female wingless, body ½–⅝" (12–16 mm). Male's wings dark gray with tan and black mottling; antennae feathery. Female is gray-brown and has threadlike antennae and very broad abdomen, which is conspicuous until eggs are laid. Caterpillar, to 1¼" (33 mm), has a bright red head, pale brown body with yellow and black stripes, and pencils of black hair and white tufts on abdomen.

Habitat: Deciduous and mixed forests.

Range: East of the Rocky Mountains.

Food: Caterpillar eats foliage of a great variety of trees and shrubs.

Flight: Throughout the year, depending upon the location.

The female moth dies soon after laying a single mass of eggs, which overwinter. Caterpillars pupate in cocoons spun of silk and hair on bark, tree branches, or other supports. There are 2 or more generations a year. A number of closely similar species occur in the United States and Canada. At times the populations of any of these may become so large locally that these pests severely defoliate host trees.

OWLET MOTHS
(Family Noctuidae)

This is the largest family of moths, represented by about 3,000 species in North America. Most are dull-colored and have wingspans of ¾–2" (20–50 mm), but some have brightly colored wings, and a few reach 5⅞" (150 mm) in wingspan. The antennae are slender and threadlike, and the tonguelike proboscis is usually well developed. There is a prominent hearing organ, or tympanum, on each side of the thorax. The wings are folded rooflike over the body when at rest. Most caterpillars are smooth or only sparsely hairy, but some are densely hairy, resembling tiger moth caterpillars. They feed on a wide range of plants, including the foliage of trees, shrubs, herbaceous plants, and grasses. Some bore into stems and roots; others feed on decayed organic matter; still others are cannibalistic. Most species pupate in cells at or below the soil surface. Others pupate in hollow galleries bored by other larvae, or in cocoons spun among debris or leaves. Adults are usually nocturnal, but some are day-fliers. Many feed on fermenting tree sap or decaying fruit, some sip flower nectar, and others do not feed. This family includes the forester moths previously grouped in the separate Family Agaristidae.

18, 521 Eight-spotted Forester
(*Alypia octomaculata*)

Description: Wingspan 1⅛–1¼″ (30–33 mm). Wings and body velvety black with 2 large yellow spots on each fore wing and 2 white spots on each hind wing. Fore legs have brilliant orange hairlike scales. Caterpillar, to 1½″ (38 mm), is bluish white marked with orange and black narrow bands and black dots; little hair.

Habitat: Woodland edges, vineyards, and cities.

Range: Eastern half of the United States.

Food: Caterpillar feeds on Virginia creeper, grape, and Boston ivy.

Flight: Spring.

Adults fly in bright daylight and are often mistaken for butterflies. The caterpillars spin thin cocoons of silk and litter or wood chips on the ground or in tunnels cut into wood. There is 1 generation a year.

---

561 Black Witch
(*Ascalapha odorata*)

Description: Wingspan 3½–5⅞″ (90–150 mm). Wings dark brown, both pairs crossed by series of alternating light and dark undulating lines and bands; often an iridescent blue cast over wings. Females are more contrastingly marked than males.

Habitat: Tropical and subtropical forests with trees of the pea family, including acacia.

Range: South and Central America, straying far
north in the United States and southern
Canada.
Food: Caterpillar eats foliage of leguminous
trees.
Flight: Migrating north in the fall.

These moths, the largest owlets in this
country, often fly great distances in
only a few nights, hiding by day
wherever they find dense shade—even
behind shutters on a house. The
caterpillars resemble giant cut worms,
sometimes becoming pests when
numerous. Formerly assigned to genus
Erebus.

559 Alfalfa Looper
(*Autographa californica*)

Description: Wingspan 1⅛–1⅝″ (30–40 mm). Fore
wings gray with a silvery mark near
middle. Hind wings and body dull gray
to brown, paler toward base, darker
along outer margin. Caterpillar, to 1″
(25 mm), is dark olive-green with paler
head.
Habitat: Crop fields, wastelands, and open areas.
Range: Western half of the United States,
extreme southern Saskatchewan, and
British Columbia.
Food: Caterpillar feeds on alfalfa, cereals,
vegetables, garden flowers, ornamental
shrubs, and orchard trees.
Flight: Midsummer.

These moths are members of a large,
widespread genus in North America.
Most adults have a silvery mark, or
"autograph," on the fore wings. Their
caterpillars move in a looping gait and
sometimes seriously damage crops.

557 Sweetheart Underwing
(*Catocala amatrix*)

Description: Wingspan 3–3⅜″ (75–85 mm). Fore wings light gray-brown, sometimes with a darker streak from base to tip. Hind wings pinkish orange with 2 black bands and a white fringe. Caterpillar is grayish, twiglike, blending in with its environment.

Habitat: Mixed and deciduous woods.

Range: Eastern United States and southern Canada, west to Arizona.

Food: Caterpillar feeds on willow and poplar.

Flight: Late summer.

Although moths are camouflaged amazingly well at rest, they are alert to the approach of a person or bird and fly off rapidly. Nocturnal, many fly to artificial lights and by day hide on tree trunks and branches. More than 100 related species occur in North America. The Widow Underwing (*C. vidua*), wingspan 3–3⅛″ (75–80 mm), is one of many species with plain, dark brown underwings. Its caterpillars feed on hickory and walnut from southeastern Canada to Florida, west to the Great Plains. The White Underwing (*C. relicta*), wingspan 3–3⅛″ (75–80 mm), occurs across the northern United States and southern Canada, where its caterpillars attack poplar and willow. The Tiny Nymph (*C. micronympha*), wingspan 1⅝–1¾″ (40–45 mm), has yellow underwings; its caterpillars are found on oaks throughout eastern North America.

556 Locust Underwing
(*Euparthenos nubilus*)

Description: Wingspan 2½″ (65 mm). Fore wings and thorax above mottled gray-brown and white in a distinctive pattern. Hind wings black, with several undulating bands of orange-yellow, the outermost usually broken into series of dots. Caterpillar is mottled dark brown, twiglike, blending in with its environment.

Habitat: Deciduous forests.

Range: Eastern North America, west to Arizona.

Food: Caterpillar eats black locust foliage.

Flight: May–August.

This moth can be quite common in areas where black locust grows abundantly, although the moths are rarely seen, except when they fly to artificial lights.

---

526 Hebrew
(*Polygrammate hebraeicum*)

Description: Wingspan ⅞–1" (23–26 mm). Fore wings and thorax above white, marked with numerous irregular black lines and 1 black dot near center of wing. Hind wings, abdomen, and undersurfaces pale grayish brown, darker brown around wing margins. Caterpillar is variously colored, often red or green with yellow dusting and 2 lengthwise yellow stripes.

Habitat: Moist woods.

Range: Eastern half of the United States and southern Canada.

Food: Caterpillar eats sourgum leaves.

Flight: July–August.

Hübner named this moth the "Hebrew" because its curving black lines and dots reminded him of a Hebrew letter. There is 1 generation a year.

Bees, Ants, Wasps, Sawflies, and Kin
(Order Hymenoptera)

Many of our most familiar insects belong to this large order, which has over 108,000 species worldwide, including 17,100 in North America. Adults are hard-bodied, active insects, usually possessing 2 pairs of membranous wings that generally have large cells and few veins. In some groups, however, adults are wingless. All adults have chewing mouthparts; in bees and some wasps certain parts are also modified into tonguelike structures for drinking liquids. Bees, ants, and wasps have a constriction, or "waist," called the pedicel between the base of the abdomen and the thorax. Sawflies and their kin do not have a "waist"—the abdomen is broadly connected to the thorax. Females of most species have a well-developed ovipositor, which in wasps, some bees, and some ants is modified into a stinger. Despite their similar appearance, sawflies and horntails do not sting.

Metamorphosis is complete. Larvae have a well-developed head with chewing mouthparts; most larvae in the order are legless, but sawfly larvae have legs on the thorax and several pairs of abdominal prolegs. In bees and a few wasps the larvae feed on pollen or nectar. Most wasps lay eggs in tunnels, cavities, or nests of mud or paper, and adult females bring the larvae insect prey to eat. Sawflies usually lay eggs in plants, and the larvae live in stems or leaves or feed externally on foliage. In braconids, ichneumons, and some others, the larvae are parasites of spiders and spider egg sacs, and other insects. Most species are solitary, but ants and some bees and wasps have a complex social organization with sterile female workers and fertile males and females. The order is economically important;

although sawflies often damage trees and crops, parasitic species are useful in the control of insect pests, and bees and many wasps are valuable as pollinators of crops and wild plants.

CIMBICID SAWFLIES
(Family Cimbicidae)

Cimbicids are large, stocky sawflies, ¾–1″ (20–25 mm) long, with distinctly clubbed antennae. As in other sawflies, the thorax and abdomen are broadly joined and not separated by a slender "waist," or pedicel. This feature distinguishes the sawflies from hornets and bumble bees. Adults sip water and nectar. The larvae feed on the foliage of willows, elms, and other trees and pupate in silken cocoons, either in the ground or attached to twigs.

450 **Rusty Willow Sawfly**
(*Cimbex rubida*)

Description: ¾″ (18–20 mm). Stout, almost parallel-sided; head squarish in front. Antennae clubbed. *Rust-colored with black markings. Wings mottled metallic blue to smoky brown.*
   Habitat: River margins and lowland woods.
    Range: Coastal California to Sierra Nevada Mountains.
     Food: Adult drinks nectar. Larva feeds on willow.
Life Cycle: Female attaches cylindrical eggs singly or in clusters to willow foliage. Larvae spin webs on branches, drag leaf fragments into these shelters, and feed. They drop to soil to pupate and overwinter under ground litter. They may pupate for 2 or more years.

This species is distinguished only by inconspicuous details from the Western

Willow Sawfly (*C. pacifica*), 1″ (25 mm), which is found in Colorado, and from California to Washington. The most common cimbicid sawfly is the Elm Sawfly (*C. americana*), ¾″ (18–20 mm), which is steel-blue with about 4 spots on its sides and smoky-brown wings. Its larvae feed on elm, willow, and other broad-leaved trees from Newfoundland south to North Carolina, and from Colorado to British Columbia and Alaska.

COMMON SAWFLIES
(Family Tenthredinidae)

This is the largest family of sawflies. Its members are black or brown, ⅛–¾″ (3–20 mm) long. Like other sawflies, they lack the slender "waist," or pedicel, between the thorax and abdomen, which are instead broadly joined. They have long, threadlike antennae composed of 7–10 segments; in other sawflies, the antennae have fewer than 7 or more than 10 segments, and are often clubbed or comblike. The larvae of most species resemble small caterpillars and feed on the foliage of trees and shrubs. A few live in galls or make mines in leaves.

481 **Northeastern Sawfly**
(*Tenthredo originalis*)

Description: ⅜″ (9–11 mm). *Elongate, almost parallel-sided. Black with yellow bands* on each abdominal segment and yellow markings on upper lip, base of mandibles, scutellum, and upper leg. Tibiae yellow with black ring at tip of hind pair; tarsi yellow. Antennae 8-segmented. Wings smoky, veins red.

Habitat: Meadows or forest glades near rapid streams.

Range: New England and temperate regions of
         Canada.

Food: Adult eats little. Larva feeds on foliage
        of streamside trees and shrubs,
        especially willow.

Life Cycle: Female uses sawlike ovipositor to cut
              slits through bark of twigs, into which
              translucent eggs are wedged. Larvae
              feed on tender young leaves. They
              descend to soil and spin silk cocoons.
              1 or more generations, depending on
              length of warm season.

Larvae often feed in groups. Each curls
up on a leaf as it feeds, forming an "S."

HORNTAILS
(Family Siricidae)

Horntails are wasplike, brownish or
black insects, ¾–1⅝" (20–40 mm)
long, with cylindrical bodies. Unlike
wasps, bees, and ants, their thorax and
abdomen are broadly joined together,
not separated by a narrow waist. At the
tip of the abdomen, on its upper
surface, there is a triangular plate,
elongated into a spine in some species.
In females the abdomen also bears a
long, slender ovipositor at the tip,
which is used for drilling into the stems
of plants or the wood of dead and dying
trees, where the eggs are laid. The
caterpillarlike larvae tunnel in wood;
after the larval and pupal periods, the
adults emerge through conspicuous
holes. Adults feed on nectar and water.

474 Pigeon Horntail
(*Tremex columba*)

Description: 1–1½" (25–38 mm). *Cylindrical.* Dark
               red to black, abdomen marked with
               yellowish crossbands. Wings dusky to
               yellowish. *Female has blunt ovipositor ¼*

as long as body, yellowish. Both sexes have horny, spearlike plate at tip of abdomen.

Habitat: Hardwood and mixed forests.

Range: Eastern North America.

Food: Adult drinks nectar. Larva eats fungus-infected wood of elm, beech, maple, oak, and other deciduous trees.

Life Cycle: Female uses ovipositor to bore through bark into wood, depositing 1 slender egg in each hole. Eggs are covered with fungal spores from a special pocket in female's abdomen. As embryos prepare to hatch, fungi begin to grow and soften wood. Larvae tunnel into infected wood, making cylindrical passageways into side branches. They feed for up to 2 years, then pupate under bark in cocoons made of silk and wood chips. Adults are active in fall.

After depositing the last egg, the female often dies without removing its ovipositor from the wood. The dead female becomes food for some insectivorous animal. Some ichneumon species, which parasitize and kill horntail larvae, are helpful in biological control of Pigeon Horntails. There are about 22 *Tremex* species worldwide but only 1 in North America.

477 **Smoky Horntails**
(*Urocerus* spp.)

Description: 1–1⅜″ (25–35 mm), males smaller than females. Ovipositor ⅛–⅜″ (3–9 mm) long. Mostly metallic or dull blue-black with *reddish to orange or yellow-brown markings on abdomen, antennae, and tarsi.* Wings usually clouded with brown at tips and middle, but sometimes clear.

Habitat: Mixed and coniferous forests.

Range: Mostly western United States and Canada, from California and Texas to Alaska and Alberta; also a few species in the eastern United States.

Food: Adult drinks nectar. Larva eats wood.

Life Cycle: Eggs are thrust into wood of dead trees or felled timber. Larvae tunnel into sap and heartwood, later prepare pupal chambers under bark or in crevices, where they overwinter. Adults emerge in late spring.

These horntails are prevalent where timber has been left on the ground after timber-cutting operations. They can be controlled by burning wastes and infected wood.

STEM SAWFLIES
(Family Cephidae)

These small, wasplike sawflies are usually black, 3/8–5/8″ (9–15 mm) long. The thorax and abdomen are broadly joined together, rather than being separated by a narrow "waist." They differ from other sawflies in having a slender abdomen that is often compressed at the sides. Adults visit flowers and drink nectar. The larvae of most species bore into the stems of grasses, shrubs, and trees.

472   Raspberry Horntail
(*Hartigia cressoni*)

Description: 1/2–5/8″ (12–15 mm). *Slender,* broadest at eyes. Male mostly black, spotted with yellow on abdomen above and with 3 yellow rings around rear 1/3. *Female is bright yellow with black ring on abdominal segment 4;* black on femora, front of thorax, and below abdomen. Antennae black with orange band. Wings dusky.

Habitat: Foothills and highlands.

Range: Nevada and California, north to Oregon and Montana.

Food: Adult drinks nectar. Larva feeds on rosaceous plants, particularly roses, blackberries, raspberries, and loganberries.

Life Cycle: White eggs are inserted into tender twigs of host plant, inducing swelling and discoloration. Larvae emerge a few days later and begin to feed, chewing a spiral path around twig and eventually killing it. They later tunnel downward into pith, often to the roots. Larvae overwinter, pupate in silk-lined cells in tunnels. 2 generations a year.

Wild and cultivated plants are often seriously infested by Raspberry Horntails.

BRACONIDS
(Family Braconidae)

Braconids are small, usually black parasites, 1/16–5/8″ (2–15 mm) long. They are similar to ichneumons but are usually smaller, stockier, and have only 1 recurrent vein rather than 2 in the wings. The antennae are long but are never marked with white or yellow, as are those of many ichneumons. Adults drink water and nectar. The larvae are internal parasites of other insects. More than 1 larva may develop in a single host, pupating in silken cocoons attached to the host's body.

484 Braconid Wasps
(*Apanteles* spp.)

Description: 1/16–1/8″ (2–3 mm). *Antennae threadlike, about as long as body.* Black, usually with yellow spots on legs. Wings clear with few veins and black spot (stigma) on front margin.

Habitat: Open areas on plants where hornworm caterpillars feed.

Range: Throughout North America.
Food: Adult drinks nectar. Larva is internal parasite of the caterpillars of hornworm moths, gypsy moths, and other moths.
Life Cycle: Eggs are laid on host's skin. Larvae burrow inside host, which at first continues to develop almost normally but eventually dies. Braconid larvae stop feeding and cut holes in host's skin in order to reach the outside. They pupate inside white cocoons spun on the host's skin. Caterpillar often dies before adult wasps emerge from cocoons.

Most larvae are gregarious, feeding and maturing together. A few species are solitary, each larva developing on a different host. Braconid wasps are regarded as beneficial because they reduce the numbers of plant-eating insects.

ICHNEUMONS
(Family Ichneumonidae)

This very large family of parasitic wasps varies greatly in color and in size, measuring $1/8-3''$ (3–75 mm) long. In some species the abdomen is long and slender, becoming thicker at the tip, and in females, often ending in a trailing, threadlike ovipositor. In most species, females do not sting. The antennae are long and constantly in motion; in many species the middle segments are white or yellow, making the moving antennae conspicuous. Ichneumons are usually larger and more slender than braconid wasps, and have 2 recurrent veins in the wings, rather than 1. Adults drink nectar and water. The larvae are parasites of a wide variety of other insects and spiders and are important in controlling insect populations. Members of this family are sometimes called ichneumonflies.

465 Giant Ichneumons
(*Megarhyssa* spp.)

Description: Male 1–1½″ (25–38 mm), female 1⅜–3″ (35–75 mm) with long ovipositor 2–4⅜″ (50–110 mm). Body pale to dark brown with *extensive yellow V-shaped markings bordered with black on sides of abdomen.* Female has more brown spots on yellow legs than male. Wings smoky.

Habitat: Deciduous and mixed forests with dead and dying broad-leaved trees and logs.

Range: Throughout North America, except plains and deserts.

Food: Adult may not eat. Larva is internal parasite of Pigeon Horntail larvae and related borers.

Life Cycle: Mated female flies from tree to tree, pressing its long antennae against the bark to detect vibrations made by horntail larvae in wood. Female curls ovipositor up over abdomen, curving it down to enter bark at right angle. Sharp tips cut progressively deeper until they reach larval tunnels. Female inserts a very slender egg into each horntail tunnel. Each ichneumon larva attacks horntail host, causing its death, but not before ichneumon larva is fully grown.

Best known is the Eastern Giant Ichneumon (*M. macrurus macrurus*), which has extensive brown markings on its wings and ranges from Florida to Mexico. A northern subspecies, the Lunar Giant Ichneumon (*M. m. icterosticta*), has clear wings and occurs from Quebec and Nova Scotia to Georgia, west to South Dakota and Texas. The Western Giant Ichneumon (*M. nortoni*) is black with red and yellow spots and mostly yellow legs. It is found mostly in mountains from California to Alaska, also in Colorado, Utah, and Nevada, and in the East, in Georgia. Its larvae are internal parasites of siricid horntails in coniferous trees.

445 Short-tailed Ichneumons
(*Ophion* spp.)

Description: ⅜–¾" (10–19 mm). *Abdomen long,
compressed on sides.* Body pale yellow to
reddish brown. *Antennae and legs long,*
pale. Ovipositor of female barely visible
at tip of abdomen. Wings clear.

Habitat: Forest canopies and shrubby fields.

Range: Worldwide.

Food: Adult drinks nectar. Larva eats internal
tissues of caterpillars of tiger moths,
noctuids, giant silkworm moths, and
satyrids.

Life Cycle: Female hunts for active caterpillar of
appropriate species and lays 1 egg on its
body, often high on the thorax where
the caterpillar cannot reach. Ichneumon
larva burrows into host, eventually
killing it but usually after caterpillar
spins its cocoon. Ichneumon larva then
pupates inside host's remains. Adults
are active May–August. 1 generation a
year.

Ichneumons fly to artificial lights at
night. They are often seen emerging
from hosts' cocoons. This genus
contains some of the most common
large North American ichneumons, but
only a specialist can identify the
different species.

453 Red-tailed Ichneumon
(*Scambus hispae*)

Description: ⅝–¾" (16–19 mm). Elongate. *Most of
abdomen brownish red;* 1st segment black;
rest of body black except for gray fore
and middle tibiae and gray tarsi.
Wings dusky.

Habitat: Meadows and forest edges.

Range: California to Alaska.

Food: Adult drinks nectar. Larva feeds on
caterpillars of small moths, in buds,
leaf rolls, leaf mines, galls, and fruit.

Life Cycle: Female actively searches foliage for

caterpillar, lays 1 egg on its back, then moves on to lay more eggs. Larva parasitizes host and may pupate without moving from site. Adults are active in August.

In late summer, large numbers of these ichneumons are seen in forests and wooded areas. They are easily recognized by their conspicuous red abdomen.

TORYMID WASPS
(Family Torymidae)

Torymids are tiny, elongate, usually metallic-green wasps, $\frac{1}{16}$–$\frac{1}{8}$" (2–4 mm) long. The hind coxae are enlarged, and the female of most species has an ovipositor that is as long as the body. The larvae live as parasites of many kinds of insects, including the larvae of gall insects, beetles, butterflies, and moths. Some species are parasites of mantid eggs, and a few feed on seeds.

463 California Torymus
(*Torymus californicus*)

Description: $\frac{1}{8}$" (3–4 mm). Ovipositor $\frac{1}{4}$" (6 mm). *Head, thorax and abdomen coppery;* male has blue 1st abdominal segment. Antennae dark with paler 1st segment. Coxae dark, metallic, legs beyond brownish red. Ovipositor blackish. Wings clear, veins grayish.
Habitat: Forest edges and brushy areas.
Range: California and Oregon.
Food: Adult drinks nectar. Larva is internal parasite of gall wasp on oak trees.
Life Cycle: Little known.

Adult wasps are sometimes seen emerging from galls induced by host.

CHALCIDS
(Family Chalcididae)

Chalcids are small, dark-colored, wasplike insects, usually $\frac{1}{16}-\frac{3}{8}''$ (2–8 mm) long. Most species are black with bright yellow legs or markings on the body. The antennae are short and elbowed. The enlarged hind femora bear teeth underneath, and the hind coxae are larger than the fore coxae. The wings have very few veins and are held flat over the abdomen when at rest. The female's short ovipositor arises from the underside of the abdomen rather than from its tip. Adults of many species can jump; others curl up and feign death if disturbed. Most adults feed on nectar or honeydew secreted by other insects. Some female chalcids stalk female insects and feed on the body liquids the prey produces after laying eggs. Chalcid larvae are internal parasites of other insects, including other internal parasites, such as ichneumons and tachinid flies.

469 Golden-yellow Chalcid
(Spilochalcis mariae)

Description: $\frac{1}{8}-\frac{3}{8}''$ (3–8 mm). *Enlarged hind femora.* Golden-yellow with black markings on head and thorax, and black ring on each abdominal segment. Antennae elbowed, clubbed, black with 1st segment yellow. Wings smoky.

Habitat: Mixed and deciduous forests.

Range: New York to Florida, west to Illinois, Colorado, and California.

Food: Adult drinks water and nectar. Larva is internal parasite in caterpillar or pupa of saturniid moths.

Life Cycle: Female uses ovipositor to thrust 1 or 2 eggs into skin of half-grown caterpillar. Larvae penetrate and feed, usually not killing caterpillar before it pupates. 1 or 2 eggs may produce many larvae by hypermetamorphosis, a form of asexual reproduction. 12 or more chalcid wasps often emerge from a single host's cocoon.

Because they can produce so many larvae from just 1 egg, chalcid wasps are often regarded as more important for biological control of insect pests than braconids or ichneumons.

GALL WASPS
(Family Cynipidae)

These minute, humpbacked wasps are usually glossy black or brown and ⅟₁₆–⅜" (2–8 mm) long. The abdomen is short and flattened from side to side. The wings have few veins. The antennae are long and not elbowed, as they are in the chalcids. Some adults drink water and nectar, others may not feed. Females of most species lay eggs in plants, each species of wasp using a particular plant species. The plant then forms a swelling, or gall, in which the larvae feed on plant tissues. The galls are often distinctive in appearance and provide the best means for identifying the species. A few cynipid larvae are parasites of other insects. Some are parasites in fly puparia; others are internal parasites of braconids, which are themselves internal parasites in the abdomens of plant lice.

448 California Oak Gall Wasps
(*Andricus quercuscalifornicus*)

Description: ⅛–¼" (3–5 mm). Brown or reddish brown. *Compressed on sides.* Short "waist." Antennae threadlike. Legs stout. Wings transparent, yellowish.
Habitat: Hardwood forests and city parks.
Range: Southern California to Washington.

Food: Adult may not eat. Larva feeds on soft
tissues inside large galls on oak twigs.
Life Cycle: Eggs are inserted into soft young twigs.
Larvae tunnel inward, causing tree to
form hard galls, known as "oak
apples." Galls are green at first, then
turn red or brown, growing to 4" in
width. Each gall encloses few to many
larvae in separate central chambers.
Larvae chew exit tunnels before
pupating. Adults are active from
October in the North to February in
the South.

This is the largest and best-known gall
wasp in the West. Sometimes other
insects, such as Filbertworm (*Melissopus
latiferreanus*) caterpillars, come to live in
the large galls. Infested twigs swollen
with galls break off soon after the galls
turn dark brown.

449 Live Oak Gall Wasp
(*Callirhytis quercuspomiformis*)

Description: ⅛–¼" (3–5 mm). Brown to reddish
brown. *Abdomen compressed on sides.*
Short "waist." Antennae threadlike.
Legs stout. Wings transparent.
Habitat: Hardwood forests and city parks.
Range: California.
Food: Adult may drink nectar. Larva feeds on
soft tissues inside galls on live oak
twigs and leaves.
Life Cycle: These gall wasps have 2 different
generations a year. The 1st consists
of females that reproduce
parthenogenetically, or asexually.
The 2nd generation consists of males
and females that reproduce sexually.
The asexual generation produces stem
galls on 1 oak species and the sexual
generation produces leaf galls on
another oak species. Galls grow to
the size of small apples. Fully grown
larvae pupate in central chambers
inside the galls.

The Live Oak Gall Wasp usually does little damage to the trees it inhabits. Formerly included in genus *Andricus,* it is now separately classified on the basis of minute anatomical details.

PELECINIDS
(Family Pelecinidae)

This family contains a single North American species easily recognized by the long and slender abdomen of the glossy black female, to 2″ (50 mm) long. The male, to about ⅝″ (15 mm) long, is very rare; it has a shorter abdomen with a thickened tip. Despite the formidable appearance of this species, it cannot sting. Adults drink nectar and water. The larvae are internal parasites of May beetle larvae and other soil-inhabiting beetles.

464 American Pelecinid
(*Pelecinus polyturator*)

Description: Male ½–⅝″ (12–15 mm), female 1¾–2″ (45–50 mm). Glossy black. Antennae almost as long as body in male, or extending ½ way along 1st abdominal segment in female. Legs and fore wings well developed. Hind wings ⅓ length of fore pair. Male abdomen enlarged at rear. *Female has extremely long slender abdomen.*

Habitat: Crop fields, woods, and suburban gardens.

Range: Throughout North America.

Food: Adult drinks nectar. Larva is internal parasite of May beetle larvae and perhaps of other scarabs.

Life Cycle: Female shoves its abdomen deep into soil to detect host larvae below, then lays eggs one at a time, each on a separate host. Pelecinid larvae burrow into hosts, killing them. They scavenge

on remains, then pupate there. Adults
are active August–September.

Males are extraordinarily rare. This is
the only pelecinid wasp species in the
United States. Because of the wide
range of this species, it is listed under
many scientific names.

TIPHIID WASPS
(Family Tiphiidae)

Tiphiids are medium-sized wasps, ⅜–
1⅜″ (10–35 mm) or more long. Some
are black and yellow; others are all
black. The abdomen often has a strong
constriction between segments—it
appears somewhat cylindrical, swelling
and later slimming toward the rear end.
Males of some species characteristically
bear an upward-curving spine near the
tip of the abdomen. The larvae are
parasites of beetle larvae.

475 Five-banded Tiphiid Wasp
(*Myzinum quinquecinctum*)

Description: 1⅛–1⅜″ (30–35 mm). Black, streaked
on head and thorax with yellow.
Abdomen has yellow bands around each
segment. Antennae black. Legs yellow.
Wings dusky brown, translucent.
Habitat: Meadows, fields, and lawns.
Range: Throughout the United States and
Canada, except in the West.
Food: Adult drinks nectar. Larva is parasite of
May beetle larvae.
Life Cycle: Female lays eggs on May beetle larvae
in soil, depositing 1 egg on each host.
Wasp larvae penetrate hosts, feed first
on nonessential tissues, later eat
essential organs, killing host. Pupae
overwinter in soil. Adults emerge in
early summer. There is 1 generation a
year.

Because this wasp preys on beetles, it is considered to be highly beneficial. It, in turn, is parasitized by velvet-ants.

VELVET-ANTS
(Family Mutillidae)

Despite their name, velvet-ants are densely hairy wasps, ¼–1" (6–25 mm) long. Most are brightly patterned in red, yellow, or orange. Males are fully winged, but females lack wings and differ from true ants in having only a slight constriction, or pedicel, between the abdomen and thorax and in having antennae that are not elbowed. Adults feed on nectar and water. Females lay eggs inside a host insect's cocoon or puparium. The larvae of most species are parasites of the larvae and pupae of other wasps and bees, although a few parasitize flies and beetles. In colder latitudes there is 1 generation a year. Females can sting painfully.

327 Thistledown Velvet-ant
"Gray Velvet-ant"
(*Dasymutilla gloriosa*)

Description: ½–⅝" (13–16 mm). *Antlike,* with only slight constriction (pedicel) between thorax and abdomen. Antennae beadlike. *Black, with loose covering of long white hair.* Male winged, female wingless.

Habitat: Arid and semiarid open lands, on the ground, and in low vegetation. Males sometimes are found on flowers.

Range: Utah, Nevada, California, and Texas into Mexico.

Food: Adult drinks nectar. Larva is external parasite of sand wasp larvae.

Life Cycle: Female actively searches for burrows dug by sand wasps, which are stocked with flies as food for developing wasp

larvae. Eggs are dropped in wasp nest.
Velvet-ant larvae feed on wasp larvae
and food brought by female wasp.
Usually they kill wasp larvae, then
pupate in host's larval chambers.

Female velvet-ants defend themselves
from wasps, ants, and people by
inflicting a painful sting.

326   Red Velvet-ant
(*Dasymutilla magnifica*)

Description: Male ¾" (18 mm), female ⅞" (21 mm).
Antlike, with only slight constriction
(pedicel) between thorax and abdomen.
Head, thorax, legs, and pedicel dark
wine-red to black with black hair.
*Male's abdomen dark wine-red with red
hair* on segments 3–8 and black hair on
rest. *Female's abdomen burnt-orange to
yellow-orange.* Male winged, female
wingless.

Habitat: Arid and semiarid open lands.

Range: Texas to California and Nevada, south
to Mexico.

Food: Adult drinks nectar. Larval food
unknown.

Life Cycle: Unknown.

These hairy wasps scurry across the
ground much like true ants.

325   Cow Killer
(*Dasymutilla occidentalis*)

Description: ⅝–1" (15–25 mm). *Antlike,* with only
slight constriction (pedicel) between
thorax and abdomen. Antennae
beadlike. *Thorax and abdomen red above,
covered with short erect red hair.* Body
below and head black. Males winged,
females wingless.

Habitat: Meadows, forest edges, and clover
fields.

Range: New York to Florida and Gulf states, west to Texas.

Food: Adult drinks nectar. Larva feeds on bumble bee larvae.

Life Cycle: Female searches for bumble bee nests and drops 1 egg beside each brood chamber. Cow Killer larvae invade brood chambers, feed on bee larvae, kill them, and scavenge on their remains. Cow Killer larvae pupate in victim's brood chambers.

Adult Cow Killers can run quickly and fight ferociously. They get their name from their painful sting—so severe that many people claim it could kill a cow.

SCOLIID WASPS
(Family Scoliidae)

Scoliids are large, hairy, robust wasps, ½–1⅛″ (13–30 mm) long. Most are dark brown or black with yellow or orange markings on the abdomen. Males differ from male velvet-ants in being larger, less hairy, and darker in color. Unlike the wingless velvet-ant females, scoliid females are winged. In both sexes the wing membrane beyond the closed cells has numerous lengthwise wrinkles. Adults visit flowers, feeding on nectar. Scoliid wasp larvae are external parasites of the larvae of scarab beetles. The female burrows in soil or wood debris, in search of beetle larvae. On finding one, the wasp stings it and digs around it, forming a small chamber. The wasp then deposits an egg on the host. If disturbed, females can sting painfully.

480 Scarab-hunter Wasp
(*Campsomeris pilipes*)

Description: 1–1⅛″ (26–30 mm). Head, antennae, thorax, and legs jet-black; sometimes with white hair on thorax. *Abdomen largely yellow with black banding resembling hornet's.* Wings dark with many lengthwise wrinkles near outer margin.

Habitat: Meadows and forest edges.

Range: Colorado west to Oregon and California, south to Texas and Mexico.

Food: Adult drinks nectar. Larva feeds on scarab beetle larvae.

Life Cycle: Female searches for beetle larvae, which is food source used to provision brood cells in subterranean burrows. Larvae feed, then pupate in cell.

This wasp is easily mistaken for a yellow jacket wasp or paper wasp but lacks the yellow marks on its head and thorax. The stinger, used to anesthetize prey, is an effective weapon in self-defense.

457 Digger Wasp
(*Scolia dubia*)

Description: ½–¾″ (13–18 mm); wingspan 1″ (26 mm). Hairy. Black with *yellow or reddish-orange markings* on abdomen. Wing blackish with many parallel, longitudinal wrinkles near outer margin.

Habitat: Gardens and meadows with flowers.

Range: Massachusetts to Florida, west to Arizona and California.

Food: Adult may take juices from Green June Beetles. Larva feeds on larvae of Green June Beetles.

Life Cycle: Female digs into ground in search of beetle larvae, sometimes tunneling a few feet deep. Female stings beetle larva, subduing it, then digs cell around body and lays 1 egg on its back.

Larva feeds on host. Wasp larva spins
cocoon and overwinters as pupa. 1
generation in the North, more in the
South.

Adults are usually found on flower
clusters or scampering over the ground
in early morning searching for beetle
larvae. Mating males and females
perform a mating dance, flying low in
an "S" or figure-8.

ANTS
(Family Formicidae)

These familiar insects, less than $\frac{1}{16}$–1″
(1–25 mm) long, are mostly black,
brown, or reddish. They have a
complex social structure usually
consisting of a wingless worker caste
composed entirely of sterile females and
a reproductive caste made up of
winged, fertile males and females. But
some species do not have a worker
caste, and some reproductives do not
have wings. Ants have a slender
"waist," or pedicel, of 1 or 2 beadlike
or scalelike segments between the
thorax and abdomen. They differ from
wasps in having distinctly elbowed
antennae. Ants live in colonies in
underground tunnels or in galleries in
dead wood. From time to time, winged
males and females emerge from the nest
and perform a brief mating flight. After
mating, the males die, and the females
lose their wings and return to the
ground to start a new colony. Workers
gather food, maintain and defend the
nest, and tend eggs, larvae, and pupae.
Most species are predators or
scavengers, but a few harvest seeds,
visit clusters of aphids to eat their sweet
secretions, raise fungus for food in
small underground gardens, or eat
leaves cut from plants. Some species
produce eggs, which are eaten by the

queen and workers. When disturbed,
most ants are capable of "biting" or
"stinging" people.

322 Spine-waisted Ants
(*Aphaenogaster* spp.)

Description: ⅜" (8–10 mm), depending on caste.
Reddish black with *2 sharp diverging
spines on rear of thorax.* 1-segmented
"waist" (pedicel) between thorax and
abdomen. Antennae and legs long.

Habitat: Arid plateaus at elevations of 2,690–
5,380' (820–1,640 m).

Range: Texas to California, south to Mexico.

Food: Small insects.

Life Cycle: Queen excavates brood chamber in soil
and lays a few eggs. Queen regurgitates
food to larvae until they can spin
cocoons and transform to adults. Then
workers hunt for food, expand galleries,
feed queen, and tend eggs and young.

The sharp spines on these ants may be
nature's way of protecting their narrow
"waist," or pedicel, from attackers.
Formerly included in the genus
Novomessor.

312 Leafcutting Ant
(*Atta texana*)

Description: ¹⁄₁₆–½" (1.5–13 mm), depending on
caste. *Light to dark reddish brown.* 2-
segmented "waist" (pedicel) between
thorax and abdomen. *Sharp spines* on
waist and backward from head.
Antennae 11-segmented without
distinct club. Legs very long.

Habitat: Woods with well-drained soil.

Range: Louisiana and Texas.

Food: Fungus from tended compost heap in
nest.

Life Cycle: Founding queen (mated female) digs
first nest, brings pellet of fungus and

fertilizes it with her own wastes. Eggs are laid in nest. Larvae are fed by queen until workers mature and assume duties. Next-to-smallest workers travel by day to find fresh leaves, cut leaves into fragments, and carry these to nest. Other workers deepen and expand nest, often to 20'. Others build up compost heap to raise more fungus. Large sterile soldiers fight off other ants and guard nest.

Large mounds of earth are piled up at the entrance to these ants' underground nests. When workers remove leaves, they often strip wild and cultivated plants. The ants sometimes invade houses, where they gather seeds and cereals. These ants communicate with sounds made by rubbing a file located behind the pedicel, or "waist," against the abdomen. This noise is clearly audible to people.

310 Texas Carpenter Ant
(*Camponotus festinatus*)

Description: ¾–1" (18–24 mm), depending on caste. *Brownish yellow*, sometimes banded. 1-segmented "waist" (pedicel) between thorax and abdomen. Antennae 12-segmented, elbowed.

Habitat: Deadwood of upright or fallen trees, timber, utility poles, and on soil surface or under stones and cow dung.

Range: Colorado and Texas west to California, south to Mexico.

Food: Other insects, honeydew, juice from rotting fruit, and sweets.

Life Cycle: Queen (mated female) begins nest in dead wood and tends 1st brood. As females mature, they extend galleries, tending eggs and young and hunting for food, which they regurgitate to feed queen. Large colony of thousands includes soldiers and workers of various sizes.

Battalions of ants searching for food sometimes invade houses at dusk or dark. Flying adults may come to artificial lights and pass barriers to reach dry wood suitable for constructing galleries.

318   Black Carpenter Ant
(*Camponotus pennsylvanicus*)

Description: ¼–½″ (6–12 mm), depending on caste; workers mostly smaller, queen large. Black. Thorax and 1-segmented "waist" (pedicel) between thorax and abdomen often brown. *Enlarged part of abdomen bears long yellow to grayish hair.* Eyes well developed. Antennae elbowed, with long 1st segment and 11 shorter segments.

Habitat: Deadwood of tree trunks, felled logs, timber, poles, and houses.

Range: Quebec to Florida, west to Texas, north to North Dakota.

Food: Other insects, honeydew, fruit juices, grains of sugar, or other sweets.

Life Cycle: Nests are built in cavities excavated in dead wood. Brood chambers accommodate eggs, larvae, and pupae inside silken pupal cases.

This ant bites but cannot sting. Unlike termites, carpenter ants do not eat wood but tunnel inside. They often cause considerable damage. The Giant Carpenter Ant (*C. laevigatus*), to ⅝″ (15 mm), is shiny black to reddish brown; it is found in forests of the Cascade, Rocky, and Sierra Nevada mountains. The Red Carpenter Ant (*C. ferrugineus*), ¼–½″ (6–12 mm), resembles the Black Carpenter Ant and is common from New York to Georgia and west to Kansas.

314 Crater-nest Ants
(*Conomyrma* spp.)

Description: ⅜–½" (10–12 mm), depending on caste. *Glossy reddish black,* sometimes appears metallic. Short 1-segmented "waist" (pedicel) between thorax and abdomen. Antennae and legs long. Well-developed compound eyes and jaws.

Habitat: Open sandy and gravelly areas that are sunny most of the day.

Range: West of the Mississippi River, south of the California-Oregon border.

Food: Other insects.

Life Cycle: Nests are built in soil. Queen tends 1st brood. Later, female workers extend galleries, tend eggs and young, and hunt for food, which they regurgitate to feed queen. Colonies are small to medium-sized.

These ants build craterlike mounds around the entrance to the nest, suggesting their common name. Workers are active, aggressive predators, ready to subdue victims with a foul-smelling fluid secreted from anal glands.

315 Texas Shed-builder Ant
"Acrobat Ant"
(*Cremastogaster lineolata*)

Description: Worker ⅛" (3–4 mm). *Abdomen heart-shaped.* 2-segmented "waist" (pedicel) between thorax and abdomen. Thorax reddish. Abdomen black.

Habit: Under stones, boards, dry logs, and stumps.

Range: Texas.

Food: Honeydew, dead and live insects.

Life Cycle: Colony builds subterranean nest under objects or excavates multiple chambers in dry wood. Queen tends 1st brood, then female workers extend galleries, tend eggs and young, and hunt for food, which they regurgitate to feed queen. Large colony of thousands includes many soldiers, males, and unmated female workers (in season).

These ants build shedlike shields of chewed vegetable matter over aphids and other insects, which they tend for their sweet secretion of honeydew. Workers hold their sharply pointed abdomens up and forward over their bodies, making these ants appear shorter than they are. The ants tumble frequently and hence are also known as "Acrobat Ants."

316   Red Ant
(*Formica* spp.)

Description: Worker ⅛–½" (3–12 mm). Rusty red to brownish black or black. "Waist" (pedicel) between thorax and abdomen 1-segmented.

Habitat: Wooded slopes at high elevations.

Range: Utah and Nevada to mountains of California, north to British Columbia and Alberta.

Food: Honeydew from aphids, nectar from flowers, and other insects.

Life Cycle: Nests are often constructed close to small trees and shrubs. Entranceways are littered with pine needles, leaf fragments, and bits of stems. Queen tends 1st brood, then female workers care for eggs and young, feeding queen regurgitated food.

Workers often stay near aphids or transport them to more convenient sites, where honeydew can be collected repeatedly. The Allegheny Mound Ant (*F. exsectoides*), same size, is rusty red on its head and thorax, blackish brown on the legs and abdomen. It occurs throughout the eastern United States and adjacent areas of Canada, west to Wisconsin, Iowa, Kansas, Colorado, and New Mexico.

446 Legionary Ant
(*Labidus coecus*)

Description: Female workers ⅛–⅜″ (3–10 mm), winged male ¼–¾″ (6–19 mm), queen ¾″ (19 mm). *Shiny reddish brown,* dull on rear ½ of thorax. Body and legs often covered with red-blond hair. Head broader than thorax. Antennae short. 2-segmented "waist" (pedicel) between thorax and abdomen. Only males have well-developed compound eyes; queens and workers lack eyes or have rudimentary eyes.

Habitat: Open land with little full shade.

Range: Arkansas to Louisiana, west to Texas and Mexico, north to Oklahoma.

Food: Mostly other insects, including caterpillars and fly pupae on soil.

Life Cycle: Workers build temporary nests under stones and in fallen trees and stumps. At 15- to 25-day intervals, workers construct covered tunnels of leaf fragments, through which the colony moves to new location by day or night. Timing of move is synchronized with development of larvae that can be carried in jaws of workers.

Counterpart of the tropical Army Ants of South America and the Driver Ants of Africa, Legionary Ant workers can bite and sting severely. They have been known to kill chickens.

321 Little Black Ant
(*Monomorium minimum*)

Description: Worker 1/16″ (1.5–2 mm). Slender, smooth, with sparse body hair. *Shiny black to dark brown.* 1-segmented "waist" (pedicel) between thorax and abdomen. Antennae 12-segmented, 1st long, last 3 form club.

Habitat: Forest edges and houses.

Range: Throughout North America, except Pacific Northwest.

Food: Sweet substances, meat fragments, cooked vegetables, other human food.

Life Cycle: Nests are constructed below ground, raising small craters around opening at the surface, or in rotting wood. Queen feeds 1st brood, then workers take over, tending young and feeding queen.

One of the most common ants in homes, this insect is active day and night and is often seen carrying particles of food many feet back to its nest. Because there are usually no winged females, these ants do not have nuptial flights. The Pharaoh's Ant (*M. pharaonis*), same size, is pale brown to reddish and more hairy. It nests in woodwork and partitions in Florida, where it is a troublesome pest.

323 Arid Lands Honey Ants
(*Myrmecocystus* spp.)

Description: Worker ⅜″ (8–10 mm), dark brownish red; replete worker ⅜–½″ (10–12 mm), paler; reproductives ¼–⅝″ (5–16 mm), male black, female brownish black. Reproductive (winged) males and females have well-developed thorax, smaller and flattened head, short jaws. Workers and repletes have much larger head and stronger mandibles, smaller thorax constricted between middle and hind legs, and a more obvious slender 1-segmented "waist" (pedicel) between thorax and abdomen.

Habitat: Arid plains and deserts.

Range: New Mexico to northern Mexico, north to Utah.

Food: Honeydew from sucking bugs, nectar from flowers, and some other plant juices.

Life Cycle: Nests are built in arid soil, where eggs are laid and larvae develop and pupate. In some species, workers gather food and pass it to repletes—workers that store food for months or years in swollen abdomens. They hang head-up by their claws from the ceiling of the nest and regurgitate food droplets to the colony. There are rarely more than a few hundred ants in a colony.

Some species are active by day, others at night. Certain species are strictly predatory and carnivorous, while others have a replete caste that only takes in sweet liquids of plant origin. Identification of the 20 different species is difficult even for specialists.

311 Big-headed Ant
(*Pheidole megacephala*)

Description: Worker ⅛″ (3–4 mm); soldier ⅛–¼″ (4–5 mm). Yellow-amber to dark brown. *Head very large, bigger than abdomen;* front surface of head rough, dull; back smooth, shiny. Antennae slightly clubbed. Thorax with 2 sharp spines. 2-segmented "waist" (pedicel).

Habitat: Fields and gardens.

Range: Florida.

Food: Mostly seeds and grains; also other insects, except species that secrete honeydew, such as aphids.

Life Cycle: Newly mated queen prepares 1st nest under a stone or stick; 1st generation of female workers enlarges nest, excavating chambers to store seeds. They tend eggs and feed larvae and queen. Most colonies contain fewer than 300 ants.

This ant regularly attacks the Fire Ant. The common Large-headed Ant (*P. bicarinata*) has workers 1/16–⅛″ (1–4 mm) and soldiers ⅛″ (4 mm); it is yellowish to dark brown and is found on beaches, deserts, and mountains from New York to Florida, west to Arizona, Nevada, and California.

317 Rough Harvester Ant
(*Pogonomyrmex rugosus*)

Description: ¼–½" (6–13 mm), depending on caste. Reddish brown. 2-segmented "waist" (pedicel) between thorax and abdomen. Winged female and wingless queen larger and darker than workers; winged male smaller.

Habitat: Lowlands, especially cultivated fields and relatively bare areas, and sandy areas near roads.

Range: Southwestern states.

Food: Seeds and grains.

Life Cycle: Mated female, with help from mate, digs small chamber to conceal clusters of milk-white capsule-shaped eggs and later larvae and pupae. First workers to emerge enlarge nest; nest opening may be level with ground or protected by conical crater of small pebbles. Separate chambers are dug to shelter eggs, developing larvae, and pupae. Ants swarm April–October.

Workers, active only by day, can bite and sting painfully. They can severely damage crops by cutting down plants and creating large barren areas. The larger Red Harvester Ant (*P. barbatus*) builds gravel mounds that are scooped out at the center; it also destroys plants and occurs in Louisiana and the West. The smaller California Harvester Ant (*P. californicus*), ¼" (5–6 mm), is pale yellowish red. It nests in sand or fine gravel from Texas to California but does not damage plants.

309 Honey Ant
(*Prenolepis imparis*)

Description: Worker ¹⁄₁₆–⅛" (1–3 mm). Glossy black to reddish brown. *Abdomen swollen, sometimes pale with reddish-brown bands.* 1-segmented "waist" (pedicel) between thorax and abdomen. Legs reddish brown. Winged adults similar size and color.

Habitat: Shady damp soil with considerable clay.

Range: Wisconsin and Ontario south to Florida, west to California, and north to Oregon.

Food: Dead insects, earthworms and small arthropods, small seeds, nectar and honeydew, ripened or decaying fruit, and secretions from galls.

Life Cycle: Nests are prepared in soil by mated queen. 1st generation workers extend chambers and food stores and tend queen, eggs, and larvae. Specialized workers, called repletes, serve as storage vessels—their distended abdomens are filled with honey supplied by other workers, and later they regurgitate food for the colony. Males and females overwinter in nest. Inconspicuous most of the year, they swarm especially in March. Colonies are small to medium-sized.

Foraging mostly at night, these ants watch over aphids and scale insects as sources of honeydew. They carry the honeydew to underground chambers to be stored by repletes. Sometimes workers forage at freezing temperatures.

313 Fire Ant
(*Solenopsis geminata*)

Description: Worker $\frac{1}{16}$–$\frac{1}{4}$" (1–6 mm), 2 or more worker castes of different sizes. Dull yellow to red or black. *Head large, jaws incurved* and usually lacking teeth. 2-segmented "waist" (pedicel) between thorax and abdomen. Fine hair mostly on head and abdomen. Legs long.

Habitat: Fields, woodlands, and open areas, in dry to moist soil.

Range: Florida and Gulf states to Pacific Coast, north to British Columbia.

Food: Other insects, seeds, poultry, fruits, honeydew, vegetables, and flowers.

Life Cycle: Females excavate nest close to shrubs
for protection from burrowing ant-
eating animals, spreading large mound
of waste earth. Sometimes nests are
built in rotting logs or under stones.
Mated queen tends 1st generation eggs
and larvae, then 1st generation female
workers take care of eggs, larvae, and
queen.

Reports of devastating battalions of Fir
Ants are well known in the South and
Southwest. But although these ants
often damage young plants, they rarely
destroy established crops. Some people
even consider this species a beneficial
predator of insect pests. Fire Ants sting
and bite, producing a burning sensation
like fire, inspiring the common name.
Other members of this common genus
are found throughout North America
but the range of individual species is
more restricted.

VESPID WASPS
(Family Vespidae)

Vespids are medium-sized or large
wasps, ⅜–1⅛" (10–30 mm) long,
with distinctly notched eyes and a very
long discal cell in the fore wing. Unlike
spider wasps and sphecids, when these
wasps fold their wings lengthwise over
the abdomen at rest, the wings appear
pleated. Many are dull brown or black,
but others are banded with yellow or
white. Some vespids are solitary and
construct cells of mud or dig mud-lined
underground tunnels for their larvae.
Many other species live in colonies, lay
their eggs in combs of cells made of
paper, and show some degree of social
behavior. Social organization ranges
from groups of cooperating fertile
females, as in the paper wasps, to the
caste system of yellow jackets and
hornets, in which there is a single

fertile queen and a large population of
smaller female workers that do not lay
eggs. All of these wasps sting painfully.

467 Potter Wasp
(*Eumenes fraternus*)

Description: ⅝–¾" (15–20 mm). Compound eyes
prominent. *2-segmented "waist"* (pedicel)
between thorax and abdomen. Abdomen
beyond abruptly wider, tapering to tip.
Black with yellow on face and across
thorax at front and rear; yellow dots on
pedicel and abdomen near base; yellow
ring at tip of abdomen. Wings smoky
with dark spot (stigma).

Habitat: Woodland edges and fields with
shrubby undergrowth.

Range: Ontario and the Atlantic seaboard west
to Texas, north to Nebraska and
Minnesota.

Food: Adult drinks nectar. Larva feeds on
caterpillars and sawfly larvae.

Life Cycle: Female builds squat spherical chamber
of mud on a twig or branch. 1 egg is
attached to wall or placed near flared
opening. Anesthetized caterpillar or
sawfly larva is placed inside as food for
wasp larva. Each chamber is sealed with
mud pellet before the next is built and
provisioned.

The provision chamber may be up to
½" (12 mm) wide. It is totally resistant
to rain until the young wasp cuts an
exit hole from the inside. Often several
of these potlike chambers appear in
rows on twigs. Some entomologists
place this species in the Family
Eumenidae.

470 Polybiine Paper Wasp
(*Mischocyttarus flavitarsus*)

Description: ⅝″ (15–17 mm). "Waist" (pedicel)
between thorax and abdomen. Black,
boldly marked with clear bright yellow.
Thorax large. Wings translucent
amber.

Habitat: Meadows, fields, and sandy areas, on
flowers.

Range: Nebraska, Colorado, and Idaho, west to
California, north to British Columbia,
south to Mexico.

Food: Adult drinks nectar. Larva feeds on
caterpillars pre-chewed by attendant
female.

Life Cycle: Usually 1 female builds a paper nest
but sometimes several work together,
then 1 female becomes dominant by
eating the eggs laid by others
(including those of daughters). Cells
remain open, allowing adults to feed
grublike larvae until they grow to full
size. Then larvae seal cells and pupate
inside.

Primarily tropical, these wasps are
active in warm or hot weather, often
visiting streams to get water.

442 Paper Wasps
(*Polistes* spp.)

Description: ½–1″ (13–25 mm). Slender, *hornetlike.*
Short 1-segmented "waist" (pedicel)
between thorax and abdomen. Upper
portion of head pointed, never notched,
as in hornets and yellow jackets. Head
and body *mostly reddish brown to black
with yellow rings and reddish areas on
abdomen.* Male's face pale, with antennal
tips hooked; female has brown face.
Wings amber to reddish brown.

Habitat: Meadows, fields, and gardens on
flowers, and near buildings.

Range: Throughout North America.

Food: Adult drinks nectar and juices from

crushed and rotting fruits. Larva feeds on insects pre-chewed by adults.

Life Cycle: In spring several females work together to construct uncovered paperlike, hanging nest of wood pulp and saliva. One female becomes dominant queen. 1st few generations in summer are all females, cared for as larvae by unmated female workers. Unfertilized eggs produce fertile males. Only mated young queens overwinter under leaf litter and in stone walls. Old queens, workers, and larvae die.

Paper Wasps are much more tolerant of people and minor disturbances than are hornets and yellow jackets. The Northern Paper Wasp (*P. fuscatus*), ⅝–⅞" (15–21 mm), is dark reddish brown and yellow with its first abdominal segment banded yellow. It is common from British Columbia to the Canadian Maritime Provinces, south to West Virginia. The Southwestern Texas Paper Wasp (*P. apachus*), ½–⅝" (13–15 mm), is reddish brown with yellow markings. It occurs in Texas, New Mexico, southern California, and Mexico.

444 Giant Hornet
(*Vespa crabro germana*)

Description: ¾–1⅛" (18–30 mm). Short "waist" (pedicel) between thorax and abdomen. Head, antennae, thorax, legs, and 1st abdominal segment reddish brown. Back of head and sides of thorax sometimes have yellow stripe. Rest of *abdomen bright yellow with dark crossbands and small spots*. Wings amber.

Habitat: Forests and towns.

Range: Southern Massachusetts to Georgia, west to Indiana.

Food: Adult eats other insects and drinks nectar. Larva feeds on insects pre-chewed by adults.

Life Cycle: A covered, tan-colored paper nest is
built in a hollow tree, under porch
floor, or in an outbuilding. 1st
generation is all female workers, which
feed later generations. In late summer
unfertilized eggs produce males that
mate and then die.

Introduced to America in the mid-
1800s, this hornet is common locally
around the western limits of its range.
It defends its nest from intruders but
otherwise avoids confrontations when
possible.

486 Yellow Jackets
(*Vespula* spp.)

Description: ½–⅝″ (12–16 mm). Body stout,
slightly wider than head. Abdomen
narrow where attached to thorax with
short "waist" (pedicel). 1st antennal
segment yellow, 2nd and subsequent
segments black. *Head, thorax, and
abdomen black and yellow or white.* Wings
smoky.

Habitat: Meadows and edges of forested land,
usually nesting in ground or at ground
level in stumps and fallen logs.

Range: Throughout North America; various
species more localized.

Food: Adult eats nectar. Larva feeds on insects
pre-chewed by adults.

Life Cycle: In spring mated female constructs small
nest and daily brings food to larvae
until 1st brood matures and females
serve as workers, extending nest and
tending young. In late summer males
develop from unfertilized eggs and
mate. When cold weather begins, all
die except mated females, which
overwinter among litter and in soil.

Yellow jackets can be pests at picnics,
and they will carry off bits of food.
Females sting repeatedly at the least
provocation. If the nest can be found

and its opening covered at night with a transparent bowl set firmly into the ground, adults will be confused by their inability to escape and seek food in daylight; they will not dig a new escape hole and will soon starve to death. The Western Yellow Jacket (*V. pennsylvanica*) and Eastern Yellow Jacket (*V. maculifrons*) are similarly colored, except the first antennal segment of the latter is all black.

478 Sandhills Hornet
(*Vespula arenaria*)

Description: ⅝–¾″ (16–20 mm). Short "waist" (pedicel) between thorax and abdomen. Black with *bright yellow on sides of head, thorax, legs, and across each abdominal segment*. Wings smoky.

Habitat: Sandy country.

Range: Northern states and Canada from Newfoundland to Alaska.

Food: Adult sips nectar. Larva eats food pre-chewed by adults.

Life Cycle: Female builds paperlike hanging nest of wood pulp and saliva. Each horizontal comb is usually turned up at the edges so that seen from above it appears concave. Larvae are fed by workers.

Globular wasp nests are often seen in spring hanging under roof eaves or built in shrubbery close to the ground. If the nests are disturbed, the wasps sting viciously.

483 Bald-faced Hornet
(*Vespula maculata*)

Description: ⅝–¾″ (16–20 mm). Head much shorter than wide; neck and "waist" (pedicel) about equally constricted. *Black and white patterns* on face, thorax, abdomen, and 1st antennal segment.

Wings smoky.

Habitat: Meadows, forest edges, and lawns.

Range: Throughout North America.

Food: Adult drinks nectar, fruit juices, and perhaps eats other insects. Larva feeds on insects pre-chewed by adults.

Life Cycle: In spring female chews wood to build small, pendant nests out of gray pulp. The 1st generation includes only female workers. They bring food several times a day to larvae. Larvae close their own cells. The nest is always constructed in the open and consists of many layers of cells that are covered on the outside, with the doorway at the bottom. In late summer males mature from unfertilized eggs and mate. They die along with older queens, workers, and young; only young mated females overwinter in soil or among litter.

Adults are extremely protective of the nest and will sting repeatedly if disturbed.

SPIDER WASPS
(Family Pompilidae)

These long-legged, medium-sized or large wasps, ⅜–2″ (10–50 mm) long, are usually seen running on the ground, nervously flicking their dark wings. Many species are glossy blue-black; others are marked with red or yellow. Spider wasps fold the wings lengthwise over the back at rest. They are distinguished from vespid wasps in having a short discal cell in the fore wing. They differ from sphecid wasps in having a wide pronotum that extends back to the base of the wings. Adults may be found at flowers, feeding on nectar. Females prey on spiders as food for their larvae. Most species dig underground burrows in which the eggs are laid on the spider food. A few species are nest parasites, laying their eggs in the provisioned nests of other spider wasps, or on spiders, which are being transported back to the burrow by nonparasitic wasps. These larvae feed on both the spiders intended for the host's larvae and on the host's larvae themselves. Adults can sting.

---

460 Blue-black Spider Wasps
(*Anoplius* spp.)

Description: ½–¾" (12–20 mm). Short "waist" (pedicel) between thorax and abdomen. *Very long hind legs.* Black to metallic blue, often with orange spot or band on abdomen. Wings iridescent blue-black.

Habitat: Bare ground or on soil between sparse grasses.

Range: Western United States.

Food: Adult drinks nectar. Larva feeds on spiders, particularly wolf spiders and funnel web weavers, less often on jumping spiders.

Life Cycle: Female seeks out spider and, curling its abdomen underneath the spider, it stings the spider below its legs. The wasp then drags anesthetized spider to site, where a nest cell can be prepared from mud. Several spiders may be needed to provision cell completely. An egg is laid, and cell is closed with mud door. The wasp larva feeds on spiders, then pupates in nest cell.

The female busily runs about on bare ground with her wings twitching, searching for a suitable spider. She is usually so intent on its activities that she can be observed closely.

---

458 Tarantula Hawk
(*Hemipepsis* spp.)

Description: ½–¾" (12–20 mm). Short "waist" (pedicel) between thorax and abdomen.

Velvety black. *Wings reddish to orange,* darker and less transparent at tip and base.

Habitat: Dry hillsides and rolling arid plains.

Range: California and Mexico.

Food: Adult drinks nectar. Larva feeds on tarantulas and trapdoor spiders.

Life Cycle: Female stings spider between legs, immobilizing it. The female quickly digs a burial chamber, drags the spider inside, lays an egg, and closes burrow. Wasp larva feeds on spider, eventually killing it.

Tarantula hawks are primarily tropical, but several large species are found in the Southwest.

SPHECID WASPS
(Family Sphecidae)

Sphecids comprise a large, diverse family of solitary hunting wasps, ⅜–2⅛″ (10–55 mm) long. They may be solid black or brown, or patterned with white, yellow, or red. They differ from spider and vespid wasps because their short, collarlike pronotum has a knoblike lobe that does not extend back to the base of the wings. They differ from vespid wasps in having a short discal cell in the fore wing and unpleated wings that are not folded over the abdomen when at rest. Some adults visit flowers for nectar; others drink aphid honeydew or body fluids of prey. Breeding females hunt for many kinds of insects and spiders, which they use to provision their nests. Many species nest in underground tunnels affixed to rocks, houses, and abandoned buildings. Some species make cells of mud. A few sphecids are nest parasites, removing the egg from the nest of another sphecid species and replacing it with an egg of their own. All of these wasps can sting painfully.

455 Thread-waisted Wasps
(*Ammophila* spp.)

Description: ⅝–2⅛″ (16–55 mm). Head, thorax, and legs grayish black. *Hind legs much longer than others.* Slender "waist" (pedicel) between abdomen and thorax; rest of abdomen swollen, partly orange or reddish. Wings smoky.

Habitat: Open areas, on the ground or occasionally on clustered flowers.

Range: Throughout United States and southern Canada.

Food: Adult feeds on nectar and small insects. Larva eats caterpillars, chiefly hairless ones, and sawfly larvae.

Life Cycle: Female digs short burrow in sand or light soil, enlarges terminal chamber to receive 1–11 immobilized insect prey; an egg is laid on first. Larva feeds initially on nonessential tissues, later eats indiscriminately, killing host. It pupates close to remains of host. Adult emerges in midsummer or later.

Some species have more than one nest at a time. After the first caterpillar is placed inside, the burrow is temporarily sealed. The wasp remembers the various locations of different nests and returns to each with prey for the larvae.

476 Eastern Sand Wasp
(*Bembix americana spinolae*)

Description: ½–⅝″ (13–16 mm). Short "waist" (pedicel) between thorax and abdomen. Head, thorax, and most of abdomen black. *Abdomen has greenish-yellow bands interrupted by black along midline.* Short white hair on sides of thorax and over back. Femora black, most of tibiae and all tarsi are pale greenish or yellow. Wings are clear with pale brownish veins.

Habitat: Sandy meadows, lakeshores, and beaches.

Range: Throughout North America, except the Pacific Coast.

Food: Adult drinks nectar. Larva eats flies of various kinds.

Life Cycle: Female digs sloping burrows with terminal cells below the ground surface. Female then catches fly, paralyzes it with stinger, and transports it to cell. Female lays 1 egg, then closes burrow with a sand door. Wasp later brings further flies to larva, provisioning it for about 5 days, opening and reclosing door each time. When full grown, larva spins cocoon in cell, while female constructs another nest nearby. Adults are active in summer.

Females often dig multiple burrows but raise young in only 1 or 2, perhaps using partly closed extra burrows to distract would-be predators and parasites. The similar Western Sand Wasp (*B. a. comata*), same size, occurs on the Pacific Coast.

461 Steel-blue Cricket Hunter
(*Chlorion aerarium*)

Description: ½–⅝″ (13–16 mm). 1-segmented "waist" (pedicel) between thorax and abdomen. *Dark metallic steel-blue.* Antennae and legs black. *Wings dark blue.*

Habitat: Meadows with nearby sandy areas.

Range: Throughout the United States.

Food: Adult drinks nectar. Larva feeds on crickets and grasshoppers.

Life Cycle: Female digs a downward-slanting burrow, tossing out quantities of sand and pebbles with hind legs. Female drags prey inside, attaches 1 egg, then backs out and closes the burrow. Larva feeds, later pupates in burrow, emerging as adult the following summer.

Because of its blue color and behavior, this wasp is often confused with the

Blue Mud Dauber (*Chalybion californicum*), same size, which lays its eggs in the provisioned nests of other mud daubers. However, the Steel-blue Cricket Hunter has a shorter pedicel than the Blue Mud Dauber.

462 Purplish-blue Cricket Hunter
(*Chlorion cyaneum*)

Description: 1–1⅛″ (25–29 mm). *Large.* 1-segmented "waist" (pedicel) between thorax and abdomen. *Blue-black or metallic green.* Front tarsi bear 1 tooth. Wings blackish to blue-black.

Habitat: Wet meadows and shores of streams or ponds.

Range: Texas, New Mexico, Arizona, and northern Mexico.

Food: Adult may take juices from crickets. Larva feeds on crickets.

Life Cycle: Female hunts cricket, using stinger to anesthetize victim, which is then dragged or carried to nest site. Female prepares a deep tunnel in muddy soil, pushes the cricket inside, and deposits 1 egg. The wasp larva chews into cricket's body, eats internal organs, and eventually pupates in cell beside remains of cricket.

This wasp is uncommon within its fair-sized range.

452 Florida Hunting Wasp
(*Palmodes dimidiatus*)

Description: ¾–⅞″ (18–22 mm). Head, thorax, and legs blue-black. *Abdominal segments 2, 3, and front ¼ of 4 brownish red;* segment 3 has a dark middle blotch above; rest of abdomen blue-black, including 1-segmented "waist" (pedicel) between thorax and abdomen. Wings blackish to blue-black.

Habitat: Fields and woods.
Range: Throughout North America, except the Northwest.
Food: Adult eats nectar. Larva feeds on camel crickets and long-horned grasshoppers.
Life Cycle: Female excavates a short burrow ending in a somewhat expanded single chamber and then temporarily closes entrance. Female hunts prey and brings anesthetized victim to burrow, then opens entrance and stuffs prey inside. Female lays 1 egg and then reseals entrance.

The similar Western Hunting Wasp (*P. laeviventris*) preys upon the Mormon Cricket, which is especially common around the Great Basin.

466 Black-and-yellow Mud Dauber
(*Sceliphron caementarium*)

Description: 1–1⅛″ (25–30 mm). Slender. Cylindrical 1-segmented "waist" (pedicel) between thorax and abdomen. Black with *large yellow area on prothorax,* yellow markings on thorax, pedicel, and 1st abdominal segment. Legs mostly yellow. Wings brown to black.
Habitat: Meadows, cliffs, and settled areas, where nests are found under rocks, overhanging cliffs, or overhanging roofs of buildings.
Range: Throughout North America.
Food: Adult drinks nectar. Larva feeds on spiders.
Life Cycle: Using its mandibles, female shapes small masses of moist mud into balls and makes joined tubular cells. Into each cell female stuffs 1 paralyzed spider, immobilized by venom, then lays 1 egg on spider and closes cell with mud. Additional cells are built parallel to the 1st. Larvae grow to ⅜–½″ (10–14 mm) long, then spin semi-transparent reddish-brown cocoons in the cells.

Males are rarely seen before midsummer and often visit flowers for nectar in late summer and autumn.

---

479 Cicada Killer
(*Sphecius speciosus*)

Description: 1⅛–1⅝" (30–40 mm). Short "waist" (pedicel) between thorax and abdomen. Black, *marked with yellow* across thorax, on sides above, and on first 3 abdominal segments. Legs yellowish; middle tibiae have 2 spurs at tip. Wings dusky.

Habitat: Forest edges and city parks.

Range: Throughout North America.

Food: Adult drinks nectar. Larva feeds on cicadas.

Life Cycle: Several females work together to build nest of branching tunnels in light clay to sandy soil, making 2 or 3 cells at end. Front legs are used for digging, hind legs for kicking out dirt. Nest entrance is usually left open, while females hunt cicadas one at a time. Each victim is stung and carried back to nest. 1–2 cicadas are placed in each cell; 1 egg is laid on last one. Adults are seen July–August.

Because of its large size, this common wasp is sometimes called the "Giant Cicada Killer."

---

456 Great Golden Digger Wasp
(*Sphex ichneumoneus*)

Description: ⅝–⅞" (15–23 mm). Short "waist" (pedicel) between thorax and abdomen. Head and thorax black with short golden hair. *Abdomen black with reddish orange on 1st segment.* Antennae and coxae black; most of legs reddish orange. Wings orangish to amber.

Habitat: Meadows with bare sandy areas.

Range: United States and southern Canada.
Food: Adult drinks nectar, possibly some juices from provisions obtained for larvae. Larva feeds on true crickets, camel crickets, and long-horned grasshoppers.
Life Cycle: Female digs almost vertical burrow in hard packed soil, preparing several tunnels that radiate from a central entryway. Then female places anesthetized prey in each of the 2–7 cells and lays 1 egg on each victim. Adults are active July–August. 1 generation a year.

These wasps sometimes construct tunnels between flagstones in a garden path or terrace.

BEES
(Superfamily Apoidea)

Bees form a large group of insects that are specialized for feeding at flowers and gathering honey and pollen. More than 3,500 species occur in North America. Bees, ⅛–1″ (4–25 mm) long, may be black, brown, or banded with white, yellow, or orange. In many species the tongue is long and pointed, adapted for probing into flowers. All bees are covered with branched or feathery hair but some have more hair than others. When a bee visits a flower, pollen sticks to the hair. Most female bees have a pollen-collecting apparatus; males do not collect pollen and lack this structure. In most species the pollen is combed into a special pollen basket or brush, which is usually located on the the hind leg. In leafcutting bees, the pollen is carried in a brush of hair on the underside of the abdomen. A few species, as well as parasitic bees, have no pollen basket. Most bees are solitary—each female constructs a nesting tunnel underground or in a plant stem or wood, then stocks the brood cells with pollen and nectar for the larvae. Eggs are laid on pollen balls inside the tunnel. Honey Bees and bumble bees are social—they live in colonies consisting of a fertile queen, sterile female workers, and males, or drones. They are the only bees to produce and store honey. The parasitic bees lay eggs in the nests of other bee species; their larvae eat the pollen and honey intended for the host's larvae. Most bees can sting, but only the social species do so readily in defense of the colony. Bees are important in the pollination of many plants, including commercial crops. The families of bees are distinguished by structural details that are often difficult to see, including the tongue structure and length, wing venation, and placement of the pollen-collecting apparatus. Representative species in the major bee families are therefore included here without separate family descriptions.

Bee Wings

Colletidae

Colletes spp.
FW: 2nd recurrent
vein S-shaped;
3 submarginal cells.
HW: jugal lobe as
long as or longer
than submedian cell.

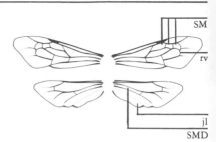

Colletidae

Hylaeus spp.
HW: jugal lobe as
long as or longer
than submedian cell.

Halictidae

Agapostemon spp.
FW: arched basal vein.
HW: jugal lobe longer
than submedian cell.

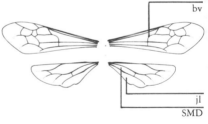

Halictidae

Nomia spp.
FW: arched basal
vein.
HW: jugal lobe
longer than
submedian cell.

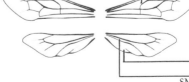

Andrenidae

FW: 2 or 3
submarginal cells.
HW: jugal lobe
longer than
submedian cell.

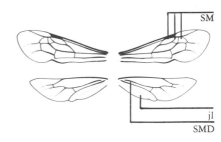

Melittidae

HW: jugal lobe
shorter than
submedian cell.

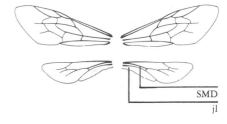

Megachilidae

FW: 2 submarginal
cells.
HW: jugal lobe
smaller than
submedian cell.

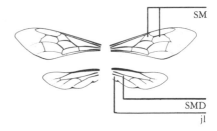

Anthophoridae

Anthophora spp.
HW: jugal lobe
shorter than
submedian cell.

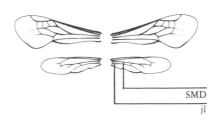

Anthophoridae

Nomada spp.
HW: jugal lobe
shorter than
submedian cell.

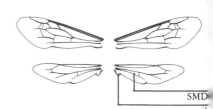

SMD
jl

Anthophoridae

Xylocopa spp.
FW: 2nd submarginal
cell triangular.
HW: small jugal lobe.

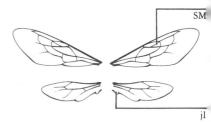

SM
jl

Apidae

Apis spp.
FW: elongate
marginal cell.
HW: small jugal lobe.

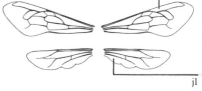

MC
jl

Apidae

Bombus spp.
HW: no jugal lobe.

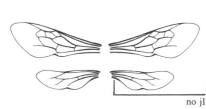

no jl

492 Plasterer Bee
(*Colletes fulgidus*)
Colletid Family (Colletidae)

Description: ⅜–½" (10–12 mm). Black. *Densely covered with long hair,* usually brown but whitish to yellowish in bands on abdomen. Tongue short, slightly notched at tip. Pollen brush on hind tibia. Wings clear.

Habitat: Meadows, gardens, and lawns.

Range: West Coast.

Food: Adult drinks nectar. Larva feeds on nectar and pollen.

Life Cycle: Female digs nest burrow a few inches deep with branching tunnels at bottom, containing individual cells. Each cell is provisioned with nectar and pollen, then 1 egg is added and cell is sealed. Burrows appear as low mounds of dirt. Adults are seen late summer–fall, midsummer in the North.

The Plasterer Bee lines its underground chambers with a thin, delicate, cellophane-like coating of saliva, suggesting its common name. Sometimes pollen sticks to hair on the thorax and abdomen as well as to the pollen brushes. Related North American species have minute differences in color, structure, and behavior. One way colletid bees are distinguished from other bee families is by differences in wing venation.

468 Yellow-faced Bees
(*Hylaeus* spp.)
Colletid Family (Colletidae)

Description: ¼" (5–6 mm). Slender; *abdomen wasplike with almost no hair.* Black with yellow markings on face, pronotum, and tibiae. Wings clear to smoky, veins often reddish brown. No pollen-collecting apparatus.

Habitat: Meadows and abandoned crop fields.

Range: Throughout North America.
Food: Adult drinks nectar. Larva feeds on nectar and pollen.
Life Cycle: Most species prepare brood cells in pith of stems, such as sumac. Some construct cells in soil using old tunnels or burrows. Cells are lined with silky secretions and provisioned with nectar and small amounts of pollen.

These bees do not have external pollen-collecting apparatuses but carry pollen and nectar in the stomach. The different species in this genus can be identified only by an expert.

517   Virescent Green Metallic Bee
(*Agapostemon virescens*)
Halictid Family (Halictidae)

Description: ⅜–½" (10–12 mm). *Thorax metallic green.* Male's abdomen ringed with short yellowish erect hair; female's abdomen ringed with white hair. Antennae and legs black. Wings smoky brown. Pollen brush on hind tibia.
Habitat: Meadows and gardens with sandy soil.
Range: Quebec and Maine to Florida, west to Texas; also Oregon to British Columbia.
Food: Adult drinks nectar. Larva feeds on nectar and pollen.
Life Cycle: Founding female digs branching burrow system, provisions cells at end of each branch with pollen ball moistened with nectar, until young females mature and help in expanding colony. Common gallery with single entrance shared by gregarious bees; 1 bee usually guards the portal by placing its head level with soil surface, as a plug. Adults fly April–October.

These bees usually nest in bare soil or in vertical banks, such as old gravel pits. Adults returning with loads of pollen stuck to the outer surfaces of hind legs are given the right of way over any bees emerging from the nest. But if an awkward entrance accidentally causes nest opening to crumble, bees inside restore passageway as they emerge. One way halictid bees are distinguished from other bees is by differences in wing venation.

518 Augochlora Green Metallic Bees
(*Augochlora* spp.)
Halictid Family (Halictidae)

Description: ⅜" (8–10 mm). Head, thorax, and abdomen *metallic green with short, erect whitish hair*. Antennae black, bare. Legs brown with short hair. Wings brown. Pollen brush on hind tibia.

Habitat: Meadows and gardens.

Range: Quebec to Florida, west to Texas and Mexico, north to Minnesota.

Food: Adult drinks nectar. Larva feeds on nectar and pollen.

Life Cycle: Female digs nest of many branching burrows in dead wood or uses pre-existing burrows of other insects. Female supplies each cell with pollen ball and nectar, and lays an egg on each ball. Larvae or pupae overwinter. Adults emerge in spring.

These bees often can be seen visiting flowers, actively crawling among stamens to reach the nectar. Returning bees carry impressive pollen loads on their hind legs as they enter the portal of the underground nest.

510 Alkali Bee
(*Nomia nevadensis angelesia*)
Halictid Family (Halictidae)

Description: ⅜" (8–10 mm). Head and thorax black, patterned with *short yellowish hair*. Abdomen reddish brown, each segment blackish and ringed with fuzzy yellow hair. Legs brown, covered with short yellowish hair. Wings smoky. Pollen brush on hind tibia.

Habitat: Meadows and gardens.

Range: Nevada, Oregon, California, and Mexico.

Food: Adult drinks nectar. Larva feeds on nectar and pollen of many flowers.

Life Cycle: Females are gregarious but excavate separate nests in ground with many cells. Full-grown larvae overwinter in brood cells and pupate in early spring. Adults fly July–August.

Because these bees like to nest in clay or alkaline soils, they are known as Alkali Bees. They are such important pollinators that, where alfalfa is grown for seed, artificial nests have been constructed to draw more bees.

503 Mining Bees
(*Andrena* spp.)
Andrenid Family (Andrenidae)

Description: ⅝" (15 mm). *Black with long hair.* Head almost as large as thorax; most species have short, pointed tongue. Abdomen usually longer than head and thorax combined. Wings smoky to dark. Pollen brush on most of hind leg.

Habitat: Meadows and forest edges.

Range: Throughout North America.

Food: Adult drinks nectar. Larva feeds on nectar and pollen.

Life Cycle: Female digs long branching tunnel in soil, prepares a brood cell at the end of each branch, and stocks cells with pollen balls and nectar. 1 egg is laid on pollen ball in each cell, then cell is sealed. Larvae develop rapidly and pupate in cells. 1 generation a year.

Mining bees are important pollinators, visiting many different plants. Many are solitary, but some are social or communal. One way these bees are distinguished from members of other bee families is by wing venation.

---

491 Willow Mining Bee
(Andrena salicifloris)
Andrenid Family (Andrenidae)

Description: ½″ (12–14 mm). Black or dark brown. *Sparsely coated with grayish hair* on head, thorax, legs, and abdomen; hair on abdomen appears as rings. Wings smoky. Pollen brush on most of hind leg.

Habitat: Arid and alpine lands.

Range: Colorado to California, north to British Columbia.

Food: Adult drinks nectar. Larva feeds on nectar and pollen.

Life Cycle: Mated female excavates long branching tunnel with a brood cell at the end of each branch. After each cell is provisioned with pollen ball and nectar, 1 egg is laid on pollen ball, and doorway is closed. Larvae develop rapidly inside, and pupate, but adults do not emerge until spring.

These bees emerge early in spring in mountain areas, collecting pollen from catkins of mountain willows. Their body hair often appears golden-yellow because of adhering pollen.

---

499 Nevada Mining Bee
(Andrena transnigra)
Andrenid Family (Andrenidae)

Description: ⅜–½″ (10–12 mm). Tongue short, pointed. Black with *yellow hair on thorax and femora;* black hair on tibiae and abdomen. Wings brownish, veins black. Pollen brush on most of hind leg.

Habitat: Arid lands.

Range: Colorado and Wyoming to California, north to British Columbia.

Food: Adult drinks nectar. Larva feeds on nectar and pollen.

Life Cycle: Mated female excavates long branching tunnel with a brood cell at the end of each branch, stocks each cell with pollen ball and nectar, then lays an egg inside. Provisions are left only once, then the cell is sealed. Larvae develop rapidly and pupate. Adults emerge in spring.

These mining bees collect huge quantities of pollen, which they store in subterranean brood cells for their larvae to eat.

494 **Clarkia Bee**
(*Hesperapis clarkia*)
Melittid Family (Melittidae)

Description: ¼–⅜" (6–10 mm). Black with *mosslike, dense whitish hair above and a few longer pale fine hairs.* Abdomen flattened. Antennae short. Pollen brush on hind tibia. Wings clear with black spot (stigma).

Habitat: Meadows, fields, and gardens on clarkia flowers, a western member of the evening primrose family.

Range: New Mexico, Arizona, and California.

Food: Adult drinks nectar. Larva feeds on nectar and pollen; each species has its one preferred flower genus.

Life Cycle: Unknown.

This recently discovered species is closely related to the similarly robust Creosote Bush Bee (*H. arida*) of Arizona and California and the Hugelia Bee (*H. rufipes*), of California. The Elegant Hesperapis (*H. elegantula*) occurs in New Mexico and has milky wings with a pale stigma. These melittid bees are distinguished from other bees by wing venation and details of the labial palps.

482 Faithful Leafcutting Bee
(*Megachile fidelis*)
Megachilid Family (Megachilidae)

Description: ⅜–½" (10–12 mm). *Black, covered by long golden hair* on face, sides of head behind compound eyes, sides of thorax below wings, rear border of each abdominal segment, and on legs. Abdomen with yellow bands. Wings blackish. *Pollen brush under abdomen.*

Habitat: Meadows and gardens.

Range: Montana and South Dakota to New Mexico, west to California, north to Oregon.

Food: Adult drinks nectar of many flowers, especially composites. Larva feeds on nectar and pollen.

Life Cycle: Working collectively, female bees cut circular or oval leaf fragments and use them to line cells constructed inside soil or rotting wood. Each cell is stocked with nectar and pollen, then 1 egg is sealed inside. Larvae grow rapidly, pupate inside chambers, overwinter, and emerge in spring.

Although leafcutting bee larvae may mature at different times, each adult does not emerge from the brood cell until all chambers between it and the end of the tunnel have been vacated. The megachilid bees are distinguished from other bees by wing venation and other structural details.

490 Western Leafcutting Bee
(*Megachile perihirta*)
Megachilid Family (Megachilidae)

Description: ⅜–½" (10–12 mm). Black with *long pale whitish-yellow hair,* particularly below thorax and abdomen. *Abdomen mostly bare, each segment narrowly fringed with whitish hair.* Wings clear, veins black. *Pollen brush below abdominal segments 2–5 is bright red.*

Habitat: Meadows and orchards.

Range: Nebraska to Texas and Mexico, west to California, north to British Columbia and Alberta.

Food: Adult drinks nectar of many flowers, especially composites. Larva feeds on nectar and pollen.

Life Cycle: Several bees work together to dig burrows in sandy or gravelly soil or in rotting wood and plant stems. They construct a series of cells and line them with fragments cut from leaves. Each cell is provisioned with pollen and nectar, then 1 egg is laid inside. Adults are seen July—August. 1 generation a year.

A leafcutting bee can snip off a leaf fragment in less than a minute, using its sharp mandibles. But because it pollinates alfalfa and other crops, it is considered highly beneficial rather than harmful. Some related species use resins from plants to cement together materials for the nest.

498 Mason Bees
(*Osmia* spp.)
Megachilid Family (Megachilidae)

Description: ⅜–½" (10–14 mm). Black with *long black hair on thorax and sides of head.* Tongue long; mandibles prominent, sharp. *Female has pollen brush below abdominal segments 2–3.* Legs black. Wings clear to brownish.

Habitat: Meadows and forest edges.

Range: Throughout North America; every state and Canadian province has a dozen or more distinct species, each showing more limited distribution.

Food: Adult drinks nectar. Larva feeds on nectar and pollen.

Life Cycle: Female constructs small nest cells of clay, individually or in clusters, that are attached to twigs or stones or built into cavities of wood. Each cell is provisioned with pollen and nectar, then 1 egg is laid inside. Larvae spin tough cocoons before pupating. Adults are seen April–June. 1 generation a year.

Mason bees convert clay into a cementlike material. Some species include plant fragments in their nest construction. Others build inside empty snail shells, and still others line each nest with snips of flower petals.

493 California Leafcutting Bee
(*Trachusa perdita*)
Megachilid Family (Megachilidae)

Description: Male ⅜–½″ (10–14 mm); female ½″ (12–14 mm). *Dark gray with whitish bands of hair* on abdominal segments 1–5. *Pollen brush under abdomen.*
Habitat: Semiarid hillsides with shrubby growth.
Range: California.
Food: Adult drinks nectar. Larva feeds on nectar and pollen.
Life Cycle: Female digs slanting, curved burrows into hillside, sometimes constructing the last, large chamber almost under entrance hole. Leaf fragments cut from buckthorn shrubs are used to construct brood cells. Cells are filled with pollen and nectar, then 1 egg is left in each.

The only other North American species known in this genus also occurs in California.

502 Digger Bees
(*Anthophora* spp.)
Anthophorid Family (Anthophoridae)

Description: ⅝″ (15–17 mm). Tongue very long. Black, *densely covered with short yellow hair* on head, thorax, and 1st abdominal segment. Legs black, covered by short black hair; tarsi brownish or yellowish. Wings clear, smoky at tip; very small spot in fore wings. Pollen brush on hind tibia.

Habitat: Meadows and gardens.

Range: Most of North America.

Food: Adult drinks nectar. Larva feeds on nectar and pollen.

Life Cycle: Nest is constructed in clay or sand bank. Entrance is concealed by a downslanted chimney made of mud. The chimney and brood cells at ends of inner branching tunnels are thinly lined with mud. Each cell contains mixture of honey and pollen plus 1 egg. Larvae feed, overwinter, and pupate in cell. Adults emerge in late spring.

Digger bees often nest together in large numbers. They are sometimes called flower-loving bees, because they visit such a wide variety of flowers. Anthophorid bees are distinguished from other bee families mostly by differences in wing venation. Within this family, the digger bees are identified by the triangular platelike area on the female's abdominal tip and other structural details.

471 Western Cuckoo Bee
(*Nomada edwardsi*)
Anthophorid Family (Anthophoridae)

Description: ⅜–½" (10–14 mm). Slender. Tongue long. Thorax black with yellow markings. *Abdomen yellow with brown crossbands.* Legs yellowish brown with *no pollen baskets.* Wings clear to brownish.

Habitat: Meadows and gardens.

Range: California to British Columbia; also Nevada.

Food: Adult drinks nectar. Larva feeds on honey and pollen stored in brood cells of mining bees.

Life Cycle: Females emerge early in spring, enter nests of other bees that prepare brood cells in the ground, and deposit eggs next to the food supply. The Cuckoo Bee's eggs hatch before those of the host. Cuckoo Bee larvae deplete host's food supply and cause host to die. Cuckoo Bee larvae pupate and overwinter in the host's cells.

Like the European Cuckoobird, this bee lays its eggs in another species' nest, inspiring its common name "Cuckoo Bee."

497 California Carpenter Bee
(*Xylocopa californica*)
Anthophorid Family (Anthophoridae)

Description: ¾–1" (20–25 mm). Robust, resembling a bumble bee but abdomen short-haired and bare in places. Black with greenish or bluish reflections. Male's face has yellowish-white center. *Male's pronotum bears white, orange, or yellow hair; abdominal segment 1 covered with white hair.* Pollen brush of short stiff hair on hind tibia.

Habitat: Forests and adjacent meadows.

Range: Utah, Nevada, and California, north to Oregon, south to Mexico.

Food: Adult drinks nectar. Larva feeds on nectar and pollen, often obtained by biting through base of flower.

Life Cycle: Female chews tunnel as deep as a foot into dry wood of dead trees, or lumber, or wood of houses. Female makes linear series of unlined cells and provisions each with pollen and nectar before laying 1 egg in each. Cells are divided by disklike partition built of cemented wood chips. Adults emerge in late summer, each waiting in line toward end of tunnel for its turn to leave.

The related Eastern Carpenter Bee (*X. virginica*), same size, is black or metallic blue-black and has short hair on its abdomen or none at all. It is found throughout eastern North America.

511 Honey Bee
(*Apis mellifera*)
Apid Family (Apidae)

Description: Male drone ⅝″ (15–17 mm); queen ¾″ (18–20 mm); sterile female worker ⅜–⅝″ (10–15 mm). Drone more robust with largest compound eyes; queen elongate with smallest compound eyes and larger abdomen; worker smallest. All *mostly reddish brown and black with paler, usually orange-yellow rings on abdomen.* Head, antennae, legs almost black with short, pale erect hair densest on thorax, least on abdomen. Wings translucent. Pollen basket on hind tibia.

Habitat: Hives in hollow trees and hives kept by beekeepers. Workers visit flowers of many kinds in meadows, open woods, and gardens.

Range: Worldwide.

Food: Adult drinks nectar and eats honey. Larva feeds on honey and royal jelly, a white paste secreted by workers.

Life Cycle: Complex social behavior centers on maintaining queen for full lifespan, usually 2 or 3 years, sometimes up to 5. Queen lays eggs at intervals, producing a colony of 60,000–80,000 workers, which collect, produce, and distribute honey and maintain hive. Workers feed royal jelly to queen continuously and to all larvae for first 3 days; then only queen larvae continue eating royal jelly while other larvae are fed bee bread, a mixture of honey and pollen.
By passing food mixed with saliva to one another, members of hive have chemical bond. New queens are produced in late spring and early

summer; old queen then departs with a swarm of workers to found new colony. About a day later the first new queen emerges, kills other new queens, and sets out for a few days of orientation flights. In 3–16 days queen again leaves hive to mate, sometimes mating with several drones before returning to hive. Drones die after mating; unmated drones are denied food and die.

Settlers brought the Honey Bee to North America in the 17th century. Today these bees are used to pollinate crops and produce honey. They are frequently seen swarming around tree limbs. Honey Bees are distinguished from bumble bees and bees in other families mostly by wing venation.

505 Golden Northern Bumble Bee
(*Bombus fervidus*)
Apid Family (Apidae)

Description: Male drone ⅜–⅝″ (10–15 mm); workers ½–¾″ (13–18 mm); spring queen ¾–⅞″ (18–23 mm). *Robust, hairy.* Face and head mostly blackish. Black band between wings. Female is yellow on most of thorax and abdominal segments 1–4, black on 5–6; male is yellow on segments 1–5, black on 6–7. Wings smoky. Pollen basket on hind tibia.

Habitat: Clearings in forests, roadsides, and open areas.

Range: Quebec and New Brunswick south to Georgia, west to California, and north to British Columbia.

Food: Adult drinks nectar and eats honey. Larva feeds on honey.

Life Cycle: Queen overwinters until early spring, enters opening in soil to build honeypots and brood cells. Small workers develop first. With warmer weather, new honeypots and brood cells are constructed, producing larger adults. Only young mated females (new queens) overwinter; the rest of the colony, including old queen, dies. Adults fly May–September.

The similar Golden-orange Bumble Bee (*B. borealis*), same size, is orangish-yellow and has more black near the legs. It occurs from the Yukon to Nova Scotia, south to Georgia; also in Michigan, Kansas, Montana, and British Columbia. The larger American Bumble Bee (*B. pennsylvanicus*) is black behind the wings with yellow on abdominal segments 1–3. It is found in the United States and southern Canada. Bumble bees are distinguished from Honey Bees and bees in other families by wing venation and other details.

506 Red-tailed Bumble Bee
(*Bombus ternarius*)
Apid Family (Apidae)

Description: Male drone ⅜–½" (9–14 mm); worker ⅜–½" (9–14 mm); spring queen ½–¾" (14–18 mm). *Robust, hairy.* Thorax yellow on sides with broad black band between the wings. *Abdominal segments above: 2–3 red;* 1 and 4 yellow; 5–6 black. Wings smoky. Pollen baskets on hind tibiae.

Habitat: Woods and open fields.

Range: Nova Scotia south to Georgia; also Michigan, Kansas, Montana, and British Columbia.

Food: Adult sips nectar. Larva eats honey.

Life Cycle: In early spring queen enters opening in soil to build honeypots and brood cells. Small workers develop first, visit flowers for nectar, and construct new brood cells. With warmer weather larger adults develop. Only young mated females overwinter.

Unlike the Honey Bee worker, a bumble bee can sting many times.

ARACHNIDS
(Class Arachnida)

Arachnids comprise the largest non-insect class of arthropod animals. They include 11 orders, with over 75,000 named species worldwide, including more than 4,000 species in North America. The first arachnids appeared some 350 million years ago. Arachnids differ from insects in lacking antennae and wings and in having 8 rather than 6 legs. They have a pair of jawlike or fang-bearing appendages, called chelicerae, in front of the mouth, and a pair of leglike pedipalps between the chelicerae and the first pair of walking legs. Unlike insects, which have 3 body segments, the bodies of most arachnids have 2 distinct parts—the cephalothorax (a combined head and thorax), and the abdomen. 8 orders of arachnids commonly found in North America are included here. The other 3 orders—short-tailed whipscorpions (Schizomida), micro-whipscorpions (Palpigradi), and ricinulids (Ricinulei) —are extremely rare and thus beyond the scope of this book.

Spiders
(Order Araneae)

Spiders are the largest group of
arachnids. There are more than 35,000
named species worldwide, including
about 3,000 in North America, but
probably most spider species are still
awaiting identification. These familiar
predators live almost everywhere—on
the ground, under rocks, among
grasses, on plants, in tree branches, in
underground caves, and even on the
water. They are easily recognized by the
4 pairs of 7-segmented legs. Like all
arachnids, spiders have a cephalothorax
and abdomen. But unlike scorpions,
mites, and daddy-long-legs, the
spider's cephalothorax and abdomen are
separated by a "waist," or pedicel. The
top of the cephalothorax is protected by
a shieldlike covering, called the
carapace. Most species have 8 simple
eyes, although some have less and a few
species have none. Often the number
and arrangement of eyes are important
in identifying the different families.
Below the eyes are 2 small jaws, called
chelicerae, that end in fangs. Venom is
produced in glands and empties
through a duct in the fangs. This
venom is used to paralyze or kill prey.
Then the spider crushes the victim by
rubbing the chelicerae against each
other and against the enlarged bases of
the pedipalps, located before the first
legs. There are usually 6 fingerlike silk
glands, called spinnerets, located
beneath the abdomen just in front of
the anus. But not all spiders spin webs.
Some live in burrows, which they line
with silk, while others have no retreat
at all. All young spiders and, in some
species, adult males release long silken
strands, which they use like a parachute
to ride the wind to other areas. This
process is called ballooning.
Most spiders lay eggs in silken sacs.

Some place the egg sac in the web,
others attach it to twigs or leaves, and
still others carry the sac around until
the eggs hatch. The young, called
spiderlings, resemble adults and are
often cannibalistic. All spiders are
voracious predators. Most feed on
insects, although a few large species
prey on small vertebrate animals.
Spiders are considered highly beneficial
because they help keep the burgeoning
insect population in check. Few spiders
bite people and the venom of most is
harmless.

FOLDING TRAPDOOR SPIDERS
(Family Antrodiaetidae)

Members of this small family live in
tubelike burrows in the ground. Many
close the entrance with a trapdoor that
folds at the midline—this feature
distinguishes them from trapdoor
spiders, which never construct folding
trapdoors. Mostly large, they are ¼–
1⅛" (6–28 mm) long. Like trapdoor
spiders, they have a rakelike row of
spines on each jaw, or chelicera, which
they use for digging, and 3 claws on
each foot. But unlike trapdoor spiders,
folding trapdoor spiders have hard
plates on the abdomen. They are
frequently seen after heavy rains, either
searching for insect or millipede prey or
repairing a broken trapdoor.

649   **Turret Spider**
      (*Atypoides riversi*)

Description: Male ½–⅝" (13–16 mm), female ⅝–
               ¾" (16–18 mm). *Cephalothorax greenish
               to brown. Abdomen purplish brown.* Legs
               dark chestnut-brown. Male's chelicerae
               have heavy spinelike extensions covered
               with hair and bristles; female's

chelicerae without extensions. Male's abdomen above has 3 hard plates; female's abdomen has 1 plate. Both sexes have 6 spinnerets.

Habitat: Slopes in pine and deciduous forests, sometimes stream banks.

Range: Northern California.

Food: Small insects, mostly ants.

Life Cycle: Spider constructs silk-lined burrow in soil and by elongating the silk lining builds a turretlike opening, which often incorporates debris. Egg sac is deposited halfway down burrow in summer. Spiderlings leave burrow the following summer. Males wander November—February, especially after heavy rains.

This spider is named for the turretlike projection above its burrow. Unlike other members of its family, it does not make a trapdoor and keeps the burrow open all day. At night when the spider forages for food, it folds the end of the turret together, closing it temporarily.

TARANTULAS
(Family Theraphosidae)

Tarantulas are large, hairy spiders, 1⅜" (35 mm) or more long, with a legspan up to 5⅞" (150 mm). Although some South American species have deadly venom, the bite of North American tarantulas is not any more dangerous to people than a wasp or bee sting. These spiders have 8 closely grouped eyes; the large middle pair are circular with 3 eyes on each side. Each leg has 2 claws at the tip and a tuft of hair underneath. There are microscopic bristles on the abdomen, which break off easily and can irritate the skin of small mice and other prey. Males have longer legs and are more active than females. Males are short-lived and do not molt after they mature at a few years of age; females continue to molt and may live 20–35 years in captivity. Nocturnal, tarantulas hide in dark cavities or burrows during the day, emerging to hunt by touch in darkness. Most line the top of the burrow with silk, but do not spin webs to catch prey. They were formerly called banana spiders, because tropical species arrived in cargoes of fruit. About 30 species are found in the southern United States.

647   **Desert Tarantula**
(*Aphonopelma chalcodes*)

Description: Male 2–2½" (50–65 mm), female 2–2¾" (50–70 mm); female legspan to 4" (100 mm). Body heavy, hairy. *Cephalothorax gray to dark brown. Abdomen brownish black.* Iridescent hair forms pad below tip of each leg.

Habitat: Desert soil.

Range: Arizona, New Mexico, and southern California.

Food: Insects, lizards, and other small animals.

Life Cycle: Male tries to maintain contact with female. If female moves away, male aggressively pursues desired mate. Eggs are concealed in some natural cavity. All spiderlings resemble females at first. After last molt, male emerges with distinctive pedipalps and more slender and relatively larger legs. Female continues to molt after reaching maturity and may live to 20 years.

The male spider wanders in the dim light after sunset or near dawn searching for a mate, then hides by day in abandoned holes or under stones. These spiders are reluctant to attack people. Usually the venom is no more poisonous than that of bees.

TRAPDOOR SPIDERS
(Family Ctenizidae)

These rather large spiders, ⅜–1¼″ (10–33 mm) long, have a rakelike row of spines on each jaw, or chelicera, either where the fang arises or on a lobe. They use this rake to dig burrows in the ground. Their 8 eyes are more or less closely grouped, arranged with 3 eyes on each side and 2 eyes in the middle. This eye arrangement is similar to that of the folding trapdoor spiders and tarantulas. There are 3 microscopic claws on each foot. Trapdoor spiders dig tubelike burrows in the ground, which they seal with a hinged lid or trapdoor. Some species construct wafer-thin lids of silk, while others make snugly fitting, thick, corklike doors of earth and debris. The spider holds the trap shut with its fangs until it senses vibrations made by passing prey. Then the spider rushes out, seizes the victim, and drags it into the burrow. Males wander on the ground seeking females, usually after rains, but females rarely emerge except for prey.

650 **Californian Trapdoor Spider**
(*Bothriocyrtum californicum*)

Description: Male ¾–1″ (18–26 mm), female 1⅛–1¼″ (28–33 mm). Cephalothorax almost as wide as long. *Abdomen egg-shaped, without lengthwise grooves.* Jaws (chelicerae) black. Cephalothorax blackish brown; male's darker than female's. Legs darker than cephalothorax. Abdomen brown to yellow-gray; male's paler than female's.

Habitat: Sunny hillsides and sloping areas of hard soil, including almost vertical cliffs into which nearly horizontal tunnels are cut as much as 8″ deep.

Range: Southern California.

Food: Insects, millipedes, and other spiders.

Life Cycle: Spiderlings leave burrow of female after spring rains to dig own burrows. Each year an extra layer of webbing is added to trapdoor, which has thick corklike lid. Several layers indicate continued use. Males wander in search of females November–February, usually after heavy rains.

The related Ravine Trapdoor Spider (*Cyclosmia truncata*), ¾–1″ (18–26 mm), is brown with a black back and has a bluntly squared abdomen with lengthwise grooves. It constructs burrows with thick silken lids on steeply sloped banks of sandy soil covered with moist leaf litter. This species is found in Tennessee, northwestern Georgia, and Alabama, and may live more than 12 years.

DICTYNID SPIDERS
(Family Dictynidae)

Spiders in this large, widespread family possess a sievelike spinning plate, called the cribellum, in front of the spinnerets. They are distinguished from other spiders that also have cribella by the arrangement of their eyes and by the 3 microscopic claws on each foot. Some spiders have 8 eyes in 2 rows—6 white eyes and 2 dark eyes. Other spiders have only the 6 white eyes. Several dictynids have large jaws, or chelicerae, and fairly big poison glands; but many males have modified chelicerae, used in mating. These spiders are seldom more than ¼″ (5 mm) long. They spin irregular webs in crevices, under leaves, and at the tips of plant stems. Some species are social. There are more than 150 dictynid species in North America.

687 **Branch-tip Spiders**
(*Dictyna* spp.)

Description: Male ¹⁄₁₆–⅛″ (2–3 mm), female ¹⁄₁₆–
¼″ (2–5 mm). Cephalothorax yellowish
to brown or reddish brown. *Abdomen*
brown to yellowish or gray, *usually with
dark markings*. Male is darker than
female.

Habitat: Woodland edges and fields on shrubs
and stiff herbs.

Range: Throughout North America.

Food: Small insects.

Web: Sticky irregular mesh net.

Life Cycle: Female produces eggs in sacs in
summer, hanging 1 sac or more near
the center of web. Both male and
female remain nearby until after
spiderlings disperse.

Often the female and spiderlings
remain in the web until autumn, when
they seek sheltered sites in which to
overwinter. Related species conceal
their webs in crevices of bark or indoors
in corners of windows.

SPITTING SPIDERS
(Family Scytodidae)

Members of this family have large
spitting glands within the
cephalothorax, which produce a sticky
secretion that is squirted on prey. With
good aim, spitting spiders can spit
about ¾″ (20 mm). Most are small
spiders, rarely more than ⅜″ (9 mm)
long. They have 6 eyes arranged in 3
groups. Most of these spiders live under
trash or stones on soil in the Southwest.
They are also frequently found in
homes.

675 Spitting Spiders
(*Scytodes* spp.)

Description: Male ⅛–¼" (4–6 mm), female ⅛–⅜"
(4–9 mm). Pale yellow to dark
chestnut-brown; sometimes mottled
with black or brown on cephalothorax
and long legs. Cephalothorax higher at
rear than at eyes.

Habitat: Woods and fields under leaf litter,
trash, or stones; also in cellars and
closets.

Range: Mostly southwestern states; some
species also in the East.

Food: Small insects.

Life Cycle: Female carries egg sac in pincerlike
chelicerae until eggs hatch and
spiderlings disperse.

These spiders spray prey with a sticky
secretion. When the spider bites, it
introduces digestive agents which help
liquefy the body contents.

VIOLIN SPIDERS
(Family Loxoscelidae)

Members of this small family, which
includes only 1 North American genus,
are known for their poisonous venom.
Most of these medium-sized spiders,
¼–⅜" (6–11 mm) long, are brownish
or yellowish. They have 6 eyes in 3
pairs. The cephalothorax is rather flat
above and has a conspicuous,
lengthwise furrow on the midline at the
rear third. Each foot has 2 claws. These
spiders spin small, irregular webs under
bark and stones. Their venom is
especially poisonous to people; those
bitten often become ill and find that
the wound does not heal quickly.

659 Desert Loxosceles
(*Loxosceles deserta*)

Description: Male ¼" (6 mm), female ¼–⅜" (6–
8 mm). *Cephalothorax rather plain, pale
orange-yellow or sand-colored,* coated with
sparse short hair. Legs pale near body,
darker beyond. Abdomen pale brownish
gray. Legs long; 1st pair 5½ times
body length in female, 7½ times body
length in male.

Habitat: Arid lands.

Range: Arizona, Nevada, Utah, and California.

Food: Small insects.

Web: Small, loose, irregular web, sometimes
spun under stone or log.

Life Cycle: Spider often spins open retreat of silk
below a stone. Disk-shaped egg mass,
whose diameter is greater than length
of spider, is attached to the silken
shelter.

Venom of this species is significantly
less dangerous to people than that of
most violin spiders. A faint indication
of a dark, violin-shaped mark is
sometimes present on the abdomen, but
never enough to warrant giving this
species the common name "Violin
Spider." For many years this species
had been known as *L. unicolor,* which
emphasized the plain body coloring.

671 Violin Spider
"Brown Recluse Spider"
(Loxosceles reclusa)

Description: Male ¼" (6 mm), female ⅜" (11 mm).
*Cephalothorax orange-yellow with dark
violin pattern.* Bases of legs orange-
yellow, rest of legs grayish to dark
brown. Abdomen grayish to dark
brown with no obvious pattern.

Habitat: Outdoors in sheltered corners among
loose debris; indoors on the floor and
behind furniture in houses and
outbuildings.

Range: Kansas and Missouri to Texas, west to
California.
Food: Small insects.
Web: Loose irregular strands.
Life Cycle: Eggs overwinter in a loose sac hung in
web and guarded by female until she
dies. Spiderlings disperse in spring and
weave loose webs of sticky silk.

This spider sometimes takes shelter in
clothing or a folded towel and bites
when disturbed. The wound commonly
develops a crust and a surrounding red
zone. The crust falls off, leaving a deep
crater, which often does not heal for
several months.

COMB-FOOTED SPIDERS
(Family Theridiidae)

These spiders have long, slender legs
with 3 claws on each tarsus, and usually
8, or sometimes 6, eyes. They are
named for the 6–10 inconspicuous
comblike bristles on the hind tarsi of
most species. Comb-footed spiders,
$\frac{1}{16}$–$\frac{1}{2}''$ (1.5–14 mm) long, spin
irregular webs and use their combs to
fling silk strands over any captive that
gets caught in the web. Usually the
swathed victim is hauled to a rest site
on the web, injected with venom, and
later eaten. The egg case is suspended
near the resting site, where the female
can guard it. If the egg case drops, the
female retrieves it and refastens it in
place. This is one of the largest families
of spiders, having more than 200
species in North America.

689 American House Spider
(*Achaearanea tepidariorum*)

Description: Male $\frac{1}{8}$–$\frac{1}{4}''$ (4–5 mm), female $\frac{1}{4}''$ (5–
6 mm). Yellowish brown. *Abdomen*

streaked and splotched with black and gray on sides. Males's legs orangish; female's yellow with black bands.

Habitat: Sheltered corners of houses, barns, and other buildings.

Range: Throughout the United States and Canada.

Food: Insects.

Web: Irregular, made of sticky strands. Webs catch dust as well as prey and are known as cobwebs.

Life Cycle: Female spins pear-shaped brownish silken cocoon around egg mass and hangs it in web. Adult females can live more than a year.

This spider emerges in darkness to produce webs beneath ceilings or in window frames, taking advantage of every angle to set sticky strands where insects may get caught. Then the spider lies in wait on the part of the web that has an extra layer of silk. If large prey, such as a camel cricket, gets ensnarled, the spider throws more silk on it and then pulls it up into the web. This spider was formerly assigned to genus *Theridion*.

651, 688 Black Widow Spider
(*Latrodectus mactans*)

Description: Male ⅛″ (3–4 mm), female ⅜″ (8–10 mm). Black. *Male's abdomen elongate with white and red markings on sides. Female's abdomen almost spherical, usually with red hourglass mark below* or with 2 transverse red marks separated by black. Legs of male much longer in proportion to body than those of female. Spiderling is orange, brown, and white, gaining more black at each molt.

Habitat: Among fallen branches and under objects of many kinds, including furniture, outhouse seats, and trash.

Range: Massachusetts to Florida, west to

California, Texas, Oklahoma, and
Kansas; most common in the South.

Food: Insects.

Web: Irregular mesh with a funnel-shaped
retreat, built in sheltered spots.

Life Cycle: Female rarely leaves web, stays close to
egg mass, biting defensively if
disturbed. Pear-shaped egg sac, ⅜–½"
(8–12 mm) wide, is pale brown.
Female stores sperm, producing more
egg sacs without mating. Spiderlings
disperse soon after hatching. Some
females live more than 3 years.

Of all spiders, the Black Widow is the
most feared—the female's venom is
especially poisonous to people. Despite
its reputation, this spider often
attempts to escape rather than bite,
unless it is guarding an egg mass.
Males do not bite. After mating, the
female often eats the male, earning the
name "widow."

SHEET-WEB WEAVERS
(Family Linyphiidae)

Most spiders in this family have long,
slender legs with strong bristles, and 3
claws on each tarsus. Some species have
stridulating ridges or a toothlike cusp
on the outside margin of the chelicerae.
Usually there are 8 eyes arranged in 2
parallel rows. Most spiders are small,
less than ¹⁄₁₆–⅜" (2–8 mm) long.
Many species spin flat, sheetlike or
dome-shaped webs. They cling beneath
this web until an insect gets entangled
on the surface, then pull the insect
downward through the web.
Supposedly the sheet protects the spider
from predators. Some species build a
second sheet under the first, which
prevents predators from attacking from
below. Hundreds of species are found in
North America, but few are widely
distributed. Only a specialist can

distinguish between members of this
family and comb-footed spiders in the
Family Theridiidae.

669 Hammock Spider
(*Pityohyphantes costatus*)

Description: Male ¼" (5–6 mm), female ¼" (5–
7 mm). Cephalothorax white to light
brown with narrow black lines at edges
of carapace and broad, forked line
extending from middle of eyes. *Abdomen
has dark brown herringbone pattern on
midline.*

Habitat: Woods, fields, shrubby areas, and near
buildings.

Range: New England to North Carolina, west
to the Pacific Coast and across southern
Canada.

Food: Small or medium-sized insects.

Web: Large sheet with barrier section above
it, or sometimes a silken tent.

Life Cycle: Spider stays concealed in corner of sheet
or hides in retreat. Leaves that fall into
web are often intertwined into the
retreat. White silken egg sac is
attached to web. Spiderlings or adults
overwinter under loose bark or under
stones. Spiders mate mid-April to mid-
May. Adults are seen until June.

Hammock Spiders get their name from
the hammocklike sheet web they
construct between fence posts,
buildings, or on the lower branches of
trees and shrubs.

ORB WEAVERS
(Family Araneidae)

Orb weavers comprise a huge family of
spiders with several hundred North
American species. These spiders vary
greatly in shape, color, and size,
measuring ¹⁄₁₆–1⅛" (2–28 mm) long.

They have 8 eyes arranged in 2
horizontal rows of 4 eyes each. Their
jaws, or chelicerae, usually have a small
bump, or boss, in the outer margin,
although some species lack this boss.
The male is commonly much smaller
than the female; males of some species
bear special spurs of clasping spines on
their legs. Most orb weavers spin
spiraling orb webs on support lines that
radiate outward from the center. The
plane of the web may be vertical,
horizontal, or slanting. Although
uloborids, ray spiders, and
tetragnathids also build orb webs, each
family has a different type of spiral
thread, and usually each species adds its
own characteristic pattern. The male
orb weaver often spins its own orb web
in an outlying portion of the female's
web. Many orb weavers replace the
entire web daily, spinning a new web
in the early evening in about an hour.
Members of this family exhibit different
degrees of sociality—some are found in
woods, others inhabit caves and dark
places, still others spin webs in grasses.
The family was formerly known as the
Argiopidae, which included ray spiders
and tetragnathids.

643 Orb Weavers
(*Araneus* spp.)

Description: Male ¼" (6 mm), female ⅜–¾" (10–
20 mm). *Abdomen bulbous.* Brown to
orange, with distinctive pattern for
each species. Legs long, yellowish
brown, sometimes ringed with black.
Habitat: Among tall grasses and shrubbery.
Range: Throughout North America; individual
species more restricted.
Food: Insects.
Web: Spiraling orb with nonsticky radiating
support lines and sticky spiral strands
in vertical plane.
Life Cycle: Spider usually hangs head downward

near center of web, or remains at a nearby resting site connected to the web by a signal line. Egg sac is attached to plant near this retreat or on foliage nearby. Spiderlings disperse after hatching.

Each night the old web is replaced with a new one, spun in complete darkness by touch alone.

644 Barn Spider
(*Araneus cavaticus*)

Description: Male ⅜–¾" (10–20 mm), female ½–⅞" (13–22 mm). Male's body and legs and female's cephalothorax and legs dark brown with dense whitish hair appearing ash-gray. Legs have some paler yellowish-brown bands. *Female's abdomen pale or yellow along midline and on each side of a dark brown mark with toothlike side projections at center;* a similar dark brown mark low on each side of abdomen. Abdomen below has brown median band bordered on each side by somewhat broken yellowish stripe.

Habitat: Barns, caves, mine openings, and overhanging cliffs.

Range: Eastern United States and adjacent areas of Canada, southwest to Alabama and Texas.

Food: Insects.

Web: Large orb stretched in shaded areas.

Life Cycle: Spider spins web at night and stands in it, but generally moves to retreat above web by day. Silk line goes down to web and signals the arrival of prey in daylight. Egg mass is attached to some shady support near web. Young disperse soon after hatching.

This widespread species is usually found in shady locations, sometimes spinning its web on the sides of cliffs.

645 Garden Spider
"Cross Spider"
(*Araneus diadematus*)

Description: Male ¼–½" (6–13 mm), female ¼–¾" (6–19 mm). Body brownish orange, legs darker because of brown encircling bands. *Abdomen above bears a median row of diamond-shaped silvery spots,* some smaller dots and dashes, and a series of short dark brown bands.

Habitat: City and suburban gardens between houses and shrubs.

Range: Boston, Massachusetts, to the Great Lakes.

Food: Flying and jumping insects.

Web: Symmetrical orb within 5- or 6-sided frame; to 20" (51 cm) across.

Life Cycle: Spider usually rests head downward at center of web. Female attaches egg mass to a leaf, twig, or other support at side of web. Spiderlings usually stay in egg sac until 1st molt.

One of the larger orb weavers, this spider was introduced from Europe. It eats the remains of the web made the previous night and spins a new web each night. The markings on the abdomen sometimes give the impression of a vertical white line crossed by a conspicuous dark line, inspiring the alternative name, "Cross Spider."

691 Marbled Orb Weaver
(*Araneus marmoreus*)

Description: Male ¼–⅜" (6–9 mm), female ⅜–¾" (9–19 mm). Abdomen oval. *Reddish yellow to yellow.* Cephalothorax marked by dark lines along sides and midline above. *Abdomen bears a marbled pattern of brown to purple,* isolating several orange-yellow spots on and near midline.

Habitat: Tall grass areas of meadows and among shrubs.

Range: Throughout the United States, north to Alaska.

Food: Insects.

Web: Spiraling orb, often irregular, built on low shrubs and grasses.

Life Cycle: Spider rests near orb in a retreat made of leaves or, if web is on tree, under bark. A signal strand extends from center of orb to the retreat, carrying warning vibrations to spider if an insect gets caught in web. In autumn many orange eggs are laid in a flattened cocoonlike egg sac of loose white silk, which is attached to a leaf near retreat. Eggs may overwinter or hatch in a few days and the spiderlings overwinter, depending on latitude and weather.

This spider tends to be inconspicuous, hiding in its retreat. It drops to the ground if approached.

690 Shamrock Spider
(*Araneus trifolium*)

Description: Male ⅛–¼" (4–6 mm), female ⅜–¾" (9–19 mm). *Female's cephalothorax white to green-gray with a broad black stripe on midline above and a narrower black stripe on each side;* male's cephalothorax brownish yellow, unmarked. Female's abdomen gray, pale green, greenish brown, purplish red, or white; always with white spots and black dots. Male's abdomen white or yellow, unmarked.

Habitat: Tall grass areas of meadows, pastures, or woodland edges.

Range: Throughout the United States and adjacent areas of Canada.

Food: Insects.

Web: Spiraling orb.

Life Cycle: Spider waits at the side or on top of web spun during the night between tall grass stems. It is often hidden in a tent made by tying leaves together. A signal strand extends from center of web to spider's retreat. In autumn female produces egg mass in or near the retreat and dies. Eggs or spiderlings overwinter, dispersing on warm winter days or in spring to spin their own webs far from one another.

Incredible numbers of Shamrock Spider webs are often conspicuous with dew at sunrise. They produce a new web every night, usually in the same site as the previous web, which the spider has eaten strand by strand. Formerly assigned to genus *Epeira*.

680 Six-spotted Orb Weaver
(*Araniella displicata*)

Description: Male ⅛–¼" (4–5 mm), female ⅛–⅜" (4–9 mm). Cephalothorax and legs brownish yellow, unmarked. *Globular abdomen brownish yellow, pink, or white, with 3 pairs of circular black spots* toward sides in rear ½; each spot surrounded by a pale ring.

Habitat: Meadows and pastures among tall grasses and on bushes.

Range: Throughout most of North America.

Food: Insects.

Web: Unusually small inconspicuous webs spun between plants near the ground.

Life Cycle: In autumn egg masses are attached to plant near web. Eggs overwinter or hatch and spiderlings overwinter. Adults die. When snow is deep, more young survive to disperse in early summer. 1 generation a year.

Both sexes may have an irregular silvery area above near the front half of the abdomen. Formerly assigned to genera *Araneus* or *Epeira*.

699 Silver Argiope
(*Argiope argentata*)

Description: Male ⅛–¼″ (4–5 mm), female ½–⅝″ (12–16 mm). *Silvery short hair on* upper surface *of female's cephalothorax and 1st abdominal segment.* Most of *abdomen* black to brownish yellow *with silver spots.* Underneath also black to yellow-brown. Legs blackish brown to yellow with 2 pale bands and black hair.

Habitat: Fields and gardens.

Range: Southern Florida to southern California.

Food: Insects.

Web: Spiraling orb with zigzag cross strands forming X-shaped mark at center, measuring to 32″ (81 cm) across.

Life Cycle: Female rests head down at center of orb web. Main spiral of sticky strands begins just beyond reach of outstretched legs. In autumn female attaches sac containing several hundred eggs to a leaf or branch just beyond orb web and dies. Spiderlings soon hatch and disperse, each spinning a new web every night after eating the old web. Spiderlings overwinter and resume spinning in spring. Few females survive to maturity, but many males survive, eventually spinning little orb webs in outlying parts of the webs made by prospective mates. The male twitches the web of female to learn when it is safe to approach. Male is often eaten by female.

Primarily a spider found in tropical regions of the New World, this species is able to survive frost only when very young and seldom is found in the North.

697, 698 Black-and-yellow Argiope
(*Argiope aurantia*)

Description: Male ¼–⅜″ (5–9 mm), female ¾–1⅛″ (19–28 mm). Cephalothorax has short silvery hair. Legs black with reddish or yellow bands near body. *Abdomen egg-shaped, conspicuously marked with yellow or orange on black.*

Habitat: Among shrubbery, tall plants, and flowers in meadows and gardens.

Range: Throughout the United States and southern Canada; not common in the Rocky Mountains and Canadian Great Basin area.

Food: Small flying insects.

Web: Spiraling vertical orb radiating out from the center.

Life Cycle: Female fills spherical egg sac, up to 1″ (25 mm) wide, with tough brown papery cover. Female attaches it to one side of web close to resting position, then dies. Eggs hatch in autumn, young overwinter in sac, then disperse in spring. Male builds web in outlying part of female's web, making a white zigzag band vertically across the middle.

This spider seems to prefer sunny sites with little or no wind. It drops to ground and hides if disturbed.

702 Crablike Spiny Orb Weaver
(*Gasteracantha elipsoides*)

Description: Male ¹⁄₁₆–¹⁄₈″ (2–3 mm), female ³⁄₈″ (8–10 mm). *Abdomen broad, hard, spiny with 2 sharp points on each side and 2 at rear.* Cephalothorax reddish black. Abdomen pale to orange or yellow with reddish-black oval spots above and reddish spines; below black with small yellow spots.

Habitat: Woodland edges and shrubby gardens.

Range: North Carolina to Florida, west to California.

Food: Small insects.

Web: Vertical orb with few or no spiral strands near center.

Life Cycle: Spider stands head downward near center of web, which is spun anew each night. Egg mass is produced late in year, then female dies. Eggs or spiderlings overwinter. Spiderlings disperse and begin web weaving in spring.

The male is smaller than the female and seldom noticed. Formerly known as *G. cancriformis.*

686 Bola Spider
(*Mastophora bisaccata*)

Description: Male ⅟₁₆″ (2 mm), female ½″ (12–13 mm). Abdomen bulbous. Medium brown or greenish mottled with paler color and with *2 shiny greenish-brown protuberances.*

Habitat: Shrubby meadows, woodland edges, and gardens.

Range: Eastern United States, Texas, and Gulf states.

Food: Moths.

Life Cycle: Hard, globular egg sac has long silken extension and is encircled with a series of irregular points. Sac is placed on tree branches. Spiderlings emerge in June. Males molt once after they emerge; females are same size as males when they emerge but molt many times.

This spider does not spin a web, but produces a dangling silken line with a globule at the end that resembles the South American bola. Supposedly the 2 protuberances on the spider attract male moths. The spider waits for the moth to approach, then throws its bola at the moth, usually snaring its wings. The spider drops down on a line spun from its spinnerets and eats the entrapped moth.

700, 701   **Arrow-shaped Micrathena**
            (*Micrathena sagittata*)

Description: Male ⅛–¼″ (4–5 mm), female ⅜″ (8–
              9 mm). Cephalothorax small, dark
              reddish brown. *Abdomen triangular,
              mottled red and yellow above,* widest at
              rear. Female has 2 dark red, long
              diverging spines and 2 pairs of dark
              red, shorter spiny tubercles with black
              tips on abdomen. Male lacks spines.
    Habitat: Woodland edges, shrubby meadows,
              and gardens.
      Range: Eastern United States, west to Texas
              and Nebraska.
       Food: Small insects.
        Web: Regular web woven in vertical plane.
              Spiral sticky strands leave a central hole
              through which spider can move easily
              from one side to other. Sometimes a
              short zigzag strengthening band is
              woven immediately above the hole.
 Life Cycle: In autumn female deposits mass of eggs
              on leaf near edge of web and soon dies.
              Eggs or spiderlings overwinter, young
              disperse in spring.

              These small spiders are often found in
              the woods in autumn, spinning snares
              for prey. Spiderlings have longer
              abdomens and their spines are short and
              blunt. Formerly known as *Acrosoma
              spinea.*

696   **Golden-silk Spider**
      "Calico Spider"
      (*Nephila clavipes*)

Description: Male ⅛″ (4 mm), female ⅞–1″ (22–
              25 mm). Female's cephalothorax pale
              gray with 3 black spots on each side;
              legs dark with brownish bands and
              *conspicuous tufts of black hair on 1st and
              last pairs of legs.* Female's abdomen
              brownish green, spotted with white in
              irregular pattern. Male's body color
              drabber; legs also have tufts of black

hair. In both sexes abdomen is 2½–3 times as long as it is broad.

Habitat: Shaded woodlands and swamps.

Range: Southeastern United States.

Food: Flying insects.

Web: Strong, slightly inclined orb with notchlike support lines. Web may measure 2–3′ (1 m) across.

Life Cycle: Female attaches elongated egg mass to undersurface of leaf and then rests nearby. Spiderlings disperse, each to make web elsewhere. At first, they build only ⅔ of a web, leaving the top somewhat irregular across from principal support line.

During the day the spider hangs head downward from the underside of the web near the meshlike center or hub. The spider repairs the webbing each day, replacing half but never the whole web at one time.

LARGE-JAWED ORB WEAVERS
(Family Tetragnathidae)

Tetragnathid orb weavers are easily recognized by their unusually large, powerful jaws, or chelicerae. Like orb weavers, they have 8 eyes and 3 claws on each tarsus. They are ⅛–⅜″ (3–9 mm) long. Many species spin orb webs although in some species only spiderlings produce webs. About 25 species are found in North America.

693 Mabel Orchard Spider
(*Leucauge mabelae*)

Description: Male ⅛″ (4 mm), female ¼″ (5–7 mm). Cephalothorax and legs yellowish to brown. *Abdomen yellowish brown with 8 evenly spaced silvery stripes and 3 pink to orange spots*. Rear femora have double row of hair.

Habitat: Woodland edges and shrubby
meadows.

Range: New England to Florida, west to Texas
and Nebraska.

Food: Small insects.

Web: Almost horizontal orb with widely
spaced spiral adhesive strands in shrubs
or trees.

Life Cycle: Female attaches egg mass to leaves near
web woven in shrubs or trees.
Spiderlings soon disperse to start webs.

This species hangs below its web until
prey is detected, or waits on a stem
nearby with 1 leg in contact with a web
strand. Orchard spiders differ from
other large-jawed spiders in possessing
a small plate underneath that protects
the opening of the female sex organs.

692 Venusta Orchard Spider
(*Leucauge venusta*)

Description: Male ⅛″ (3–4 mm), female ¼–⅜″ (5–
8 mm). Cephalothorax yellowish green,
striped with brown along sides.
*Abdomen silvery above with dark stripes,
sides yellow with red spot near tip and
red spot underneath.*

Habitat: Woodland edges and shrubby
meadows.

Range: Maine to Florida, west to Nebraska and
Texas.

Food: Small insects.

Web: Almost horizontal orb web with widely
spaced, spiral adhesive strands usually
in small shrubs or trees.

Life Cycle: Egg mass is attached to leaves and
twigs near web. Spiderlings disperse
and spin own webs.

This spider clings below its web or to a
nearby twig until prey blunders into
the web and shakes it.

695 Elongate Long-jawed Orb Weaver
(*Tetragnatha elongata*)

Description: Male ¼" (5–7 mm); female ⅜" (9 mm). *Long, diverging jaws* (chelicerae) with many teeth. Pale yellowish brown with slightly darker lengthwise stripes on abdomen. *Leg pairs 1, 2, and 4 twice as long as body,* held extended while spider clings against a twig or grass blade.

Habitat: Meadows and marshes, usually near running or standing water.

Range: Throughout North America, more common in the East.

Food: Insects.

Web: Small orb tilted (not vertical) with few radii and a central hole. Web is covered with strings of beadlike silk.

Life Cycle: Egg mass is affixed to a leaf near web but females do not guard eggs. Spiderlings disperse to spin own webs.

This spider waits for prey at the central hole of web. It can dodge through the web to either side. Inconspicuous details, including structure of the male genitalia, help specialists identify this species and the 16 others in the genus.

694 Long-jawed Orb Weaver
(*Tetragnatha laboriosa*)

Description: Male ¼" (5 mm), female ¼" (6 mm). *Cephalothorax pale yellow. Abdomen silvery* striped with dark gray below; about 3 times length of cephalothorax. *Male's chelicerae unusually short,* about ⅔ as long as cephalothorax. Female's chelicerae held almost vertically.

Habitat: Shrubby meadows and woodland edges.

Range: Throughout United States and southern Canada.

Food: Small insects.

Web: Small orb web spun between branches of shrubs, generally not quite vertical and sometimes almost horizontal. Inner spiral small with central hole, from which all cross strands have been removed.

Life Cycle: Egg mass in silken cocoon is attached to nearby plant. Spiderlings quickly disperse and spin own webs.

This spider stands at one side of the web, keeping its legs on a radial support line to detect the arrival of any prey. It drops from the vicinity of its web at the slightest disturbance. Related species have the same general size and coloring but differ in the proportions of the chelicerae and legs, and in the arrangement of the eyes.

FUNNEL WEB WEAVERS
(Family Agelenidae)

Funnel web weavers are small to medium-sized spiders, $\frac{1}{16}$–$\frac{3}{4}''$ (1–20 mm) long, that are often found in homes and fields. Most of these spiders have 8 eyes, arranged horizontally in 2 parallel rows, but some have 4–8 eyes in curved rows. Some cave species have no eyes. In many species, the hind pair of spinnerets is twice the length of the other 2 pairs. All members of the family have long, thin legs with 3 claws on each tarsus. Funnel web weavers spin sheet webs of nonadhesive silk. There is a characteristic funnel extending off from the center to one edge, where the spider hides, and a 3-dimensional barrier web over the top. When a flying insect hits the barrier, it falls into the sheet below. The spider then rushes out of the funnel, bites its insect prey, drags it back to the funnel, and feeds. Many funnel web weavers live in leaf litter. Some are often found near ants. There are over 400 species in North America.

670 Grass Spiders
(*Agelenopsis* spp.)

Description: Male ⅝″ (15 mm), female ¾″ (20 mm).
Cephalothorax yellowish with pale center stripe flanked by 2 dark bands and light bands at edges. Abdomen may have light stripe down middle, darker bands with gray at edge. Large rear spinnerets. 8 eyes arranged in 2 arches.

Habitat: Grassy areas and low shrubs, stone fences, and buildings.

Range: Throughout North America.

Food: Insects.

Web: Horizontal sheet with funnel extending from center to 1 edge. No sticky fibers but many long telegraph fibers across to funnel opening.

Life Cycle: Eggs overwinter in convex egg sac, which is usually found outside the web, sometimes under bark on a nearby tree. Often sac is still among legs of female spider that produced it in autumn, then died. Spiderlings disperse in spring, build small webs remote from one another. Survivors construct bigger webs by late summer, often in same place where thin inconspicuous webs were produced earlier.

These quick-running spiders depend on speed to capture their prey. The many different species can be distinguished only by a specialist.

NURSERY WEB SPIDERS
(Family Pisauridae)

These large spiders, ¼–1″ (7–26 mm) long, resemble wolf spiders. Their 8 eyes are arranged in 2 rows, the front row having 4 eyes in a straight line and the back row with 4 eyes in a U-shape. There are 3 claws on each tarsus. Members of this family do not build webs to catch prey, but use silk to construct a special nest or nursery web

for the young. The female carries a
spherical egg sac beneath its body,
holding it in its fangs, until hatching
time approaches. Then the female
builds a nursery web held together with
silk and suspends the egg sac among
the strands. She rests nearby until
almost all the spiderlings have
dispersed. The most spectacular nursery
web spiders run over the surface of
ponds and streams and, if pursued,
sometimes even go underwater. They
occasionally capture tadpoles and small
fishes near the surface, but usually prey
on insects.

663　Brownish-gray Fishing Spider
(*Dolomedes tenebrosus*)

Description: Male ¼–½" (7–12 mm), female ⅝–1"
(15–26 mm). Female legspan over 3"
(75 mm). *Body and legs brownish gray,
marked extensively with blackish brown.*

Habitat: Woods, often at a distance from any
stream or pond.

Range: Virginia north to New England and
adjacent areas of Canada.

Food: Small insects.

Life Cycle: Egg mass is laid in part of egg sac
produced previously, then rest of egg
mass is made. Female carries it in
chelicerae even when running across
open water. Egg mass is discarded
when spiderlings emerge. Spider hunts
widely, finding prey on water surface,
in water, and on land.

When disturbed near the water, this
spider may creep below the surface and
remain there motionless for 30 minutes
or longer. It occasionally seizes a small
fish or tadpole that swims by. Air
clinging to body hair appears sufficient
for the spider to breathe underwater.
Sometimes it is found some distance
from the water in wooded areas.

668 Six-spotted Fishing Spider
(*Dolomedes triton*)

Description: Male ⅜–½" (9–13 mm); female ⅝–¾"
(15–20 mm) with legspan to 2½" (64
mm). *Greenish brown with silvery-white
lengthwise stripes* along each side of
cephalothorax and abdomen. Abdomen
has 12 white spots. Undersurface paler,
except for 6 *black spots between leg bases.*

Habitat: Slow-flowing streams or ponds.

Range: East of the Rocky Mountains in the
United States and southern Canada; rare
in Rocky Mountains and Great Plains.

Food: Small insects, sometimes tadpoles and
small fish.

Life Cycle: Female sometimes carries egg sac across
open water and holds or stays close to
sac until most spiderlings have
dispersed. Egg sacs are produced June–
September, occasionally in April.

Sometimes these spiders are eaten by
fish. They are often seen scampering
over waterside plants.

642, 666 Nursery Web Spider
(*Pisaurina mira*)

Description: Male ⅜–⅝" (9–15 mm), female ½–⅝"
(12–15 mm). *Yellowish brown, sometimes
with light to dark brown band* down
middle of back; bordered narrowly with
white on abdomen.

Habitat: Old fields, meadows, and woods in tall
grass and shrubs; also houses.

Range: Ontario, Quebec, and Nova Scotia to
Florida, west to Texas, north to Kansas
and Minnesota.

Food: Small insects, occasionally larger ones.

Life Cycle: Female constructs a nursery web in
high weeds or low shrub. Egg sac is
produced June–July. Female guards
eggs until they hatch and spiderlings
disperse. Males are seen in June.

This spider wanders as a hunter over vegetation and does not build a web to catch prey. The female is often seen carrying a bulbous egg sac under the cephalothorax. In some European *Pisaurina* species, the male offers the female a courtship gift of a fly or other food.

WOLF SPIDERS
(Family Lycosidae)

Wolf spiders, ⅛–1⅜″ (3–35 mm) long, are widespread hunters, named for the Greek word "lycosa," meaning "wolf." They have 8 dark eyes of unequal size arranged in 3 rows, the first having 4 eyes. The abdomen and cephalothorax are usually as long as wide. The long legs bear 3 microscopic claws at each tip. Most wolf spiders live on the ground and hunt for prey at night. Their dark mottled colors camouflage them among dead leaves, stones, and other debris. Except for one genus, these spiders do not spin webs. Some dig burrows in the ground, others make holes under rocks, and many have no retreat at all. Males court potential mates by rhythmically waving the pedipalps. The female spins a large spherical egg sac, attaches it to her spinnerets, and drags it about after her until spiderlings hatch. Spiderlings clamber onto the female's back and are carried until they are ready to disperse. This large family has over 200 species in North America.

646   **Burrowing Wolf Spiders**
(*Geolycosa* spp.)

Description: Male ½–⅞″ (14–22 mm), female ¾–⅞″ (18–22 mm). Body and legs gray to sand-colored, *speckled with black.* Legs have many black spines.

Habitat: Sandy areas.

Range: Throughout North America; individual species more restricted.

Food: Insects.

Life Cycle: Vertical burrows up to a few feet deep are dug in sand. Spider cements sand particles for walls with silk, throwing loose sand out doorway. Female brings up egg sac and exposes it to warmth of sun in doorway.

This spider seldom emerges from its burrow, but lies in wait for insects to pass within reach. If it senses vibrations in the ground, it disappears down the burrow and waits minutes before climbing up to the doorway again.

648 Carolina Wolf Spider
(*Lycosa carolinensis*)

Description: Male ¾″ (18–20 mm), female ⅞–1⅜″ (22–35 mm). *Cephalothorax gray-brown. Abdomen may have darker stripe along midline.* Female has sparse covering of gray hair; somewhat paler in male.

Habitat: Open fields on the ground.

Range: Throughout the United States and southern Canada.

Food: Insects.

Life Cycle: Female digs a burrow 6–8″ (150–203 mm) deep, often with a high rim around the entrance. Female often produces more than 1 egg sac, which gradually darkens from satiny white to earth color before spiderlings emerge. Female guards egg sacs in burrow but does not interfere with spiderlings as they emerge and disperse.

The Carolina Wolf Spider is the largest wolf spider in North America. It hunts almost exclusively at night.

665　Forest Wolf Spider
(*Lycosa gulosa*)

Description: Male ⅜″ (10–11 mm), female ⅜–½″ (10–13 mm). Dark brown with *grayish-yellow middorsal stripe on cephalothorax and narrow grayish-yellow stripe on each side.* Male's abdomen has 2 incomplete black stripes on front ⅓. Pedipalps large, hairy.

Habitat: Woods, among litter.

Range: Maine to Georgia, west to Utah, north to southern Manitoba.

Food: Small insects.

Life Cycle: Female drags eggs in a spherical sac until they hatch. Spiderlings ride on female until able to fend for themselves.

This spider hides among litter by day, hunts at night. It makes no nest or silken shelter, although it secures a dragline before leaping upon potential prey. The light of a flashlight is reflected from its silvery eyes, making this wolf spider easy to find at night.

667　Rabid Wolf Spider
(*Lycosa rabida*)

Description: Male ½″ (13 mm), female ⅝–⅞″ (16–21 mm). Brownish yellow. *Cephalothorax has 2 lengthwise dark stripes; 1 dark stripe between 2 pale areas on abdomen.* 1st pair of legs often black or dark brown on male.

Habitat: Woods and meadows, among litter and on low foliage.

Range: Oklahoma north to Nebraska, east to Maine, south to Florida.

Food: Small insects.

Life Cycle: Female spins silken cocoon around egg mass, attaches cocoon to spinnerets, and drags it about. It darkens from shiny white to dirty brown. Spiderlings ride on female's back until ready to disperse.

This species is harmless to people, but its bite is often feared—hence its common name. According to a legend, the only way to save a victim bitten by the related European Tarantula (*L. tarentula*), male to 1" (25 mm) and female larger, is to dance the tarantella.

654, 660 Thin-legged Wolf Spiders
(*Pardosa* spp.)

Description: Male ⅛–⅜" (4–10 mm), female ¼–⅜" (5–10 mm). *Slender, long-legged.* Dark or with lengthwise dark-and-light stripes. Covered with long hair. Upper row of large eyes occupies major area on the front of the head. Spines on legs relatively long.

Habitat: Soil surface in grassy fields.

Range: Throughout North America.

Food: Insects.

Life Cycle: Female spins lens-shaped oval cocoon, attaches it to spinnerets, and drags it about. Cocoon is greenish at first but soon turns dirty gray. Spiderlings are carried on female's back.

This spider does not build a shelter. It hunts over a limited territory and often basks in the sun, keeping warm and ready for quick pursuit of potential prey. There are more than 100 species of this genus in the United States and Canada. Only a specialist can distinguish them reliably.

LYNX SPIDERS
(Family Oxyopidae)

These hunting spiders are recognized by the distinctive arrangement of their 8 eyes—6 eyes form a hexagon with 2 smaller eyes facing forward. They are usually small, measuring ⅛–⅝" (4–

16 mm) in length. Their legs have
prominent spines and 3 claws on the tip
of each tarsus. In many, the abdomen
tapers to a point. Lynx spiders do not
spin webs or build retreats. Most are
found on plants, tall grasses, and low
shrubs, where they hunt for prey. A
few live among rocks or near water, like
crab spiders. Some species use a silken
dragline to seize an unsuspecting
victim; others do not. While waiting
for prey, they often rest on the hind
legs and keep the front legs raised, in a
pose that somewhat resembles that of
mantids. They are able to jump
through vegetation onto prey with
considerable agility. Egg sacs are
attached to plants. Mostly tropical
spiders, there are fewer than
20 North American species.

641 Jumping Lynx Spider
(*Oxyopes* spp)

Description: Male ⅛–¼″ (3–5 mm), female ⅛–⅜″
(4–8 mm). *Cephalothorax yellow with 4
lengthwise vague, pale stripes;* 2 black
lines from below eyes to tip of
chelicerae. Male's abdomen gray or
black with iridescent scales; female's
pale, darker below and on sides. *Narrow
black line below each yellow femur;* legs
spiny.

Habitat: Fields in tall grasses and among
herbaceous vegetation.

Range: Throughout the United States, most
common in the East, the Rocky
Mountains, and Canadian Great Basin.

Food: Small insects.

Life Cycle: Female attaches spherical egg sac to
plant, tying several leaves together with
silk. Female guards egg sac until
spiderlings disperse, each soon hunting
on its own. In the North, egg sac,
embryos, or spiderlings overwinter.
Adults are usually seen June–
September. Usually 1 generation a year.

This hunting spider does not build a
web or nest. Active by day, it hunts
among herbaceous vegetation,
including tall grasses and low plants.
The related Texan Golden Lynx Spider
(*O. acleistus*), same size, is variably
colored: tan to golden-yellow with a
pattern of dark gray or black on the
upper body and along the sides. The
male's pedipalps are conspicuously
velvety black. It is found in Louisiana,
Texas, and Mexico.

677, 681   **Green Lynx Spider**
(*Peucetia viridans*)

Description: Male ½" (12–14 mm), female ½–⅝"
(14–16 mm). Bright leaf-green, ivory-
white, or tan. Legs yellowish with
many black spines. Cephalothorax has 6
eyes on pale area, 2 other eyes face
forward. *Abdomen marked above with 3
incomplete chevrons.*

Habitat: Fields and woods on tall grasses and
flowerheads, especially of wild
buckwheat.

Range: Southern United States and Mexico.

Food: Small insects.

Life Cycle: Female spins a silken egg sac that is
flattened on one side and sometimes has
pointed projections on the sides. Egg
sac is attached to plants, then female
rests on top of it. As soon as spiderlings
hatch, they disperse, catching whatever
tiny insects they can.

This spider uses silk to cover and secure
eggs and also as a dragline when
pouncing on prey.

SAC SPIDERS
(Family Clubionidae)

Sac spiders are hunters with somewhat stout to elongate bodies, ⅛–⅜″ (3–10 mm) in length. They have 8 eyes of similar shape that are arranged in 2 rows and have 2 claws on each tarsus. Some species have tufts of hair between the claws. These spiders spin saclike, tubular resting sites inside a rolled leaf or under bark or a stone. Many species may escape the attention of predators because they slightly resemble ants in appearance and behavior. There are 100 North American species, and more than 1,500 species worldwide.

319, 640   Ant-mimic Spiders
(*Castianeira* spp.)

Description: Male ¼–⅜″ (5–8 mm), female ¼–⅜″ (5–10 mm). *Antlike.* Orange, brown, or black. Individual species *patterned with light or dark bands, midline stripe, or spots.*

Habitat: Woods, fields, and gardens on soil, tree bark, low shrubs, and under stones.

Range: Southeastern states.

Food: Small insects.

Life Cycle: Female attaches flat, parchmentlike egg sacs with metallic luster to a stone. Eggs hatch in autumn. Spiderlings overwinter, dispersing in spring.

These spiders often live near anthills, where their antlike behavior may confuse predators. Their tubular sacs are often concealed in cracks of tree bark, between the twig and main stem, or in a rolled leaf.

WANDERING SPIDERS
(Family Ctenidae)

These hunting spiders, ¼–1⅝″ (5–40 mm) long, are mostly dull yellowish brown with blackish stripes and leg bands. They have 2 or 3 claws on each leg, and their 8 eyes are arranged in 3 rows—the first row has 2 eyes, the second 4 eyes, and the third 2 widely separated eyes. They slightly resemble wolf spiders, which also have eyes arranged in 3 rows. Some wandering spiders live on foliage; others wander on the ground either by day or night. They do not make webs. Some species live in burrows in the ground. Primarily tropical, many are accidentally imported in cargoes of bananas. They are most common in the South and the West.

672 Wandering Spiders
(*Ctenus* spp.)

Description: Male ¼–⅞″ (5–21 mm), female ⅜–1″ (8–25 mm). *Cephalothorax yellowish to orangish brown.* Legs orangish brown to brown, palest close to body. Abdomen grayish yellow to black. Body sparsely coated with fine down.

Habitat: Woods, on the ground or on foliage; also in caves.

Range: Alabama and Mississippi; southern Texas and northeastern Mexico.

Food: Insects.

Life Cycle: Egg mass is concealed among litter or attached to undersurface of a leaf. Apparently they mature at any time of year and overwinter at any stage. This spider hunts in litter, among mosses, or under stones and logs almost like a small wolf spider.

Most members of this genus are tropical and are transported into North America on plants and fruit. Identification is

based on differences in the genitalia,
spines on the legs, teeth on the
chelicerae, and color patterning.

GIANT CRAB SPIDERS
(Family Sparassidae)

Giant crab spiders are large spiders, ⅜–
1″ (10–25 mm) long, named for their
crablike legs. They somewhat resemble
crab spiders but are larger and have a
soft, 3-lobed membrane at the end of
the sixth joint of each leg, which gives
the foot an extra extension. Like crab
spiders, giant crab spiders have 2 claws
at the tip of each tarsus. Each fang folds
into a distinct groove that is armed
with teeth. The 8 eyes are arranged in 2
rows of 4 eyes each. Mostly tropical
spiders, they are found in the
Southwest and Florida. The group was
previously known as Family
Heteropodidae.

664 Huntsman Spider
(*Heteropoda venatoria*)

Description: Male ¾″ (19 mm), female 1″ (24 mm).
Body brown. *Unusually long and bristly
legs. Cephalothorax* flattened, *almost
encircled by narrow, pale brownish line;*
central area dark or blackish, sometimes
divided by a pale midline. Abdomen
pale brown with dark brown midline
stripe on 1st half.

Habitat: Woods and gardens on bark and house
walls.

Range: Florida.

Food: Cockroaches and other insects.

Life Cycle: Female carries spherical tan egg sac in
jaws until eggs hatch and spiderlings
disperse.

These spiders hide in crevices by day,
hunting in darkness. They are

appreciated in many southern homes as free exterminators of household insect pests.

658 Golden Huntsman Spider
(*Olios fasciculatus*)

Description: Male ½–⅝″ (12–17 mm), female ¾–⅞″ (19–21 mm). *Cephalothorax reddish yellow to golden* with short black hair. *Abdomen,* coated with gray hair, *bears Y-shaped black marks toward rear.*
Habitat: Shady woodlands and thickets.
Range: New Mexico and Utah, west to California.
Food: Insects.
Life Cycle: Female carries egg sac in jaws until spiderlings emerge and disperse.

This species makes no organized webs but wanders in slow search for prey.

SELENOPID CRAB SPIDERS
(Family Selenopidae)

These rather flat-bodied spiders, ⅜–⅝″ (8–17 mm) long, live under stones and bark. They are distinguished by having 6 of their 8 eyes in a single, transverse, front row. The long legs are held to the sides, and each ends in 2 claws. When disturbed, they run sidewise, crablike, and hide in a crevice. Mostly tropical spiders, there is only 1 North American genus, which occurs in the Southwest.

661, 662 Selenopid Crab Spiders
(*Selenops* spp.)

Description: Male ⅜″ (8–10 mm), female ⅜–⅝″ (10–17 mm). Body yellowish brown with *grayish to brownish mottling.* 6 of the 8 eyes in row across front of cephalothorax, remaining 2 behind on sides.

Habitat: Woods, fields, and gardens in crevices of bark or under stones.

Range: Southwestern United States.

Food: Small insects.

Life Cycle: Eggs are concealed in crevices and then left unattended. Spiderlings disperse widely, sometimes entering houses.

These spiders are active in dim light and darkness, seldom by day. They creep forward slowly to reach prey but run sideways much more rapidly, a characteristic of this genus.

CRAB SPIDERS
(Family Thomisidae)

Most spiders in this family hold their legs outstretched to the sides like crabs and can move forward, sideways, or backward. They have a short, rather broad body, $\frac{1}{16}-\frac{3}{8}''$ (1.5–10 mm) long, and 8 small eyes that are often located on raised bumps in 2 backward-curved rows of 4 eyes each. The second pair of legs is often much heavier and longer than the third or fourth pair. There are 2 claws on each tarsus. These spiders wander over the ground, climbing flowers and plants in search of prey. They do not spin web snares, retreats, or overwintering nests, but the male of some species may cover a prospective mate with loose silken webbing and tie the female down. The female produces a silken sac for eggs, usually guards it a while, but dies before spiderlings emerge. There are more than 200 North American species.

678, 679 Goldenrod Spider
"Flower Spider"
"Red-spotted Crab Spider"
(*Misumena vatia*)

Description: Male ⅛" (3–4 mm), female ¼–⅜" (5–
10 mm). *Female yellowish to white with
crimson streaks* on each side of abdomen
and a reddish-brown stain between
eyes. Female's legs pale. Male's
cephalothorax dark reddish brown with
white spot in center and in front of
eyes; abdomen white with 2 red bands.
Male has 2 pairs of reddish-brown fore
legs, 2 pairs of yellow hind legs.

Habitat: Meadows, fields, and gardens on
daisies, goldenrod, and other white or
yellow flowers.

Range: Throughout North America and
southern Canada.

Food: Flower-visiting insects.

Life Cycle: Eggs are protected in a silken sac.
Female usually dies before spiderlings
hatch and disperse.

This spider's coloring often changes to
yellow, camouflaging it on yellow
daisies and goldenrod. The smaller
Flower Spider (*Misumenops asperatus*),
male ⅛" (3 mm), female ¼" (5 mm), is
brownish to pale gray marked with 2
pink stripes on the cephalothorax, a
pattern of pink streaks and spots on the
abdomen, and pink bands around tibiae
and tarsi of first 2 pairs of legs. It is
covered with stiff short hair. This
spider lives among grasses and leaves in
the eastern United States and southern
Canada, west to Arizona, and north to
Alberta.

682 Elegant Crab Spider
(*Xysticus elegans*)

Description: Male ¼" (5–7 mm), female ⅜" (8–
10 mm). *Dull brown with pale areas* on
top of cephalothorax and on side of

abdomen; narrow crossbands on rear.

Habitat: Woods on bark and among debris on ground or on low vegetation.

Range: Maine to Georgia, west to Arizona, north to Alberta.

Food: Small insects.

Life Cycle: Female guards eggs in silken sac but usually dies before spiderlings emerge.

This spider does not use silk to wrap prey or construct a nest, although the male may lightly swaddle female in silken strands as an act of courtship. It is commonly seen running over split rail or stump fences in rural areas east of the Rocky Mountains.

683 Thrice-banded Crab Spider
(*Xysticus triguttatus*)

Description: Male ⅛–¼" (3–5 mm), female ⅛–¼" (4–6 mm). Male's cephalothorax dark brown, paler mid-back; abdomen above with alternate black and white crossbands. *Female's cephalothorax dark to yellow-brown; abdomen whitish, both speckled with black above and with 3 black broken crossbands near rear.*

Habitat: Forests on ground, under rocks, or on bark; woods, fields, and meadows on foliage and stems of low shrubs and tall grasses.

Range: Newfoundland to Georgia, west to New Mexico, north to Alberta.

Food: Small insects.

Life Cycle: Eggs in silken cocoon are attached to plants. Female stands guard but rarely lives until spiderlings emerge and disperse.

This spider waits in ambush until prey comes within striking distance. It extends its front legs, snatches prey, and holds it until it is sucked dry. No silk is used to wrap prey or construct nest, although the male may swaddle a potential mate with it. The male is so

unlike the female that it can easily be
mistaken for a different species.

PHILODROMIDS
(Family Philodromidae)

Philodromids are small, crablike
spiders, $\frac{1}{16}-\frac{3}{8}''$ (2–11 mm) long. They
have elongate bodies and slender legs
with brushes of hair on each leg and
tufts of adhesive hair between the 2
claws on each tarsus. Their 8 eyes are
arranged in 2 parallel curving rows.
These spiders live in the bark of trees,
under rocks on the ground, or in
vegetation. Most are well camouflaged.
Swift runners, they can flee from
predators or escape the fingers of a
collector. Because philodromids
somewhat resemble crab spiders, some
arachnologists include them in the
family Thomisidae.

657 Inconspicuous Crab Spiders
(*Philodromus* spp.)

Description: Female $\frac{1}{8}-\frac{3}{8}''$ (3–8 mm), male slightly
smaller. Camouflaged coloring:
brownish to reddish, yellow, or
grayish, with spots, bands, or
mottling. *Abdominal tip usually pale.*
Habitat: Woods on tree bark and adjacent
foliage, or on ground among litter.
Range: Throughout the United States and
southern Canada.
Food: Small insects.
Life Cycle: Flattened egg sac is attached to a leaf,
twig, or rock, or left under bark. It is
guarded by female until spiderlings
emerge and disperse.

These spiders are inconspicuous when
they rest on bark. Occasionally they
stray into a house.

JUMPING SPIDERS
(Family Salticidae)

This large family gets its name from the spectacular leaps the spiders make pouncing on prey. Small to medium-sized spiders, ⅛–⅝″ (3–15 mm) long, they have short legs with 2 claws at the tip of each tarsus. Their 8 eyes are arranged in 3 rows—the first row near the midline contains the largest pair, which faces forward, permitting binocular vision; the second row, 2 very small forward-facing eyes; the third row, 2 pairs of eyes facing upward and only slightly forward. These spiders have the sharpest vision of all spiders, and consequently are excellent hunters. As they leap on a victim, silk comes out from the spinnerets, creating a long, silken dragline. They do not spin webs but make little, silken shelters under bark, stones, or leaves.

676   Green Lyssomanes
(*Lyssomanes viridis*)

Description: Male ¼″ (5–7 mm), female ¼–⅜″ (7–8 mm). *Body bright green* with black around eyes and red scales on front ⅓ of cephalothorax. *8 black spots on abdomen.*

Habitat: Low bushes.

Range: North Carolina to Florida, west to Texas.

Food: Small insects.

Life Cycle: Female encloses egg mass in a silken cocoon and guards it until spiderlings emerge. Spiderlings overwinter and mature by midsummer.

Some arachnologists place this genus in the Family Lyssomanidae because lyssomanes have 4 rows of eyes instead of 3 rows and a domed cephalothorax.

653, 655 Dimorphic Jumping Spider
(*Maevia inclemens*)

Description: Male ¼" (5–7 mm), female ¼–⅜"
(7–10 mm). *Male has 2 forms*—1st has
3 low tufts of black hair on
cephalothorax; 2nd has grayish body
with red, white, and black markings.
Female resembles 2nd form with faint
V-shaped markings and paler abdomen.

Habitat: Meadows on foliage and flowers.

Range: New England to Georgia, west to
Nebraska and Oklahoma.

Food: Small insects.

Life Cycle: Male courts potential mate by waving
hairy pedipalps in rhythmic gestures.
Mated female produces mass of eggs on
a leaf and guards eggs until spiderlings
hatch and disperse.

Spiders in this genus have longer, more
slender legs than most jumping spiders,
but like others, they run forward,
backward, or sideways.

656 Metaphid Jumping Spiders
(*Metaphidippus* spp.)

Description: Male ⅛–¼" (3–5 mm), female ⅛–¼"
(3–6 mm). Brown to yellow. *Body and
legs somewhat grayish* due to covering of
dense hair. Male usually has white band
on sides of abdomen. Both sexes have
spots, bands, and chevrons.

Habitat: Meadows and woods, on foliage, tree
bark, fence posts, and tall grasses.

Range: Throughout North America; individual
species more restricted.

Life Cycle: After mating, female constructs cocoon
for eggs, attaches it to twigs, and stays
close by. Spiderlings disperse rapidly.

These spiders run freely, producing an
anchor line when they leap on potential
prey or walk about on the ground.

320 Ant-mimic Jumping Spider
(*Peckhamia picata*)

Description: Male ⅛″ (3–4 mm), female ⅛–¼″ (3–5 mm). *Antlike.* Constrictions behind rear eyes and on abdomen; thick plate on back of abdomen. Mostly reddish brown. Violet around eyes with white spots behind rear eyes. Abdomen has white spots on each side and black at rear.

Habitat: Meadows and woods, on soil or leaves.

Range: Quebec to Florida, west to Texas and Nebraska.

Food: Small insects.

Life Cycle: Egg sacs are left under stones and in crevices. Spiderlings disperse quickly.

This hunting spider, the most common in the genus, is an accomplished jumper.

652 Daring Jumping Spider
(*Phidippus audax*)

Description: Male ¼–½″ (6–13 mm), female ⅜–⅝″ (8–15 mm). *Black with a short gray or white crossband on abdomen and several gray or white spots.* Pale markings, often yellow or orange in spiderlings. Chelicerae metallic green.

Habitat: Woods and gardens on tree trunks and fallen limbs, and in ground litter.

Range: Atlantic Coast to Rocky Mountains in the United States and southern Canada.

Food: Insects.

Life Cycle: Male courts female, backing or jumping away if female approaches too rapidly. Female produces egg sac while in silken retreat, then attends it until spiderlings disperse.

This common species enters houses, hunting on windowsills and sashes.

Scorpions
(Order Scorpionida)

Familiar in the South and West, scorpions somewhat resemble miniature lobsters—they have lobsterlike pincers, but their long upcurved "tail" ends in a poisonous stinger. These medium-sized to large arachnids, 1⅝–5" (40–127 mm) long, have a compact cephalothorax that is broadly joined to a long, 12-segmented abdomen; in fact the last 5 segments of the abdomen are really the "tail." Scorpions have 2 eyes in the center of the cephalothorax and 2–5 eyes on each side. A few species are blind. The small jaws or chelicerae, have 3 segments.

Nocturnal, scorpions use their poisonous stinger to kill spiders and large insects. Females give birth to living young that resemble tiny adults. The young ride on the back of the female until they molt for the first time. Then the young become solitary and catch their own prey. They grow slowly, some taking as long as 5 years to become adults. Most scorpions are not dangerous and do not attack people. If disturbed, they will inflict a sting that can cause painful swelling, but the poison of most North American species is not lethal to people. In Egypt and other tropical and subtropical countries where scorpions sting people frequently, an antivenin has been developed. In ancient times the scorpion's sting was feared almost as much as the lion's bite. So revered was this animal that it was given a place in the zodiac. The families are distinguished by the shape of the breastplate and the number of spurs between the last 2 abdominal segments. There are more than 70 species in North America, out of 1,500 to 2,000 worldwide.

BUTHID SCORPIONS
(Family Buthidae)

Buthid scorpions, 2–2¾″ (50–70 mm)
long, have a triangular breastplate, or
sternum, between the legs and 1 or 2
spurs on each side of the last tarsal
segment of the last pair of legs. Unlike
iurid scorpions, most buthids have a
spine under the stinger. In addition to
2 pairs of eyes on top of the
cephalothorax, there are 3–5 eyes on
each side. Buthids are the largest family
of scorpions, having over 700 species
worldwide and only 5 species in North
America.

635 Centruroides Scorpions
(*Centruroides* spp.)

Description: 2–2¾″ (50–70 mm). *Dark brown to
tan,* often striped with greenish yellow
along midline above. Some species have
greenish-yellow parallel stripe on each
side of cephalothorax. Abdomen
slender, constricted at each segment,
pale or dark according to species. *Tooth
beneath venom bulb.*

Habitat: Dark crevices under bark, stones, and
litter on the ground, and on dry
abandoned dirt roads.

Range: Florida and Gulf states, west to Arizona
and Mexico.

Food: Small insects.

Life Cycle: Female keeps eggs in sac, then carries
hatchlings on back until they can fend
for themselves. Male uses pincers to
pull female on top of him.

These scorpions seize prey in pincer-
tipped pedipalps and kill them with
their stingers. Included in this genus is
the dreaded Sculptured Centroides
(*C. sculpturatus*), 2″ (50 mm), whose
sting is extremely poisonous for people
and sometimes fatal. It is found in
Arizona.

IURID SCORPIONS
(Family Iuridae)

Members of this family are 1½–5½"
(37–140 mm) long, have a broad
pentagonal breastplate, or sternum,
between the legs. Unlike buthid
scorpions, which have 2 spurs, iurids
have only 1 spur on the joint of the last
tarsal segment of each leg, and no spine
under the stinger. Iurids can also be
recognized by the prominent tooth
underneath the small jaws, or
chelicerae. There are about 25 species
worldwide, and 9 species in North
America, including the largest
scorpions in the United States.

636 Giant Desert Hairy Scorpion
(*Hadrurus arizonensis*)

Description: 5½" (140 mm) *Cephalothorax black, each
segment rimmed in pale yellow.* Abdomen,
pincerlike pedipalps, and legs pale
yellow. Undersurface pale. Abundant
erect dark brown hair on legs,
pedipalps, and abdomen.

Habitat: Deserts

Range: Southwestern United States

Food: Insects; occasionally small lizards and
snakes.

Life Cycle: Female bears live young, and carries
brood on back for 10–15 days. Young
do not feed during this period. After
shedding their first skin, they scatter to
live independently. Young grow slowly
and shed their skin several more times
before reaching maturity.

Scorpions are nocturnal and are preyed
upon by owls and bats.
Venom in the scorpion's stinger is used
to subdue struggling prey and for self-
defense. *H. arizonensis* is the largest of
9 species of *Hadrurus* inhabiting the
United States. They range north as far
as Idaho and east as far as Colorado.

Pseudoscorpions
(Order Pseudoscorpionida)

Pseudoscorpions are small, flat arachnids, usually less than ¼" (5 mm) long. Like true scorpions, they have enlarged, pincerlike pedipalps, but the short, oval abdomen does not have a tail or stinger. Some species have 2–4 simple eyes; others do not have eyes. Many species have poison glands in the pedipalps, which are used to subdue insect prey and small invertebrates. Pseudoscorpions also have silk glands, but unlike spiders, which have them at the tip of the abdomen, the duct openings are located on the jaws, or chelicerae. They use this silk to spin cocoons, in which they overwinter and molt. The male in one species produces another kind of silk from the rectum, spinning threads that direct the female to the stalked sperm packet. In most species, the female's brood pouch remains attached to the female's abdomen, and hatchlings stay in the pouch, feeding on a milklike liquid from the female's ovaries. In some species, the female abandons the pouch, and the embryos develop unattended. Often young require several years to mature to adults. Some species can live up to 4 years.

Pseudoscorpions are common in many different habitats—at the seashore, in caves, in houses between the pages of books, and in woods among mosses and under loose bark, leaves, and stones. They can maneuver with great ease, moving forward, backward, and sideways. Frequently, they latch onto large insects, especially flies and beetles, and ride them to a new location. There are over 350 North American species, and about 2,500 species known worldwide.

637   CHERNETIDS
(Family Chernetidae)

Members of this family of common
pseudoscorpions live in many different
habitats, ranging from the woods under
leaf litter or bark to deserts or homes.
These small arachnids, $\frac{1}{16}-\frac{1}{8}''$ (2–
4 mm) long, have 5-segmented legs
and are usually tan to dark brown.
Some species have 2 eyes, while others
have none. The pincerlike pedipalps
contain well-developed venom glands
and ducts in the movable part of the
pincer, but are poorly developed or
absent in the fixed part. Most species
are predators, feeding on small flies,
bark lice, butterflies and their
caterpillars, ants, mites, and small
earthworms. Some of the species are
found under the elytra of long-horned
beetles, where they prey on mites on
the beetles' bodies. Females in this
family carry the brood pouch until the
young hatch. Some females construct a
brood nest.

Daddy-long-legs
(Order Opiliones)

Although daddy-long-legs resemble spiders, they are not. They are easily recognized by the long legs of most, but a few species have short legs. The flat oval body, ⅛–¾" (4–20 mm) long, consists of a compact cephalothorax that is broadly joined to the abdomen. The slender jaws, or chelicerae, in front of the mouth have 3 segments, while the leglike pedipalps behind the mouth have 6 segments without a pincer at the tip. Stink glands are located on each side of the cephalothorax near the base of the first pair of legs. The 2 eyes are raised on low turretlike protuberances at the middle of the cephalothorax. Despite the stiltlike legs, daddy-long-legs hold the body close to the ground, keeping the middle part of the legs high. If disturbed, they furiously wave the second pair of legs in the air. Males and females in this order do not perform mating rituals, but touch each other with their legs. Unlike any other arachnid, the male daddy-long-legs has a penis between the coxae where the legs meet, which he inserts between the female's chelicerae into an oviduct. The mated female uses a long ovipositor to thrust eggs in damp soil under stones, wood, or vegetable debris. In the North most species die in autumn after the female lays eggs; in the South many females overwinter under ground litter, laying eggs the following spring. Most species in this order live about 1 year.

Nocturnal, adults and young feed on many kinds of invertebrates, including other daddy-long-legs, spiders, flies, aphids, leafhoppers, snails, and earthworms. Some scavenge dead invertebrates and bird droppings. Others eat the gills of fungi and sugary sweets left as moth bait. There are over 200 species in North America out of

3,500 worldwide. They are also called harvestmen, because the first species to be described were seen in fall at harvest time.

DADDY-LONG-LEGS
(Family Phalangiidae)

Members of this family have 4 pairs of extremely long, slender legs. Most are ⅛–½″ (4–12 mm) long and differ from other daddy-long-legs in having a microscopic claw at the tip of each pedipalp and in having the coxae of the fourth pair of legs joined near the base to the abdomen. The second pair of legs is longer than the others, and is used like antennae. Sometimes legs may be missing because they break off easily and cannot be regenerated. Adults are frequently seen in huge gatherings, standing with legs interlaced. They prey on small insects and feed on soft, decaying organic matter.

674 Eastern Daddy-long-legs
(*Leiobunum* spp.)

Description: ¼–⅜″ (6–8 mm). *Long thin legs.*
Yellowish to greenish brown with blackish stripe along midline above and on each side. Legs pale to dark. Each pedipalp ends in a microscopic claw.
Habitat: Open areas on foliage and tree trunks, or on shady walls outside buildings.
Range: East of the Rocky Mountains.
Food: Minute insects, mites, and plant juices.
Life Cycle: Female uses slender ovipositor to insert eggs individually as far as possible into soil, where they overwinter, hatching in spring. Adults most common in autumn.

Sometimes daddy-long-legs cluster in tree holes with their legs intertangled, as though seeking insulation from the winter cold. In areas where frost is frequent, few adults survive until spring. The different species are distinguishable only on the basis of inconspicuous features.

673 Brown Daddy-long-legs
(*Phalangium opilio*)

Description: ⅛–¼" (4–6 mm). *Long thin legs.* Body reddish brown. Legs dark with prominent paler coxae. Eyes on black turret; 1 eye to right, 1 to left.

Habitat: Fields on tree trunks and open ground.

Range: Throughout North America.

Food: Small insects and decaying organic matter.

Life Cycle: Female thrusts ovipositor into soil to deposit eggs. When warm weather arrives, young creep out and grow slowly. Normally they mature in summer, then mate without courtship. 1 generation a year.

On cool afternoons adults often climb trees or sides of buildings, seemingly to benefit from residual heat of the sun. A warm knothole may attract dozens of daddy-long-legs, which stand close together with legs interlaced all night.

Mites and Ticks
(Order Acarina)

Mites and ticks comprise a large group of arachnids, including about 1,000 named species of ticks and more than 30,000 known species of mites worldwide with probably at least a million more still to be identified. Minute to small, they range from less than ¹⁄₆₄″ to 1⅛″ (0.3–30 mm) long. Unlike most arachnids, their oval to elongate bodies are not separated into cephalothorax and abdomen and lack indication of segments. They have 1–4 pair of legs, each with 6–7 segments. The jaws, or chelicerae, and pedipalps differ in adaptive features from one family to another; they are reduced or merely vestigial claws. Most species lay eggs, although a few give birth to living young. Newly hatched young, called larvae, have 6 legs. They gain 2 more legs after the first molt and are known then as nymphs.

Many mites are beneficial, preying on the eggs of aphids, attacking insects on plants, and hunting roundworms in the soil. There are some mites that are external and internal parasites, particularly in the larval stages. Ticks are generally larger than mites, some species measuring up to 1⅛″ (30 mm) in length. Both adults and young are external blood-feeding parasites of birds, mammals, and reptiles. Ticks cling to their host using a dartlike anchor located just below the mouth. The anchor's outer surface has backward curving teeth, which grip so firmly that pulling on the tick is likely to sever the body from the head, which remains in the wound. Ticks sometimes carry diseases from one animal to another. Rocky Mountain spotted fever is a rickettsial infection transmitted by a widespread tick found commonly on dogs, coyotes, and many other wild animals.

SPIDER MITES
(Family Tetranychidae)

Familiar pests in gardens, orchards, and
greenhouses, spider mites have tiny,
slightly flat, oval bodies, less than $\frac{1}{64}$–
$\frac{1}{32}$" (0.3–0.8 mm) long, and piercing
mouthparts for sucking plant juices.
Females use a silk-gland opening near
the mouth to spin loose webs on foliage
and fruit, where they lay their eggs. As
cool weather arrives in autumn, they
sometimes invade homes. Most die of
the winter cold, and only eggs survive.
Members of this family damage leaf
surfaces, often introducing disease
agents and reducing the commercial
value of fruits, vegetables, and other
crops.

631 Two-spotted Spider Mites
(*Tetranychus urticae*)

Description: Less than $\frac{1}{64}$–$\frac{1}{32}$" (0.3–0.8 mm).
Flattened oval. *Pale yellow to green or
pink with dark spots* and many black
bristles on body and legs.

Habitat: Gardens, orchards, and greenhouses on
foliage and fruits.

Range: Throughout North America.

Food: Soft tissues and juices of American elm;
also house plants.

Life Cycle: Eggs laid on twigs and buds may
overwinter. In spring hatchlings
disperse onto opening foliage and flower
buds, where they penetrate tissues
using sharp mouthparts. Rapid growth
may lead to several generations before
fall. The male protects the unmated
female from another male by shooting
silk to ward off competitors.

Although they are motionless while
feeding, these mites are easily
disturbed, and they run quickly among
the loose webbing they spin as a
shelter. Mites weaken plants, making

them susceptible to other maladies. Related species on orchard trees and commercial crops are serious pests, difficult to control by environmentally safe measures.

WATER MITES
(Family Hydrachnellae)

These elongated, globular mites, $\frac{1}{16}$–$\frac{1}{8}''$ (2–3 mm) long, inhabit ponds and slow streams, creeping on immersed vegetation or swimming with peculiar gliding action through the water. In many species the 4 pairs of legs are fringed with hair. Red freshwater mites are the most conspicuous; others are green, yellow, or mottled, tending to blend in with the environment. The parasitic larvae attach themselves to the body or legs of backswimmers or other aquatic insects, and sometimes they cling to the mantle of bivalve mollusks and other pond dwellers. Others attach themselves to fully grown stoneflies or to the naiads of damselflies and dragonflies. They often hitchhike a ride on adult insects to remote bodies of water.

628 Red Freshwater Mite
(*Limnochares americana*)

Description: $\frac{1}{8}''$ (3 mm). Red. *Smooth, globular, somewhat elongated.* Legs hairy.

Habitat: Ponds and slow-flowing streams.

Range: Throughout North America.

Food: Small insects and mollusks.

Life Cycle: Eggs are dropped in water at random and settle to the bottom. Larvae creep or swim to aquatic invertebrate animals, mostly insects, on which they attach themselves as external parasites. Nymphs and adults are active predators.

Fish appear to avoid freshwater mites, although many must be eaten accidentally along with the water insects. Only a specialist can reliably identify mites.

VELVET MITES
(Family Trombidiidae)

These mites are easily recognized by their red coloring and dense, velvety hair. Adults measure $\frac{1}{64}$–$\frac{1}{8}$″ (0.5–4 mm) long. Adults eat insect eggs; larvae, known as red bugs, are parasites of insects, spiders, daddy-long-legs, and scorpions. Some cling to a host's wing even in flight. Many are found in moist woods, others in deserts.

629 Velvet Mites
(*Trombidium* spp.)

Description: ⅛″ (3–4 mm). Oval to rounded rectangle. *Bright velvety red.*
Habitat: Moist woods on vegetation and soil.
Range: Throughout North America.
Food: Adult eats certain insect eggs. Larva is external parasite of insects, spiders, daddy-long-legs, and scorpions.
Life Cycle: Eggs are scattered on soil. Larvae find hosts and begin to feed. Nymphs and adults are predacious.

These mites help keep the insect population in check, which makes them beneficial, although only of minor commercial importance.

SOFT TICKS
(Family Argasidae)

Despite their name, soft ticks have a tough, leathery skin, or integument.

They are called "soft" because they lack the hard plate on the back that is present in hard ticks. The mouthparts of soft ticks are attached at the front end of the body and do not project forward, as they do in hard ticks. Most are ⅛–¼" (4–5 mm) long. The female usually lays small masses of eggs several times a year. Ticks feed intermittently, usually on one host. Some species feed only in the larval stage, while others feed in all stages except the larval stage. Some species have more than 1 nymphal stage. Unlike hard ticks, they usually are not important carriers of diseases. Soft ticks attack birds or mammals that have nests or regular sleeping sites.

633 Mammal Soft Ticks
(*Ornithodoros* spp.)

Description: ¼" (5 mm). *Brownish. Oval, flattened, with tough pebbly skin.* Legs less than half as long as body width.

Habitat: Pastures and range lands.

Range: Throughout North America.

Food: Blood of warm-blooded animals, on which ticks feed intermittently, dropping off between meals.

Life Cycle: After feeding, mated female drops off host to ground and lays hundreds of eggs, then returns to host. Hatchlings creep from nest of host to feed, but drop off to molt and repeat process to reach another host. Eggs are usually laid in several batches a year.

In the West, ticks sometimes transmit relapsing fever to people. Like other soft ticks, these feed intermittently, showing a preference for mammals that have regular nesting areas or burrows where immature ticks can develop. They become spectacularly distended as they gorge on blood.

HARD TICKS
(Family Ixodidae)

Hard ticks get their name from the
hard plate on top of the body. They
hold their mouthparts forward, unlike
soft ticks, whose mouthparts are
attached underneath at the front. Most
hard ticks are $\frac{1}{16}-\frac{1}{8}''$ (2–4 mm) long.
They feed on 2 or 3 hosts, in contrast
to soft ticks, which parasitize only 1
host. Males of certain species do not
feed. The female lays many eggs and
dies.

632 Eastern Wood Ticks
(*Dermacentor* spp.)

Description: $\frac{1}{8}''$ (3–4 mm). Male's body pale gray
with reddish-brown spots and legs.
*Female's body reddish-brown with small
shield of black-speckled gray near head. Legs
brown; head often orange above.*

Habitat: Woodlands and shrubbery beside trails.

Range: Eastern North America.

Food: Adult feeds on larger animals and blood
of mammals, especially deer. Larva
feeds on rodents.

Life Cycle: Tick clings to plants while extending
fore legs to seize passing host. Tick
climbs on prey for a meal, dropping off
after fully engorged. If not yet mature,
tick molts and repeats process. Mature
female, if mated before last major meal,
drops many eggs, producing six-legged
larvae.

This tick can transfer disease organisms
from one host to the next. After a walk
through a field, it is wise to inspect
clothing and hair for ticks. Then the
ticks should be removed and burned or
drowned in alcohol.

630 California Black-eyed Tick
(Ixodes pacificus)

Description: ⅛" (3–4 mm). Adult flattened oval when unfed. Head, thoracic shield, and legs dark brown. Abdomen large, reddish brown. *Body almost spherical after eating,* still clinging with mouthparts. Young dark brown, sometimes mottled with paler areas.

Habitat: Fields, pastures, and ranch lands.

Range: Pacific Coast from Mexico to British Columbia.

Food: Blood of mammals, including domestic livestock and deer.

Life Cycle: Engorged female drops large number of eggs, which hatch on ground in a few days in warm weather. Hatchlings climb vegetation and latch onto passing host. They drop off after a meal, molt, and climb to another host, repeating process until full grown.

Ticks are most abundant on livestock in spring. They can transmit disease-causing organisms from one host to another.

Whipscorpions
(Order Uropygi)

Whipscorpions somewhat resemble
scorpions but have a long, whiplike tail
instead of a stinger at the end of the
abdomen. These reddish or brownish-
black arachnids, 2¾–5⅛" (70–130
mm) long, have jaws, or chelicerae,
between 2 heavy, pincerlike pedipalps
and 4 pairs of legs—the very long, thin
first pair ends in a multisegmented
antennae-like filament and is used as
feelers; the other 3 pairs are used for
walking. Nocturnal predators,
whipscorpions use their highly movable
tail to spray a defensive secretion,
which contains acetic acid and an agent
that attacks the victim's body covering.
But unlike scorpions, whipscorpions do
not have poison glands and, therefore,
are never harmful to people.
The mated female carries 20–30 eggs
in a membranous sac below the
abdomen until they hatch. The young
ride on the female's back until they
shed the first skin and are independent.
During the day whipscorpions hide
under logs and stones, or burrow into
the sand. Mostly tropical, there are
about 70 species worldwide; in North
America there is only 1 species, which
is found in the Southwest and Florida.
The order is also called Thelyphonida.
Previously whipscorpions were grouped
with Short-tailed Whipscorpions and
Tailless Whipscorpions under the Order
Pedipalpida.

VINEGARONES
(Family Thelyphonidae)

Vinegarones are large, measuring up to
3⅛" (80 mm) long, not including their
whip. Although they do not have a
stinger, the pincerlike pedipalps can
deliver a painful pinch. When
disturbed, they emit a vinegar-scented
acid from a gland at the base of the tail
whip—hence their common name.
This is the only family of whipscorpions
in North America.

634   Giant Vinegarone
"Grampus"
(*Mastigoproctus giganteus*)

Description: 3–3⅛" (75–80 mm). *Brown to black.*
Tail filament normally held curled forward
over the back. Pincers turned forward;
1st pair of legs held forward, like
feelers. 8 eyes (2 in middle, 3 on each
side of head).

Habitat: Outdoors among debris on soil, under
logs and rotting wood; indoors in
humid dark corners.

Range: Southern and southwestern United
States.

Food: Small insects.

Life Cycle: Female watches over eggs. Young ride
on female's back until the 1st molt,
when they disperse. Young are
colorless.

This formidable-looking whipscorpion
is seldom encountered, because it hides
by day and hunts in darkness.

Tailless Whipscorpions
(Order Amblypygi)

Tailless whipscorpions resemble spiders, ⅜–2″ (8–51 mm) long, with enlarged, spiny pincers, or pedipalps. Like whipscorpions, most have 8 eyes arranged in 3 groups, but some species have 6 eyes in 2 groups, and one species has no eyes. All have a shieldlike covering, or carapace. Unlike whipscorpions, the cephalothorax is unsegmented, the abdomen is joined to the thorax by a slender stalk or pedicel, and there is no tail. The first pair of legs have long, slender femora with a multisegmented, antennae-like appendage at the tip, which is about 3 times the body length. Tailless whipscorpions hunt at night, hiding under bark or stones by day. They run sideways if exposed. The female carries eggs briefly below the body in a lens-shaped egg sac that is held together with an enamel-like material. Hatchlings clamber on the female's back and are carried for 4–6 days before they disperse. Tailless whipscorpions were formerly included in the same order as Whipscorpions. There are 60 species of tailless whipscorpions worldwide, including 3 species in North America.

TAILLESS WHIPSCORPIONS
(Family Tarantulidae)

Members of this family have flattened
bodies, ⅜″ (8–11 mm) long, with
spiny pedipalps and slender, flat
antennae-like legs. The cephalothorax is
often wider than long, and the
abdomen is generally shorter and
narrower than the cephalothorax. The
abdomen and cephalothorax are
connected by a short stalk. The first
pair of legs have long filamentous or
whiplike tips. The remaining 3 pairs of
legs are held to the side, crablike.
Primarily denizens of humid tropics,
most North American species are found
in Florida and Gulf states, where they
occasionally enter houses.

---

639 Side-spotted Tailless Whipscorpions
(*Tarantula* spp.)

Description: ⅜″ (8–11 mm). Pale to dark blackish
brown, depending on time since latest
molt. Pedipalps and femora dull black;
legs beyond paler. *Abdomen has* 4 rows
of small *circular dark spots above,* which
tend to coalesce as body darkens,
leaving series of intervening pale spots.

Habitat: Woods in moist crevices, under bark or
stones; indoors on walls of shower stalls
and in damp basements.

Range: Florida.

Food: Insects and arthropods.

Life Cycle: Female carries plasticlike egg sac
attached to abdomen until young
hatch. Hatchlings ride on female's back
until the first molt after 4–6 days.
Young disperse and grow slowly.

Nocturnal, this tailless whipscorpion
hides under stones or bark during the
day. It scurries sidewise if approached.

Windscorpions
(Order Solpugida)

Windscorpions are easily recognized by the pair of large, pincerlike chelicerae on the head in front of the mouth and by the slight, waistlike constriction near the middle of the body. Unlike the broadly joined cephalothorax and abdomen of scorpions, windscorpions have 3 distinct body regions—a segmented cephalothoracic area with 2 eyes at the front margin, a 3-segmented thorax, and a 10-segmented abdomen. The chelicerae are used independently of each other to chew food—one pair holds the prey, while the other cuts it. The long, slender pedipalps do not have pincers and are used to scoop up water and bring it to the mouth. The first pair of legs are longer than the others and function in conjunction with the pedipalps as feelers. The other 3 pairs of legs are used for walking. Additionally, specialists identify this order by minute, T-shaped sensory organs on the hind pair of legs. These small arachnids, $\frac{3}{8}-2''$ (8–50 mm) long, are mostly brownish or yellowish and often hairy. Most are found in dry or desert regions, where they prey on insects and small vertebrates, including lizards. Some species live in the mountains. Females lay about 50 eggs in subterranean burrows, then stand guard over the eggs and young for up to several weeks until the young molt for the first time. They are called windscorpions because they appear to run as fast as the wind. Although most are nocturnal, these arachnids are sometimes called sunspiders after their sunny desert habitat. There are almost 120 species in North America out of 800 to 900 worldwide.

EREMOBATID WINDSCORPIONS
(Family Eremobatidae)

Most North American windscorpions
belong to this family of medium-sized
arachnids, ⅝–1¾" (15–45 mm) long.
They differ from the only other North
American family, Ammotrechnidae, in
having a straight front of the head and
1 or 2 claws on the first pair of legs,
rather than a rounded or pointed head
and no claws. Adults usually live only a
few months. All are found in the West,
particularly in the Southwest.

638   Pale Windscorpion
(*Eremobates pallipes*)

Description: Male ⅝–1" (15–26 mm), female ⅞–
1¼" (22–32 mm). Yellowish brown.
*Chelicerae large, pincerlike, held forward
close together.* Pedipalps heavy, leglike,
held like coarse antennae. 1st pair of
walking legs as long as other pairs but
much more slender.

Habitat: Arid and semiarid lands.

Range: Arizona to North Dakota and adjacent
areas of Canada.

Food: Insects and small vertebrate animals,
such as lizards.

Life Cycle: Female digs out area in soil, then hides
eggs there. Female stands guard until
they hatch. Young are primarily
nocturnal, venturing about in daylight
only when they approach adult size.
Males remain smaller than females and
their legs are longer in proportion to
body length.

These solitary windscorpions are
independent hunters from hatchling to
adult stages. Only a specialist can
distinguish the more than 100 species
in this genus.

Part III
Appendices

GLOSSARY

Abdomen The hindmost of the 3 subdivisions of an insect's body or of the 2 subdivisions of a spider's body.

Aggregated eye A cluster of simple eyes (ocelli), usually of indefinite number, not necessarily alike on 2 sides of the head, never fitted into a tight array, as are the facets (ommatidia) of a compound eye.

Antenna One of a pair of sensory appendages on each side of an insect's head.

Arista A large bristle extending from the third antennal segment in many flies.

Asexual Reproducing without the union of sperm and egg.

Beak Protruding mouthparts, usually sharp or with jaws at the tip.

Bristle A stiff hair, ordinarily arising from a cavity or socket in the body surface.

Calypter A small lobe on the hind margin near the base of the wing of some flies.

Carapace A hard covering over the dorsal surface of the body, or at least of the thorax.

Caste In social insects a specialized form of adult with a distinct role in the colony.

Caudal Pertaining to the hindmost part of the body; from the Latin word *cauda,* tail.

Cell An area of wing membrane partly or completely surrounded by veins.

Cephalothorax The first subdivision of a spider's body, combining the head and thorax.

Cercus One of a pair of sensory appendages at the posterior end of an insect's body.

Cervical shield A hard plate located above the first segment of the thorax of a caterpillar; synonym, prothoracic shield.

Chelicera One of the first pair of appendages below the cephalothorax of a spider, consisting of a heavy first segment and a sharp second segment or fang.

Chrysalis The naked pupa of a butterfly.

Claspers A pair of appendages among the male sex organs, used to hold the female while mating.

Clubbed Enlarged toward the tip, clavate.

Cocoon An enclosure, or case, of secreted fibers of silk produced by a fully grown larva as a chamber in which to pupate.

Colony An aggregation of individuals with some degree of organization into a community, normally of 1 species.

Compound eye One of the paired visual organs consisting of several or many light-sensitive units, or ommatidia, usually clustered in a radiating array with exposed lenses (facets) fitting together.

Coxa The first, or basal, segment of an insect's leg.

Cremaster A spine-studded projection at the posterior end of a pupa, frequently used for attachment.

Cribellum A sievelike spinning plate in front of the spinnerets, found in members of some spider families.

Cuticle The secreted, noncellular outer covering of an insect's body.

Dimorphism Having 2 distinctive forms of the same species.

Drone One of a caste of social bees, consisting only of reproductive males.

Elytron The thickened fore wing of earwigs and beetles, serving as protective covers for the hind wings.

Epidermis The layer of living cells underlying and secreting the insect cuticle.

Exoskeleton A supporting structure on the outside of the body, enclosing all living cells.

Eyespots Spots resembling eyes on winged insects, such as butterflies and moths.

Facet The outer surface of a single unit (ommatidium) of the compound eye, through which light enters the sensory structure.

Fang The sharp-pointed second segment of a spider's chelicera.

Femur The third segment of an insect's leg, between trochanter and tibia.

Gall An abnormal plant growth induced by the presence of an insect or other living organism (roundworm, fungus, etc.).

Generation The young that hatch from eggs produced by 1 female at 1 time.

Genitalia The external sex organs.

Gills Respiratory organs of aquatic animals which take dissolved oxygen and salts from the water.

Girdle   A loop of silk that some caterpillars spin as extra support for the body during pupal transformation.

Grub   A heavy-bodied, immature insect, usually slow-moving, with legs on its thorax but no prolegs on its abdomen, and usually with a distinct head; characteristic of beetles.

Haltere   The specialized hind wing of a fly, arising from the metathorax, supposedly aiding in balance.

Hemelytra   The thickened fore wings with membranous overlapping tips of true bugs.

Honeydew   The sugary liquid discharge of various insects, especially aphids.

Humeral   Pertaining to the front basal portion of the wing, as in the humeral angle.

Hypermeta-morphosis   Metamorphosis including more developmental stages than the usual 4.

Imago   The mature or adult stage of the insect.

Instar   A stage in development, the first instar being that between hatching of an egg and the first molt.

Mandible   One of a pair of jaws of an insect, normally used for chewing.

Margin   Each edge of a wing.

Maxilla   One of a pair of mouthparts, immediately posterior to the mandibles.

Mesonotum   The upper surface of the mesothorax.

Mesosternum   The lower surface of the mesothorax.

Mesothorax   The second or middle segment of the thorax, bearing the second pair of legs and the fore wings, if present.

Metamorphosis The transformation of an immature to a mature insect, following the feeding (nymphal and larval) stages. Said to be complete when it involves a pupal stage, or simple when it requires no pupal stage but when the immature resembles the adult except for lack of sex organs and, in some groups, wings.

Metanotum The upper surface of the metathorax.

Metasternum The lower surface of the metathorax.

Metathorax The third or last segment of the thorax, bearing the third pair of legs and the second pair of wings, if present.

Molt The shedding of the confining outer layer of the body (the cuticle, integument, or exoskeleton) to permit growth or metamorphic change.

Naiad The aquatic young of mayflies, stoneflies, dragonflies, and damselflies.

Nymph The terrestrial young of insects with simple metamorphosis.

Ocellus A simple eye with a single light-collecting lens and a cup-shaped set of light-sensitive receptor cells.

Ommatidium A single receptor unit with an exposed lens (facet) of a compound eye.

Overwinter To go through a period of dormancy during the cold season.

Oviparous Reproduction by release of eggs.

Ovipositor An egg-laying organ at the rear end of the female's abdomen.

Ovoviviparous Reproduction by release of active young, from eggs that have hatched within the body of the mother.

Palp A sensory structure associated with the mouthparts.

Parthenogenesis Asexual reproduction in which eggs develop without fertilization.

Pedicel In ants and other members of the Order Hymenoptera, the basal stalk of the abdomen. In spiders, a slender connection between the cephalothorax and abdomen. Also called petiole.

Pedipalp One of the second pair of appendages on the cephalothorax of a spider, usually leglike in a female but enlarged at the tip in a male as a special organ for transferring sperm; used by both sexes for guiding prey to the mouth.

Polymorphism Having more than 2 distinctive forms of the same species.

Postscutellum A conspicuous swelling underneath the scutellum on the thorax; characteristic of some flies.

Proboscis A prolonged set of mouthparts adapted for reaching into or piercing a food source.

Proleg A soft, fleshy, unjointed extension, usually paired, from the abdomen, providing support or attachment; characteristic of caterpillars.

Pronotum The upper surface of the prothorax.

Prosternum The lower surface of the prothorax.

Prothorax The first or most forward of the 3 thoracic segments, supporting the head and the first pair of legs, if present.

Pupa The inactive stage of insects during which the larva transforms into the adult form, completing its metamorphosis.

Puparium The hardened cuticle of the last larval stage, within which a maggot forms a pupa and metamorphoses into an adult stage.

Reproductive The male or female of a caste of social insects capable of reproduction, usually gaining wings for brief mating flights, as occurs in termites and ants.

Scutellum A triangular portion of the mesonotum, located between the bases of the fore wings and usually visible when the wings are folded; characteristic of true bugs and beetles.

Segment A subdivision of the body or of an appendage, between membranous regions that permit flexibility.

Simple eye A light-sensitive organ consisting of a convex lens bulging from the surface of the head, concentrating and guiding light rays to a cup-shaped cluster of photoreceptor cells. Also called an ocellus.

Social Living in organized communities, with division of labor and castes (at least a worker caste).

Solitary Living alone and independently, not in aggregations or communities.

Spinneret One of the 2–4 pairs of nozzlelike appendages below the rear part of the spider's abdomen, through which silk is extruded and manipulated to form the strands of a web.

Spiracle A breathing pore in the surface of the body through which gas enters and leaves the internal system of tubular tracheae.

Spur A stout, spinelike projection, usually movable, such as is present commonly toward the end of the tibia.

Stigma A thickened area of the wing membrane along the front edge near the tip, often pigmented and conspicuous in many wasps and dragonflies.

Style or stylus   A slender, elongate projection, often hollow and movable.

Subimago   The winged, flying stage of a mayfly, which precedes the final molt to a reproductive adult (imago).

Tarsus   The last or "foot" portion of the leg, consisting of 1 to 5 segments, attached basally to the tibia.

Thorax   The subdivision of the body between head and abdomen, consisting of three segments (the prothorax, mesothorax, and metathorax) and bearing whatever legs and wings are present.

Tibia   The fourth segment of the leg, linking femur to tarsus.

Trachea   A tube permitting movement of air into and out of the body, facilitating respiration in internal organs.

Trochanter   The second segment of the leg, linking coxa to femur.

Tubercle   A small, knoblike projection, often found on caterpillars.

Venom   Secretion of glands opening through pores near the tip of each cheliceral fang, serving to immobilize prey.

Vesicle   A pouch or sac, such as is found beneath the abdominal segments of jumping bristletails.

Viviparous   Giving birth to active young that have developed with no identifiable egg stage within the mother.

Worker   One of a caste of social insects, usually incapable of reproduction, that procures and distributes food or provides defense for the colony.

PICTURE CREDITS

The numbers in parentheses are plate numbers. Some photographers have pictures under agency names as well as their own. Agency names appear in boldface. Photographers hold copyright to their works.

David H. Ahrenholz (202, 238, 303, 306, 402, 432, 447, 587)

Amwest
Yva Momatiuk (172)

Animals Animals
Tom Brakefield (647) George K. Bryce (643) Bruce A. Macdonald (674) Raymond A. Mendez (79, 355, 668, 671) Robert Mitchell (319, 661) C. W. Perkins (32, 299, 665)

David C. Berta (5)
James H. Carmichael, Jr. (103, 109, 114)
Robert P. Carr (26, 80, 108, 132, 162, 163, 234, 255, 288, 312, 428, 620, 694)

Bruce Coleman, Inc.
Jen and Des Bartlett (194) James H. Carmichael, Jr. (693) Harry N. Darrow (662) E. R. Degginger (124, 209, 479, 689) Robert L. Dunne (561) John Markham (87) David T. Overcash (295) John Shaw (33) Larry Stepanowicz (38, 669, 680, 695) Larry West (186, 207, 676) Dale and Marion Zimmerman (200)

Sturgis McKeever (545)
Charles R. Meck (393, 394, 395)
Paul Miliotis (353, 361, 369)
Lorus J. and Margery Milne (6, 176, 271, 343, 377)
Robert W. Mitchell (71, 73, 83, 95, 96, 100, 107, 126, 127, 128, 129, 148, 174, 183, 185, 190, 208, 225, 226, 235, 246, 249, 262, 265, 289, 292, 296, 302, 310, 314, 315, 317, 322, 323, 374, 383, 421, 423, 433, 446, 459, 475, 509, 511, 520, 527, 530, 566, 573, 633, 634, 640, 672, 685)
C. Allan Morgan (17, 21, 25, 85, 93, 227, 257, 276, 277, 337, 396, 495, 516)

National Audubon Society
Collection/Photo Researchers, Inc.
A. W. Amber (580) N. E. Beck, Jr. (136, 150, 237, 489, 529) A. Cosmos Blank (101) John Bova (663) Ken Brate (142, 338, 372, 503, 531, 581, 589) Robert Bright (550) Stephen Collins (123) Stephen Dalton (201) Marjorie Dezell (540) P. W. Grace (44) Gilbert Grant (595) William J. Jahoda (528) A. Kalnik (644) Robert Lee (627) Charles Mann (523) Tom McHugh (636) Lincoln Nutting (645) Richard Parker (105, 111, 161, 181, 638) Noble Proctor (43, 248) Louis Quitt (46) Bucky Reeves (650) Alfred Renfro (177) J. H. Robinson (628) Harry Rogers (76, 104) Ray Simons (134) Alvin E. Staffan (117) M. T. Tanton (373) Larry West (284, 604, 691)

Daniel Otte (261, 273, 274)
Betty Randall (102, 158, 217, 228, 264, 285)
Manuel Rodriguez (442, 651, 654, 664, 683, 688)
Edward S. Ross (8, 15, 40, 45, 47, 48, 49, 50, 51, 52, 53, 54, 58, 61, 62, 68, 75, 81, 88, 92, 98, 110, 118, 119, 120, 121, 122, 130, 139, 144, 146, 152, 156, 159, 160, 164, 165,

INDEX

Numbers in boldface type refer to color plates.
Numbers in italics refer to pages. Circles
preceding common names indicate insects and
spiders illustrated in this book, making it easy for
you to keep a record of those you have seen.

STAFF

Prepared and produced by
Chanticleer Press, Inc.

Founding Publisher: Paul Steiner
Publisher: Andrew Stewart

Staff for this book:

Editor-in-Chief: Gudrun Buettner
Executive Editor: Susan Costello
Guides Editor: Susan Rayfield
Project Editor: Jane Opper
Assistant Editor: Mary Beth Brewer
Editorial Assistant: Constance Mersel
Art Director: Carol Nehring
Art Assistant: Ayn Svoboda
Production: Helga Lose, Amy Roche
Picture Library: Edward Douglas,
Dana Pomfret
Symbols: Paul Singer
Drawings: M. J. Spring

Original Series Design: Massimo Vignelli

All editorial inquiries should be
addressed to:
Chanticleer Press
568 Broadway, Suite #1005A
New York, NY 10012
(212) 941-1522

To purchase this book, or other
National Audubon Society illustrated
nature books, please contact:
Alfred A. Knopf
201 East 50th Street
New York, NY 10022
(800) 733-3000

NATIONAL AUDUBON SOCIETY FIELD GUIDE SERIES

Also available in this unique all-color, all-photographic format:

African Wildlife

Birds *(Eastern Region)*

Birds *(Western Region)*

Butterflies

Fishes, Whales, and Dolphins

Fossils

Mammals

Mushrooms

Night Sky

Reptiles and Amphibians

Rocks and Minerals

Seashells

Seashore Creatures

Trees *(Eastern Region)*

Trees *(Western Region)*

Weather

Wildflowers *(Eastern Region)*

Wildflowers *(Western Region)*

DATE DUE

MIDBOREAL MAYFLIES
(Family Ephemerellidae)

Mostly brownish mayflies, members of this family measure ¼–¾" (5–19 mm) long, excluding the 3 tail filaments. They are distinguished from other mayflies by details of wing venation. Naiads are either carrot-shaped or flattened. They have broad, flattened femora and often have tubercles on the head, thorax, and abdomen. Well-developed gills appear on abdominal segments 4–7 or 3–7, but those on segment 1 are rudimentary. The one North American genus is found near rapid streams or small clear lakes.

394 Midboreal Mayfly
(*Ephemerella subvaria*)

Description: ⅜–½" (10–13 mm) excluding 3 tail filaments. *Dark brown to reddish brown.* Legs almost black. Wings clear.
Habitat: Near small flowing streams.
Range: Nova Scotia to Maryland, west to Kansas, north to Manitoba.
Food: Adult does not eat. Naiad feeds on minute green plants and small insects.
Life Cycle: Eggs hatch a few minutes after female drops them into water. Naiads seek shelter in water, moss, or crevices and may overwinter in water, completing full growth in spring. Subimagos emerge April–May and transform to adults 22–30 hours later.

Males swarm 6–50 feet above the ground. Any female that flies into the swarm is seized by a male and the pair drops downward to mate. Touch and chemical signals confirm that the other is of the same species.

BURROWING MAYFLIES
(Family Ephemeridae)

Many of the largest and most common
North American mayflies belong to this
family. Some measure more than 1"
(25 mm) excluding the 2 or 3 tail
filaments, which may be even longer.
Adults are distinguished from mayflies
in most other families by their
4-segmented tarsi. The spindle-shaped
naiads have sharp, tusklike mandibles,
fringed gills on abdominal segments
2–7, and fore legs adapted for digging
in the muddy bottoms of rivers, lakes,
or gently flowing streams. Most North
American species are found in the East,
usually in water no more than a few
yards deep.

393 Coffin Fly
(*Ephemera guttulata*)

Description: ¾–⅞" (18–23 mm) excluding 3 tail
filaments. *Chalk-white.* Fore legs dark.
Wings clear to amber with numerous
dark veins.

Habitat: Near shallow lakes and quiet bays of
rivers with sandy or silty bottoms.

Range: Nova Scotia to Alabama and Florida,
north to Ontario.

Food: Adult does not eat. Naiad feeds on
microscopic green plants and organic
matter.

Life Cycle: Eggs hatch in 2–4 weeks. Naiads
burrow into bottom silt, feed, and
grow rapidly, reaching full size by
autumn or overwintering and
completing growth in spring.
Subimagos emerge May–August, rarely
as late as October. They transform to
adults in 48 hours or less.

Incredible numbers of these mayflies
emerge from the water, fly to lights,
and die soon after mating, sometimes
forming piles of bodies 3 feet deep.